T0138344

Large Carnivore Conservation

Large Carnivore Conservation

Integrating Science and Policy

in the North American West

Edited by Susan G. Clark and Murray B. Rutherford

THE UNIVERSITY OF CHICAGO PRESS *Chicago and London*

Susan G. Clark is the Joseph F. Cullman 3rd Adjunct Professor of Wildlife Ecology and Policy Sciences in the School of Forestry & Environmental Studies at Yale University. **Murray B. Rutherford** is associate professor in the School of Resource and Environmental Management at Simon Fraser University.

The University of Chicago Press, Chicago 60637
The University of Chicago Press, Ltd., London
© 2014 by The University of Chicago
All rights reserved. Published 2014.
Printed in the United States of America

23 22 21 20 19 18 17 16 15 14   1 2 3 4 5

ISBN-13: 978-0-226-10740-0 (cloth)
ISBN-13: 978-0-226-10754-7 (e-book)
DOI: 10.7208/chicago/9780226107547.001.0001

Library of Congress Cataloging-in-Publication Data
Large carnivore conservation : integrating science and policy in
   the North American West / edited by Susan G. Clark and
   Murray B. Rutherford.
       pages ; cm
   Includes bibliographical references and index.
   ISBN 978-0-226-10740-0 (cloth : alkaline paper) — ISBN
   978-0-226-10754-7 (e-book) 1. Carnivora—Conservation—
   North America. 2. Wildlife conservation—North America.
   3. Wildlife management—North America. 4. Top predators—
   Conservation—North America. 5. Grizzly bear—
   Conservation—North America. 6. Puma—Conservation—
   North America. 7. Wolves—Conservation—North America.
   8. Human-animal relationships—North America. I. Clark,
   Susan G., 1942– editor. II. Rutherford, Murray B., editor.
   QL737.C2L336 2014
   333.95'97—dc23

                                            2013027587

♾ This paper meets the requirements of ANSI/NISO Z39.48-1992
(Permanence of Paper).

Contents

Preface

Over the last two centuries wolves, grizzly bears, and mountain lions have been eliminated or displaced from much of their original range in North America. To avoid further loss of these large carnivores from the continent, we need to find ways to reduce conflicts between people and carnivores, so they can coexist in healthy ways on the landscape. We also must find common ground among people about carnivores and carnivore conservation, despite distrust, differing perspectives, and disagreement about our relationship with these animals and, more broadly, our responsibility to "nature." In this book we discuss and evaluate the efforts of diverse people across western North America who are trying to improve governance (decision making in the public sphere) and reduce conflict in order to live on a more sustainable basis with large carnivores and each other. The cases in this book and other hard-won experience over decades point the way for more sound conservation that serves the public interest.

Many people care deeply about conserving large carnivores, and there is broad and growing public support for more effective efforts to ensure the continued existence of these animals outside of zoos and protected areas. But living with large carnivores is not easy—people must cope with restrictions on land development, occasional losses of livestock and pets, rare instances of personal harm or even death, and other deprivations of value. These deprivations, and the typical response of systematically eliminating the carnivores involved, have made decision making about carnivores highly contentious. In a rapidly changing world, there is real doubt in the minds of some people about whether large carnivores can persist in viable wild populations in many parts of the globe. The challenge that we face is enormous to be sure, but far from insurmountable. In order to address the physical challenge and these doubts and to find more effective strategies and practical actions, we need to deepen and enlarge our understanding and insight about the problems of carnivore conservation and our own efforts to solve these problems.

Sustainable coexistence depends on having two kinds of knowledge. First is knowledge of carnivores and their environments (content knowledge). Second is knowledge about people: how they interact and how they make individual and collective choices that directly or indirectly affect carnivores (procedural knowledge). Both types of knowledge are vital, but the focus of attention of the vast majority of conservation efforts and in the literature has been on obtaining and reporting content knowledge about the animals and their ecology. In recent years this knowledge has been supplemented by research on human attitudes, values, and behavior, but the overriding emphasis remains on knowledge about carnivores. As important as the biology of a species is, such a view of conservation leaves out consideration of how this information can best be used through effective decision-making processes. This underattention to knowledge about governance has led to a plethora of scientific, management, and policy problems, resulting in intractable conflict, widespread killing of animals, and recycled intolerance and misunderstanding. Problems are perpetuated by a complex, dynamic mix of individual and institutional factors. This book examines these factors as they play out in actual cases.

Many of these problems are avoidable. To be successful in carnivore conservation, content knowledge and procedural knowledge must be integrated in practical and effective ways to make wise decisions. Fortunately, a way to integrate—a strategy and method—already exists and is detailed in this book. We use this integrative approach to focus on governance processes in large carnivore conservation, asking, "How well does decision making function to achieve coexistence and sustainability?" and "What can be done to make it more effective?" We harvest the lessons of experience to answer these questions and turn them into lessons of foresight for use by a wide audience.

We seek to contribute to and advance an ongoing dialogue about how to sustain large carnivores and make governance more effective. We believe that people need to come to grips with shortfalls and dysfunction in conservation. This observation is not intended as a criticism of any one person or program. It is as much a comment on our systems for education, management, and policy making (i.e., our institutions). Like many of our colleagues, we want to foster more practical, effective, and ameliorative conservation efforts.

The book includes eleven chapters by seventeen authors from universities, state and federal agencies, and nongovernmental organizations. Combined, they have over 150 years' experience with large carnivores in western North America and other places in the world. This is a unique assemblage of perspectives, skills, and experience.

The introductory chapter provides an overview of the biophysical and sociopolitical dimensions of large carnivore conservation in North America, discusses the failures of conventional approaches to governance in wildlife management, and describes the analytical or integrative approach and criteria used in the case studies and other chapters. We introduce the common problems of carnivore conservation and stress the importance of understanding decision-making processes and how they cause or contribute to these problems.

The introduction is followed by six case studies: mountain lions in Arizona, wolves in Wyoming, grizzly bears in Yukon, wolves in Alberta, grizzly bears in Montana, and grizzlies in Alberta. Each case study is written by one or more academic or practicing professionals who know the ecological, social, and institutional context and have the expertise in wildlife science, management, and the integrative policy sciences to be able to understand the biological and human dimensions of the case. The cases focus on individual and collective decision making, examining how these governance processes function and how they can be improved. Several cases feature innovative strategies that have been successful in improving decision making, alleviating conflict, and reducing carnivore mortality. By evaluating and comparing the cases using a stable frame of reference—an insightful and practical analytic approach—the authors identify pervasive problems of governance, highlight examples of success, and develop lessons for these and other settings.

Following the six case studies are three chapters that examine large carnivore conservation from a higher-level perspective, synthesizing lessons from the case studies and much other work. The first of these chapters discusses constitutive decision-making processes for large carnivores in North America—that is, the broad cultural context and institutions that set the constraints and incentives under which local initiatives operate (e.g., in the six cases). These factors ultimately determine whether innovations prosper and are diffused to other settings. The second of these chapters looks at the North American Model of Wildlife Conservation and what needs to be done to make it more effective and bring it into alignment with modern times. The third such chapter takes a network systems-wide view on large carnivore conservation. It uses an analytic design approach to develop better ways to assess and communicate information about the adequacy of knowledge, social interaction, and decision making in individual cases as they are ongoing. This chapter includes plexus diagrams that visually summarize key features of the cases. The concluding chapter of the book reviews our overall messages, using examples from the cases to illustrate recurring malfunctions in decision making and summarizing recommendations about how to overcome these malfunctions. An index and list of authors are provided at the end of the book.

In the end, our goal is to improve governance and promote successful coexistence between large carnivores and people. We hope this book will be widely discussed and used globally.

Acknowledgments

We thank world-renowned wildlife photographer Tom Mangelsen for use of his work. The Northern Rockies Conservation Cooperative (NRCC) in Jackson, Wyoming, contributed research funding and served as the organizational link that brought the authors together. Many of the authors are affiliated with this innovative nongovernmental entity; some have been for decades. The NRCC has a long successful history of creative contribution to conservation, leadership, and problem solving in the common interest. Both Peyton Curlee Griffin and Jason Wilmot of the NRCC deserve special acknowledgment for their help, and we would like to recognize board member Cathy Patrick for her long-term support of the NRCC and her commitment to effective conservation. Funding for copyediting and indexing was provided in part by a grant from the University Publications Fund of Simon Fraser University and in part by the NRCC. We thank Richard Reading of the Denver Zoological Foundation for his encouragement. Our universities and other organizational affiliations merit special recognition as well, and we also want to acknowledge fully the many students, colleagues, and programs that we have collaborated with or shared ideas with—too many to identify individually. Finally, we are grateful for the essential contribution of Denise Casey as overall editor and coordinator of the book manuscript. This volume has been greatly improved as a result of her dedication, skill, keen eye for detail, and friendship with all of the authors.

1

Large Carnivores, People, and Governance

SUSAN G. CLARK, MURRAY B. RUTHERFORD, AND DAVID J. MATTSON

Introduction

Large carnivore conservation involves complex practical and policy problems that severely challenge our capacity to make well-reasoned decisions for the common good. Consider the recent series of decisions about managing grey wolves (*Canis lupus*) in the Northern Rocky Mountains of the United States. In 2008 the US Fish and Wildlife Service (USFWS) removed the region's wolves from federal protection under the Endangered Species Act, but conservation groups quickly obtained a court injunction blocking this delisting. The agency revised and reissued the delisting for states other than Wyoming, but the revised delisting was soon overturned by another court decision. The US Congress then stepped in, adding a rider to a budget appropriation bill that directed the USFWS to reissue the delisting. Not surprisingly, conservation groups sued to challenge the rider. In his ruling on the challenge, Judge Donald Molloy complained of "legislative prestidigitation" and called the rider "a tearing away, an undermining, and a disrespect for the fundamental idea of the rule of law" (*Alliance for the Wild Rockies v. Salazar* 800 F. Supp. 2d 1123 [D. Mont. 2011]). He decided, however, that the sweeping language of the rider precluded him from declaring it to be unconstitutional. His decision was upheld on appeal and the states of Idaho and Montana assumed control of wolf management within their jurisdictions. Then, in the fall of 2012, the USFWS delisted wolves in Wyoming, and conservation groups promptly sued to challenge this decision as well. Whatever the eventual outcome of this latest lawsuit, it seems highly unlikely that anyone involved with these convoluted processes would consider this to be a reasonable and appropriate way to make public decisions.

Why are decisions about large carnivores so bitterly contested, and why is common ground so elusive? In this book we contend that conserving and living with large carnivores is as much a problem of people and governance—authoritative decision making in the public sphere—as it is a problem of animal ecology and behavior. Governance is essentially about who gets what, when, and how, and who gets to decide (Lasswell 1936). For large carnivore conservation, governance involves how people interact, how they make and implement decisions, and how those decisions affect carnivores, people, and the settings in which carnivores and people live together.

People and Large Carnivores

The case of wolves in the Northern Rockies may be an extreme example, but large carnivore conservation is often characterized by conflict, contested science, litigation, and dysfunctional decision making. A USFWS decision to delist grizzly bears (*Ursus arctos*) in the Yellowstone region was disputed in the courts for four years before it was overturned in November 2011 when the Ninth Circuit Court of Appeals restored the threatened status of grizzlies under the Endangered Species Act. In California the Fish and Game Department's proposal to institute sport hunting for mountain lions (*Puma concolor*) was rejected in 1990 as the result of a public referendum in which 52 percent of voters supported a prohibition on hunting. A subsequent referendum to allow sport hunting of mountain lions was rejected in 1996, this time by 58 percent of the votes. In Alberta, Canada, the government-appointed Endangered Species Conservation Committee recommended in 2002 that grizzly bears be listed as threatened in the province, but the recommendation was contested by hunters and industry, and the provincial government did not follow through with the listing until eight years later. Of course, listing grizzlies did not end the dispute, as hunters argued that the status of bears should be assessed separately for each local region rather than for the province as a whole, and conservationists argued that the new protections were still inadequate. Meanwhile, the number of grizzly bears in Alberta has declined from an estimated pre-European population of as many as 6,000 bears to less than 700 (Alberta Sustainable Resource Development and Alberta Conservation Association 2010).

LEARNING FROM EXPERIENCE IN LARGE CARNIVORE CONSERVATION

This book, then, is about more than people trying to conserve and live with large carnivores; it is also about people trying to live with other people and make decisions about the collective good in spite of diverse and sometimes conflicting perspectives, beliefs, and values. As the following chapters illustrate, people hold radically different notions about appropriate attitudes and behavior toward large predators. These notions are readily catalyzed into conflict by the symbolic potency of animals that not only can threaten human life and property, but also can engender an uncanny sense of kinship. Our case studies examine the intense conflict associated with large carnivore conservation, the diverse measures taken to alleviate this conflict, and the effectiveness of such measures.

We believe that governance in large carnivore conservation can be substantially improved and that better governance is the key to reducing conflict and ensuring that large carnivores continue to survive. We also believe that better governance will ultimately lead to greater overall civility and dignity among those involved with and affected by large carnivore conservation. Accordingly, the book's overall goal is to offer insights from practical experiences about how to improve governance processes in large carnivore conservation.

In the chapters that follow, the authors use an integrative interdisciplinary approach drawn from the policy sciences to analyze the problems of governance in large carnivore conservation. Chapters 2 through 7 are case studies of carnivore management or conservation initiatives in North America. In each case, a governmental agency, nongovernmental organization, or group of individuals has attempted to improve decision making and alleviate troubling conflict between people and carnivores, or among people, or both. The authors apply a standard set of evaluative criteria to assess the decision-making processes in these cases and develop recommendations for both the case at hand and other settings. The case studies are followed by chapters that adopt a broader focal lens, examining higher-level governance and sociopolitical factors that apply across all the cases in this book and elsewhere, and offering lessons of broader relevance. The book concludes with a summary of common themes and recommendations. Later in this chapter we describe the integrative approach and associated evaluative criteria used by our authors, but first we offer a brief introduction to the context of large carnivore conservation in North America, highlighting factors that make governance so problematic.

Our cases focus on wolves, grizzly bears, and mountain lions. These animals all weigh more, on average, than 25 kg as adults. Carnivores of this size are especially prone to endangerment because of a combination of comparatively low reproductive rates, large range sizes, and low densities (Mattson 2004). When in contact with humans, these animals tend to die at rates that exceed reproduction. One important factor contributing to this high mortality rate is that people typically have many reasons to kill large carnivores: carnivores occasionally kill or injure livestock, they prey on ungulates that might otherwise be targets for human hunters, they provoke fear for human life, and for some people they simply engender intense dislike. A substantial number of people also have a strong desire to hunt these animals, recently evidenced by the fact that in 2011, the year wolves were delisted in Montana, more than 18,000 wolf-hunting licenses were purchased (Montana Fish, Wildlife and Parks 2011). As a result of these and other factors, wolves, grizzly bears, and mountain lions have been extirpated from large parts of their former ranges, especially in the United States and the eastern and southern portions of Canada (Laliberte and Ripple 2004).

Our case settings extend along the mountainous spine of western Canada and the United States from Yukon to Arizona. The settings encompass not only a spectrum of biophysical environments, but also a variety of social, political, and jurisdictional arenas. In Yukon, relations among Aboriginal peoples, European descendents, territorial government, and federal national parks administrators create a unique dynamic. Farther south in Canada, the problem of carnivores interacting with livestock in agricultural landscapes is featured, alongside contested federal and provincial management in and around national parks that have objectives of not only protecting nature, but also providing "visitor experiences" and commercial opportunities. In the northern US Rockies, the focus is on federal management of officially designated or prospective endangered species, controversies over recovery efforts and delisting, and the struggles of western states and local communities to control or at least influence decision making. Finally, farthest south, we examine state-level management of a comparatively abundant species, the mountain lion.

SYMBOLIC CREATURES

Human cognition is shaped by language and emotion. Our consciousness is correspondingly narrative in nature and encoded in symbols. It follows that, despite being biological entities, large carnivores are almost entirely represented as symbols and emotions in the cognitive processes through which we construct meaning in the world. We respond not so much to the carnivores themselves as to our own ideas of these creatures, encapsulated in evocative narratives. Few people in western North America have to deal in an immediate physical way with large carnivores. Instead, the vast majority of people deal almost exclusively with these animals as they exist in their imaginings or in the narratives and imaginings of other people. Thus, the emotive symbolic constructs we call "wolves," or "grizzly bears," or "mountain lions" are rarely empirical. There is ample evidence to suggest that they are instead largely creations of our cultures, upbringings, personalities, neuroses, and even psychoses. Whatever immediate experiences we may have with large carnivores during a lifetime rarely penetrate and transform our symbolic constructions. More typically, we interpret our experiences in ways that fit with and reinforce our established worldviews. Consider the awe of the environmentalist who, while hiking in the backcountry, sights a pack of wolves across the valley, or the anger of the rancher who, while checking on his cattle, finds the carcass of a calf that was chased and taken down by wolves.

The large carnivores featured in this book have different symbolic profiles, albeit layered on core similarities. These similarities include the fact that all these animals eat meat, some exclusively, which perhaps engenders some degree of bond with our (mostly) meat-eating selves. Like us, they also tend to be intelligent and provide their offspring with relatively prolonged care. These traits position large carnivores and humans closely typologically, especially compared to creatures such as rodents or ungulates. This nearness can generate both empathy and fear. Many of us are drawn to our imaginings of large carnivores as strong and noble. For others, large carnivores are savage, bloodthirsty, and cruel—for the same physical reasons. Humans seem to have a particularly strong symbolic affinity for bears. Like us, bears are omnivores and intelligent in similar ways for many of the same reasons. Like us, they provide lengthy care for their offspring, and, of great symbolic import, they enact the miracle of virtual death and rebirth through the annual cycle of hibernation. This relatedness between bears and ourselves is richly expressed in the mythology of Aboriginal peoples, still vigorous in arctic and subarctic North America (Clark and Casey 1993). By contrast, wolves are more often demonized, especially among descendents of Europeans, rooted in stories and ancestral memories that trace back to Eurasian rather than North American experiences (Casey and Clark 1996). Wolves did prey on and scavenge humans in Europe, probably for reasons with little physical relevance to North America (Graves 2007). Yet the myths persist. In contrast to bears and wolves, mythologies of mountain lions are not as well developed, probably because of the reclusive and cryptic nature of these large cats (Kellert et al. 1996). However, mountain lions may still evoke fear, respect, a desire to exert dominion, and other emotional responses.

Aside from the unique ways in which people have mythologized large carnivores, these animals are also caught up in broader attitudes or perceptions of wildlife and nature. Such worldviews are particularly relevant to policy cases such as the ones we examine in this book, largely because these more encompassing perspectives touch on many facets of human relations and governance. Some researchers classify human attitudes toward nature along a single gradient, which has been given a variety of names, including intrinsic versus instrumental, protectionist versus consumptive or utilitarian, and eco- or biocentric versus anthropocentric. Although these bipolar schematics give a sense of the strong differences in perspectives that may underlie people's positions on a particular problem, they fail to capture important nuances that influence not only how people respond to large carnivores, but also how they frame policy issues. Other researchers have put forward multidimensional value typologies. Manfredo and Teel (2008) assess cultural value orientations using two scales: the "domination wildlife value orientation," grounded in an ideology of human dominion over wildlife, and the "mutualism wildlife value orientation," based on an egalitarian view of wildlife. Kellert (1996) uses a nine-part typology of human values or attitudes toward nature, which Mattson and Ruther (2012) condense into five main categories: negativistic (fearful), dominionistic/utilitarian (valuing domination and instrumental utility), scientistic/ecologistic (valuing ecological connections or wholeness and opportunities for learning), aesthetic/naturalistic (valuing beauty and naturalness), and humanistic/moralistic (seeing animals as "virtual" people and feeling moral obligation for their well-being). Thus, although scholars do not agree on a single "best" model of human values for nature and wildlife, it is clear from this research that people involved in carnivore conservation may hold quite distinct and conflicting values and attitudes concerning these animals and the environments on which they depend.

POLITICAL PERSPECTIVES

Diverse and often strongly felt worldviews regarding humans and nature, including large carnivores, give rise to correspondingly diverse and strongly felt demands regarding the state of the world and the outcomes of management (Mattson et al. 2006). For some people, large carnivores are a nuisance, a threat, or even a plague—in their view these animals do not contribute to economic progress, and their presence and activities interfere with the need to control or efficiently make use of nature. They see little or no place for large carnivores except to the extent that the animals generate income or have some other instrumental value. For others, large carnivores are essential features of pristine or healthy ecosystems and entail little real cost for humanity. For these people, large, widespread, and ecologically functional populations of large carnivores are a necessary component of the world in which they wish to live. For yet another group of people, large carnivores are simply majestic and beautiful and for that reason alone are an essential presence on the land. In concrete terms, these diverse desired conditions are incompatible. It is impossible to have few or no large carnivores at the same time and place as robust ecologically functional populations. This incompatibility alone sets the stage for conflict. Moreover, as the cases in this book show, other aspects of the identities and groupings of participants add fuel to the figurative fires. Nature-views are not independent of other human traits, including gender, race, economic status, education, age, employment, and place of residence (Kellert 1996). It is virtually inevitable that people will coalesce as groups around shared worldviews, reinforced by other aspects of identity, and, having done so, progressively harden group boundaries, demonizing "alien" others who pose a material, symbolic, or existential threat. The recipe for conflict, even hatred, is potent.

Overt political factors often mesh with these social ingredients, compounding the difficulties people have in finding common ground (Rutherford and Clark 2005). In cases where carnivores are managed under a federal mandate, there are often struggles over authority with state, provincial, or Aboriginal governments, under circumstances laden with the residue of past conflicts. Contests between state and federal governments for control of natural resources are legion in the western United States, as are similar conflicts among federal, provincial, territorial, and Aboriginal governments in Canada. In both countries, wildlife management has historically been mainly the purview of state, provincial, or territorial governments. Federal control is uncommon, except in national parks or for species at risk. For wildlife under state/provincial/territorial management, conflict with the federal government often centers on issues associated with maintaining wildlife populations that are threatened or that cross the boundaries of national parks, but underlying these issues are deeper contests over jurisdiction and power.

In the United States, grizzly bears and wolves since the 1970s have been mainly managed under the federal Endangered Species Act, which expressly took authority away from the states and tacitly rendered a judgment on the competence of state-level management. This kind of federal control is often identified by local residents as involving the imposition of alien values held by an amorphous national public residing somewhere else, usually far away (Nie 2003). State managers and many of their constituents are sometimes intensely resentful of these impositions of federal control in their traditional domain. These conflicts are also saturated with tension arising from federal managers operating almost always under a protectionist mandate, in contrast to state managers operating almost always in service of "use," or "sustainable use," particularly the sustained supply of game species. Wolves in the Northern Rocky Mountains and the Western Great Lakes have recently been delisted and returned to state management, but the federal USFWS is required to continue to monitor these populations for five years after delisting, thus sustaining the perception that state governments cannot be trusted to manage carnivores properly.

In Canada the federal government has little direct role in the management of grizzly bears, wolves, or mountain lions except on federal lands (including national parks) or under joint management agreements with Aboriginal peoples and territorial governments in the Canadian north. These carnivores are not listed under the federal Species at Risk Act, although the Committee on the Status of Endangered Wildlife in Canada has designated grizzly bears as of "special concern." Provincial governments have vehemently opposed any attempt by the federal government to expand its authority into fields the provinces perceive as properly under provincial jurisdiction, such as wildlife management (Boyd 2003).

In both countries, large carnivore management has become identified with the political polarization of environmental issues. At one time, Democrats and Republicans in the United States were not necessarily dissimilar in their stands on the environment. However, over the last twenty years party identity has become almost synonymous with a pro- or anti-environmental stance, at least as assessed by metrics such as League of Conservation Voters scores for elected officials. This has come to mean that being Republican, or otherwise politically conservative, is often synonymous with antipathy toward conservation of wolves or grizzly bears, and being a Democrat typically means the opposite. In Canada the presence of additional political parties such as the New Democratic Party and the Green Party complicates the direct association of environmental views with support for a particular political party, but those who are sympathetic to large carnivore conservation often perceive themselves to be on the left-of-center politically. Thus, in both countries positions about large carnivores are increasingly reflexive rather than reflective and are often reinforced by political loyalty.

Given this litany of forces that potentially polarize and alienate stakeholders in large carnivore conservation, it is not surprising that decision makers have had trouble finding common ground. This is the fundamental challenge of governance—to find and advance common ground among diverse and conflicting perspectives and demands.

Governance and Large Carnivore Conservation

The term "governance" is used in many ways, but a good working definition for our purposes is provided by institutional scholar Oran Young (2013, 3): "At the most general level governance is a social function centered on efforts to steer or guide the actions of human groups—from small, local associations to international society—toward the achievement of desired ends and away from outcomes regarded as undesirable." In a democracy, the ends toward which governance steers should embody the common interest, and it is assumed that the best means to identify these ends is by giving all citizens an informed and equal voice in decision making (Lasswell 1971; Brunner 2002). These ideals are captured in Dahl's (1989, 1998) criteria for democratic processes: effective participation by all, equal say and equal choice in decisions, the opportunity to learn about and understand policies and their consequences, control over how decisions are made and what is decided, and full inclusion of all adults.

Governance involves authoritative decision making, but authority is not limited to formal institutions such as Congress or Parliament, and decision making takes place across a range of scales: "Governance refers, therefore, to all processes of governing, whether undertaken by a government, market, or network, whether over a family, tribe, formal or informal organization, or territory, and whether through laws, norms, power or language" (Bevir 2012, 1). For large carnivore conservation, governance can vary from a few individuals working together on a program to deter carnivores from killing livestock, to international institutions that regulate the trade of endangered species.

Scholars have developed broad conceptual frameworks to classify and study the biophysical and sociopolitical systems involved in natural resource governance. One prominent example is the Social Ecological Systems Framework developed by Elinor Ostrom and her collaborators, which distinguishes subsystems made up of resource users, resource units, resource systems, and governance systems (Ostrom 2009, 2011). These subsystems interact with each other within a social, economic, and political setting to produce outcomes for the system and related ecosystems. The Social Ecological Systems Framework recognizes three hierarchical levels of decision making: the operational tier, where individuals make applied decisions in local settings; the collective-choice tier, where repeated collective decisions are made under established rules; and the constitutional tier, where decision makers prescribe rules that specify how other decisions will be made and who may participate (Ostrom 2011).

In order to be effective in an uncertain and rapidly changing world, governance must also include mechanisms for learning from experience. Peters (2012) argues that successful governance involves selecting goals, reconciling and coordinating multiple goals, implementing decisions to advance goals, and learning through feedback and accountability. Concepts such as adaptive management (Walters 1986; Lee 1993) and adaptive governance (Brunner et al. 2005; Brunner and Lynch 2010) explicitly incorporate appraisal, learning, and adaptation into the design of decision-making processes. A good appraisal process helps decision makers and others to identify the location and causes of error when policies or programs fall short of goals. Such policy failure may arise because of flaws in the technical design of programs, but failure may also occur because of malfunctions in the decision-making processes through which programs are developed and implemented, or political factors that determine the support and resources made available, and how success or failure will be assessed and attributed (Howlett 2012). Appraisals should also consider the possibility that the goals, objectives, and framing of policies may need to be revisited (Argyris and Schön 1978).

People take part in governance for large carnivore conservation in many ways. For example, agency officials are charged with putting laws and policies into action on the ground, adapting and applying them to particular species and situations, and, in principle, seeing that the public will is carried out. Scientists, often under the auspices of a university, government, or nongovernmental organization, conduct research to obtain basic information about the species in question, or its biophysical environment, or perhaps about the sociopolitical or human context. Environmentalists, hunters' associations, recreational users, and other interest groups may offer up goals, information, and perspectives that could otherwise be overlooked or left out of official conservation efforts. Scholars of policy and management appraise decision-making processes and outcomes. This, of course, is only a partial list; there are many other roles for participants in carnivore conservation. Effective governance must integrate all of these participants and roles to make decisions that are in the common interest (e.g., Peacock et al. 2011).

THE TRADITIONAL APPROACH TO
GOVERNANCE IN CONSERVATION

For most of the twentieth century the approach known as "scientific management" dominated governance in wildlife management, including large carnivore conservation. Although this approach has been widely criticized in recent years and management is slowly evolving in new directions, the doctrine of scientific management is still a pervasive influence on the attitudes and behavior of many people involved in carnivore conservation. Consequently, it is important to understand the main principles of this approach. In the classic paradigm of scientific management, conservation decisions and activities are dictated by biological facts, and wildlife or conservation biology professionals are considered to be the most qualified to offer up these facts. The role of scientists is to discover the facts, through observation, description, hypothesis testing, mathematics, and statistics, leading to predictions. Scientists then communicate the facts and predictions to government managers, who set the rules for management and policy. Thus, under scientific management conservation becomes an activity driven largely by a body of facts that emerge from scientific discovery, and the job of government administrators is to implement relatively straightforward rules based on unambiguous facts. Science is perceived to be the legitimate and authoritative source of the key information on which decision making should be based, and scientists are expected to be largely objective and neutral. Other sources of information and other viewpoints are considered politically inferior and given much less weight in decision making or discounted entirely (see Reisman 1995–1996).

Although the scientific management approach can be very effective in dealing with technical problems, it is inadequate for complex problems involving high uncertainty, diverse conflicting goals and values, irreversibility, and uniqueness—the types of problems that planners call "wicked" (Rittel and Webber 1973; Xiang 2013). Carnivore conservation is plagued by wicked problems. Scientific management cannot handle such problems because it is based on a series of unwarranted assumptions: that conservation is a technical problem for science and law, that the independent moral stance of science is different and superior to other views, that the methods, objectivity, and ways of knowing of science are virtually unassailable, that the nature of problems, society, and decision making are relatively straightforward, and that scientific views are more relevant than other perspectives on problems (Pielke 2007).

Rather than simply imposing rules based on scientific facts, resource management needs to combine knowledge about ecological, sociocultural, and governance systems to make choices that integrate or balance conflicting values (Chen 1989; Brunner et al. 2005). As the reintroduction of wolves into Greater Yellowstone strikingly demonstrates, science-based decisions may strongly affect peoples' values. Governance must take into account the consequences of decisions for people and institutions. Of course, decisions should be rational, but to be effective and to endure in a democratic society, decisions must also be politically acceptable and morally justified (Clark 2002).

This is not to suggest that science and scientists are unimportant in governance for carnivore conservation. On the contrary, rigorous scientific knowledge is critical. But rigorous scientific knowledge is not sufficient, and science cannot resolve value disputes. Moreover, there are normative and conceptual ambiguities involved in obtaining scientific facts that must be acknowledged and taken into account; scientists are not immune from the influences of their own values in the positions they take on conservation problems.

The many difficulties with the traditional scientific management approach to governance—"facts lead to rules"—are well recognized and frequently discussed (Brunner et al. 2005). Many scientists and other participants in carnivore conservation are cognizant of these difficulties. Increasingly, participants in carnivore conservation are graduates of contemporary schools of environmental studies or environmental sciences, where the importance of interdisciplinary knowledge and the roles of diverse values in environmental problems are emphasized (Clark et al. 2011). But, as the case studies in this book show, the doctrine of scientific management is deeply embedded in the institutions of wildlife management, and elements of this doctrine repeatedly surface in the perspectives and demands of participants. For example, advocates in conservation disputes often call for "decisions based on science rather than politics." This demand may come from hunters arguing that wolves should be delisted and returned to state management, while the same demand is being made by environmentalists arguing for continued protection. Meanwhile, managers will hold up science as a defense against accusations of poor management ("we're just following the best available science"). Social ecologist William Burch describes this phenomenon succinctly: "All resource allocation decisions are matters of political struggle rather than technical facts. . . . Resource managers, when confronted with social value decisions, will seek to convert them into technical decisions" (Grumbine

1997, 46). When disputes end up in court, as they frequently do, cases deteriorate into duels of scientific experts, with judges forced to decide value-based disputes by choosing which scientific opinion and credentials they consider to be the strongest.

AN INTEGRATIVE APPROACH TO CONSERVATION

Scientific management has a long history, but it is slowly giving way to more effective approaches that integrate knowledge and methods from multiple sources and disciplines to address conservation problems. All the authors in this book use an integrative and interdisciplinary approach to governance developed in the policy sciences (Lasswell 1971; Lasswell and McDougal 1992). This approach has been called interdisciplinary problem solving, the policy-oriented or configurative approach, or the law, science, and policy approach. It is the foundation of the concept of adaptive governance (Brunner et al. 2005). Our authors use this integrative approach to investigate, evaluate, improve, and learn from large carnivore conservation in practice. At the core of the approach is a conceptual framework for organizing inquiry about the full range of variables in the ecological and sociopolitical context for conservation and for analyzing and evaluating the governance system. The approach also includes tools for examining the analyst's own standpoint, defining problems, and developing solutions.

This integrative approach has proven to be effective in helping people find common ground and achieve on-the-ground conservation successes in a variety of settings, including in large carnivore conservation (Brunner et al. 2002, 2005; Clark 2002; Rutherford et al. 2009). It helps users to escape the assumptions and limitations of scientific management and the bounded, disciplinary, and narrow perspectives of conventional bureaucratic, scientific, legal, or advocacy approaches, yet it draws on the strengths of traditional science and practical experience. It also provides a procedurally rational, stable frame of reference in different situations and across community boundaries.

The focus is on decision making (governance) as it extends across the full range of people, perspectives, values, and organizations, and across the hierarchy of power. Governance always deals with people's values and the "value processes" of people's interactions in society. The integrative approach considers the development of conservation science and legislation, the implementation of decisions in the field, and the deliberations in courts or other institutions. Similar to the Social Ecological Systems Framework discussed earlier, the integrative approach recognizes different hierarchical levels of decision making, distinguishing between ordinary decisions and higher "constitutive" or structural levels.

From the standpoint of the integrative problem solver in conservation, the main objective is to help people make wise collective choices, taking into account differing perspectives and values. This means that decision makers must prescribe and apply policy in ways that maintain community order and at the same time achieve the best possible approximation of the community's social goals: "The continuing task of governance—in any community that respects equal rights for all—is finding common ground on policies that advance the common interest" (Brunner 2002, 8–9).

CHARACTERISTICS OF THE INTEGRATIVE APPROACH

The main elements of this integrative approach to conservation problem solving and governance are (1) clarifying observational standpoint, or how problem solvers view themselves in relation to the situation; (2) choosing a focal lens or lenses through which to make observations and understand the situation; (3) performing the intellectual tasks of problem orientation (identifying, defining, and solving problems); (4) mapping the social process that problem solvers are trying to understand and influence; (5) examining and evaluating the decision-making process and assessing the potential to influence decisions; and finally (6) clarifying and securing the common interest. In the following paragraphs we expand on these elements of the integrative approach and explain how we see them in relation to the objectives of this book (for a comprehensive explanation of the integrated approach, see Lasswell 1971; Lasswell and McDougal 1992; Clark 2002).

The first element is clarifying and establishing an observational standpoint. Like the proverbial blind men and the elephant—a metaphor about different perspectives—we must ask which of several standpoints will be most useful in examining and understanding a given situation, because different standpoints affect the situation and how it is interpreted. The observer's standpoint needs to be kept as distinct as possible from the phenomenon or process under examination so that the observer can develop appropriate evaluation criteria and as clear a picture as possible. Standpoint affects all other problem-solving tasks, including how problems are defined, what goals are advanced, and what intellectual skills are brought to bear to resolve them. Consequently, clarification of observational standpoint is an essential task, both intellectually and practically. We recommend adopting a standpoint identified with the present and future community that seeks a healthy sustainable environment and is committed to genuine democracy.

This is the standpoint adopted by the editors and authors in this book. We encourage observers, analysts, and problem solvers to establish a standpoint as free as possible from parochial interests and other cultural biases and to examine themselves for both latent and overt emotional problems, neuroses, subgroup parochialism, misplaced loyalties, and distortions arising from training and affiliation with particular professions.

The second element of the integrated approach is choosing the focal lenses of inquiry. Different lenses focus attention on different aspects of a process or situation, perhaps on a species' life history, the status of the science, government responsibility, or the quality of decision making. We all look at our environs and specific issues through conceptual frames, even though we may be unaware that we are doing so. Indeed, many people believe that their own way of seeing the world is the only one and true way and that those who see the world differently are misguided, ignorant, or malevolent. In the physical sciences, for example, different lenses permit observers to bring different features or properties of the same viewed object into sharper focus or greater prominence. Geologists, soil scientists, plant ecologists, wildlife biologists, or landscape ecologists will all see different things as a consequence of the particular "window" they are looking through. When it comes to complex social and decision-making processes, differences are also to be expected as viewers look through different lenses. Scholars in the social sciences have studied how we can become more aware of our own "window on reality" by using carefully crafted conceptual frames. In this book we adopt a focus of attention on decision making, using a conceptual framework that draws on a broad range of concepts and tools from the social sciences. We encourage others to consider this very important dimension of individual and collective problem solving.

The third element of the integrative approach is formulating the problem(s). How problems are viewed and formulated, or defined, has a direct and major impact on the outcomes of decision making. Problems are discrepancies between goals (or demanded values) and achievement of those goals through decision making. In other words, problems are differences between what people want to have happen and what they perceive is actually happening. There are several tasks involved in orienting to problems: the community's goals must be identified, trends mapped and compared with goals, underlying factors that explain trends identified, likely future events forecast, discrepancies between goals and forecasts examined, and, finally, alternatives or options invented, appraised, and selected for implementation. We adopted this understanding of problem orientation across all the chapters in this book. When combined with the other tools in the integrated approach, these tasks of problem formulation provide a comprehensive, yet economical, way of addressing problems.

The fourth element of the integrated approach is mapping the social process that is the context of problems. In addition to the focal lenses of *how* observers look at pertinent data about conservation and decision making, another key consideration is *what* observers look at. We use a scheme adapted from cultural anthropology that is designed to describe any social process comprehensively. It allows a systematic accounting of those who engage in or are affected by conservation (the *participants*), the subjective dimensions that animate them (their *perspectives*), the *situations* in which they interact, the resources or *base values* on which they draw, the ways they manipulate those resources (*strategies*), the aggregate *outcomes* of the process of social interaction, including decision making, and the long-term *effects* on values and institutions. Outcomes are portrayed in terms of changes in a comprehensive set of values. Operational indices of all these variables allow for comprehensive inquiry into the social context of any problem.

Values are a particularly important focus in mapping social process. Values are preferred events, what people want, aspire to, or cherish. The approach we take in this book involves use of a set of universal value categories borrowed from ethical philosophers and other value or normative specialists:

Respect—equality, recognition, and freedom of choice
Power—influencing, making, and enforcing community decisions
Enlightenment—gathering, processing, and disseminating information and
    knowledge
Well-being—safety, health, and comfort
Wealth—production, distribution, concentration, and consumption of goods
    and services
Skill—acquisition and use of capabilities in vocations, professions, and arts
Affection—intimacy, friendship, loyalty, and positive sentiments
Rectitude—participation in forming and applying norms of responsible
    conduct

The fifth element of the integrative approach is examining and evaluating the decision-making process. In most conventional conceptions of conservation, the term "decision" refers to a government official or judge applying rules to a particular problem or dispute in an organized situation (e.g., from a government office or courtroom). But the process of making and implementing choices involves many more activities or functions than just authorities rendering judgments from their offices. All these other activities are also part of decision making.

From the integrative standpoint, the decision process has seven functions or characteristic activities (Lasswell 1971; Clark 2002):

1. Intelligence—Initial identification of a problem; gathering, processing, and disseminating information about social choices.
2. Promotion—Investigation and debate about the nature and status of the problem and possible solutions; the process by which awareness of a discrepancy between a desired state and an existing state leads to demands for some type of intervention or regulation.
3. Prescription—Choosing a plan to address the problem, which often involves setting new rules, laws, or policies. A legislature or some other organized body may accomplish this, but often, especially in wildlife management, prescription is accomplished by government officials or takes place informally through "custom."
4. Invocation—Invoking the rules or policies in specific cases; the provisional characterization of a certain action as consistent or inconsistent with a prescription, law, or norm that has been established, often accompanied by the demand that an appropriate community institution take action.
5. Application—Applying the rules through administrative activities (conventionally called dispute resolution); organizing the facts of a dispute, specifying which norms apply and fashioning a binding resolution or adjudication of the conflict. When this takes place in a courtroom, it is called a judgment, but it also occurs in informal, unorganized situations. Invocation and application are often referred to together as implementation.
6. Appraisal—Evaluating the individual and aggregate performance of all decision functions in terms of general standards and particular community requirements.
7. Termination—Terminating or modifying the rules if they fail or no longer apply and ending existing norms and social arrangements based on them; developing transitional regimes and, where appropriate or necessary, designing compensation programs for those who have made good-faith investments expecting the old regime to continue.

The decision functions do not necessarily occur in a linear sequence as they are listed here, but may occur simultaneously and interactively in actual cases.

Any decision process and each individual function can be analyzed, understood, and managed by participants to meet high standards. The case studies in this book appraise decision processes using well-grounded standards of performance. Taken together, these standards amount to sound democratic principles for governance. For example, the intelligence activity, which is about recognizing and characterizing a problem, demands that the sources of information be dependable, that the data considered to be relevant are comprehensive yet selective, and that participants exhibit both creativity and openness in gathering, processing, and disseminating information. During promotion, which is about open debate, the process should be rational, integrative, and comprehensive as various groups advocate for their interests or preferred policies. New policies, rules, or guidelines should be effective, rational, comprehensive in resolving problems, and fair to all affected people. In addition to standards specific to each decision function, there are overall standards for effective decision making that apply to all functions, including honesty, economy, technical efficiency, involvement of loyal and skilled personnel, complementary outcomes, effective impacts, flexibility, and realism in adjusting to change, deliberateness, and responsibility.

The appendix at the end of this chapter lists the decision-making functions and their standards for effectiveness as sixty-six questions, which we used in analyzing the cases and issues in this book. Using these widely recognized, well-supported, and practical standards gave us a consistent, comprehensive, and systematic basis for evaluating large carnivore conservation efforts and drawing lessons.

The sixth characteristic of the integrative approach is clarifying and securing the common interest. All communities struggle to clarify and secure the common interest, a concept that has been debated in theory and negotiated in practice in societies since the dawn of time. The conflict and contentiousness that we see in large carnivore conservation illustrates how difficult this process can be (e.g., polar bears, Clark et al. 2008 and Peacock et al. 2011; grizzly bears, Cromley 2000), yet conserving large carnivores is clearly in the common interest.

An interest is a demand for values made by individuals or groups, a demand that is supported by the expectation that the demand will be advantageous to them (Lasswell and McDougal 1992). Special interests benefit the few at the expense of the broader community, whereas common interests are those that benefit the community as a whole (note, however, that it may be in the interest of the majority to protect minority rights). Special interests and common interests are typically in conflict in any community. A common interest is at "stake whenever people who act on their perceived interests also interact enough to form a community around an issue" (Brunner 2002,

12). Large carnivore conservation is one high-profile example of the tension between special interests and the common interest.

There is no single, simple, or objective way to determine what is in the common interest; it has to be worked out in practice. However, there are three partial tests of the common interest that can guide us—the procedural, substantive, and pragmatic tests (Brunner 2002; Steelman and DuMond 2009):

The *procedural test* asks if inclusive and responsible participation is involved in the decision-making process, if those interests not directly included are represented or reflected in the outcome, and if those making decisions are held accountable.

The *substantive test* asks whether the expectations of participants about what will be accomplished are reasonable, whether the decision-making process considers all valid and appropriate concerns, whether it ostensibly solves the problem, whether the outcomes are supported by participants, and whether those outcomes are compatible with broad societal goals (e.g., democracy, fair play, timeliness).

The *pragmatic (or practical) test* asks if the decision outcomes work in practice, if they are pragmatic, if they uphold the reasonable expectations of participants, and if decisions are adapted to deal with changing circumstances over time.

These tests are applied in each case study in this book to assess the performance of governance in serving the common interest.

Conclusion

Much of the conflict over large carnivore conservation, and even the intense emotion and passion, arises from failure to understand conservation as a problem of governance, laden with emotive symbolic constructs, conflicting perspectives and values, and highly politicized arenas of interaction. Inherent differences of perspective lead to conflict because people fail to adopt and use integrative tools to improve decision making and find common ground. We believe that the common interest goals of improved carnivore conservation and sustainable coexistence with these animals will continue to elude us unless we deploy the integrative approach just described—prioritizing common interests, orienting to problems, mapping the social and decision-making context, and applying multiple methods. This book elaborates and uses this integrative approach in case studies of large carnivore conservation from Yukon to Arizona involving mountain lions, grizzly bears, and wolves in diverse situations. These studies offer up considerable insight, individually and collectively, into how communities go about making decisions in conservation. They reveal lessons that show us how we can do much better in conservation if we so choose. For people who demand an ideology of strict compliance with "scientific facts" or with established authority, the integrative approach that we use may seem threatening, destructive, misguided, or perhaps just wrong. We hope this book will convince them of its benefits.

The integrative approach to conservation, with a sharp focus on decision making, is a powerful methodology that is well-suited to the complexity of conservation problems involving large carnivores or other issues. In situations where political and power factors are at play, where the environment (both social and biophysical) is relatively unstable and unpredictable, where consensus is low on many critical social goals, and where judgments about consequences differ, the perspectives and tools of integrated problem solving can be extremely helpful. These, or some equivalent methods of interdisciplinary adaptive governance, are essential to achieve practical gains. Our hope is that the perspectives, concepts, and tools described in this book will help readers to realize practical improvements in large carnivore conservation—a goal that we all share.

Appendix
Criteria for Evaluating Decision Processes

The following list of questions can be asked in order to evaluate how well each function of a decision-making process and the process as a whole meet accepted standards (adapted from Lasswell 1971).

INTELLIGENCE
Dependability
1. Is the information gathered for decision making accurate? Is it reliable?
2. Are estimates given of the credibility of the information and the uncertainty involved?
3. Are the best sources of information used? What is the quality of the expertise?
4. Can the sources be relied on to supply realistic statements of information? What are their loyalties and identifications?
5. Are appropriate methods and competence mobilized and used when needed to gather information?

6. Is the information accurately transmitted from the sources to the recipients?
7. Do recipients acknowledge the credibility and realism of the sources of information?

Comprehensiveness
8. Is information obtained for all relevant components of the problem (goals, trends, conditions, etc.) and the context of the problem (social process, decision process)?
9. Is information obtained from all appropriate sources and all affected people (traditional science, traditional ecological knowledge, local knowledge, social, economic, etc.)?
10. Are estimates made of the benefits, costs, and risks of each of the alternatives?

Selectivity
11. Are intelligence activities focused on the key aspects of the problem, guided by adequate problem orientation?
12. What proportion of outputs are related to problems perceived to be important by insiders in comparison with problems perceived to be important by other people who are affected or interested?

Creativity
13. Are realistic new objectives and strategies adopted to gather intelligence?
14. Are new objectives and strategies compared with older objectives and strategies? Are they better?

Openness
15. To whom is intelligence communicated?
16. Is anyone excluded from access to intelligence, and is the exclusion justified?
17. Is there general cooperation in obtaining intelligence?

PROMOTION
Rationality
18. Are all reasonable alternatives brought to the attention of decision makers?

19. Are proposed alternatives assessed and justified in terms of value indulgences or deprivations?

Integrativeness

20. Are proposed alternatives broadly supported?
21. Is the debate about alternatives bipolar (yes-no, either-or)? Can it be made more nuanced?
22. Is there coercion in the debate about alternatives? Is there a way to eliminate it?

Comprehensiveness

23. Do the alternatives proposed reflect the full range of community interests?
24. Are the views of the weak or neglected incorporated through direct participation or other means of solicitation?
25. Is thorough debate of all views encouraged before proposals are adopted or entrenched?

PRESCRIPTION

Stability of Expectation (Effective)

26. Do those affected or interested generally consider decisions to be lawful and enforceable?
27. Are decisions for which there is general support made and implemented promptly?
28. Are decisions for which there will not be continuing support avoided?
29. Before proposed decisions are made, are they brought to the attention of groups beyond those most immediately interested?
30. Do prescriptions specify *goals*, *norms* (rules of conduct), *contingencies* (the circumstances in which the norms apply), *sanctions* (to deter noncompliance and encourage compliance), and *assets* (resources for implementation)?

Rationality

31. Do decisions further the common interest rather than special interests, and do they balance inclusive interests (those that affect everyone) with exclusive interests (those that affect only some)?

Comprehensiveness

32. Are decisions appropriate for all potential situations (high, middle, and low crisis)?
33. Do sanctions exist to:
    a. deter nonconformity with prescriptions?
    b. resist acts of noncompliance?
    c. rehabilitate persons and restore damaged assets?
    d. prevent future recurrences?
    e. correct motivations and deficiencies in education that stimulate nonconformity?
    f. reconstruct institutions to encourage conformity?

INVOCATION
Timeliness

34. Are justifiable complaints by the less powerful encouraged, and are actions taken in response?
35. Are all the decisions carried out and done so consistently?

Rationality

36. Are the decisions applied fairly whenever the specified contingencies exist?
37. Are expedient processes applied where circumstances call for quick action?
38. Are more deliberative processes applied where circumstances do not require quick action?

Nonprovocativeness

39. Do initiatives impose the minimum deprivation necessary to be effective?

APPLICATION
Rationality and Realism

40. Is performance supervised and reviewed and nonconformity with prescriptions rectified?

Uniformity

41. Is there any discrimination in application?
42. Are third-party participants mobilized in order to neutralize special interests?

TERMINATION

Timeliness

43. Are obsolete programs promptly terminated when they are no longer justified?

Dependability and Comprehensiveness

44. Are the facts used in judging whether to terminate a policy dependable and comprehensive?
45. Are termination policies dependable and comprehensive?

Balance

46. Is a balance maintained between expediting or inhibiting change?

Ameliorative

47. Are valid losses arising from termination compensated?
48. Is compensation denied in circumstances where it will not be effective because strong coercive opposition to policy is likely to continue?
49. Are windfall advantages arising from termination expropriated?

APPRAISAL

Dependability and Rationality

50. Are appraisal policies and criteria generally agreed on?
51. Are the data used in evaluation dependable?
52. Are the explanatory analyses relevant and explicit?
53. Is there explicit imputation of formal responsibility for successes and failures?

Comprehensiveness and Selectivity

54. Is appraisal comprehensive and yet selective (does it use the integrated framework for assessment)?

Independence

55. Are the appraisers protected from threats or inducements?
56. Are internal appraisers supplemented by external appraisal?

Continuity

57. Is appraisal continuous rather than intermittent?

OVERALL CRITERIA

Honesty

58. Are official personnel fiscally honest?

Reputation for Honesty

59. Does the public believe that program officials are fiscally honest?

Money Economy

60. Is the process efficient in the use of money (benefit/cost, cost effectiveness)?

Technical Efficiency

61. Is the process efficient in the use of nonmonetary resources?

Loyalty and Skill of Official Personnel

62. Are official personnel committed to the overriding goals of public policy (human dignity)?

Complementarity and Effectiveness of Impact (in Decision and Social Process)

63. Does the program contribute to the mobilization of immediate and continuing political support for the political system as a whole and to the overriding goals of public policy (human dignity)?

Differentiated Structures

64. Are there differentiated structures in place for each function of decision making (e.g., an independent body responsible for appraisal)?

Flexibility and Realism in Adjustment to Change

65. How well does the system adjust to changes (e.g., shift from crisis to noncrisis and vice versa)?

Deliberateness and Responsibility

66. Are decisions made in a deliberative and responsible manner rather than in an erratic and impulsive manner?

NOTES

The perspectives here do not represent the official views of the US Geological Survey or the US Government, with which D. Mattson was affiliated at the time he contributed to this chapter.

We want to thank many people, too numerous to mention all by name, for the ideas and their description in this chapter. We are members of a growing worldwide movement to upgrade natural resource conservation, its science, management, law, and policy. We have benefitted from the support of our employing organizations and many others. Several anonymous readers provided critical reviews.

REFERENCES

Alberta Sustainable Resource Development and Alberta Conservation Association. 2010. *Status of the Grizzly Bear* (Ursus arctos) *in Alberta: Update 2010.* Wildlife Status Report 37, Update 2010. Edmonton: Alberta Sustainable Resource Development.

Argyris, C., and D. Schön. 1978. *Organizational Learning: A Theory of Action Perspective.* Reading, MA: Addison Wesley.

Bevir, M. 2012. *Governance: A Very Short Introduction.* Oxford: Oxford University Press.

Boyd, D. 2003. *Unnatural Law: Rethinking Canadian Environmental Law and Policy.* Vancouver: University of British Columbia Press.

Brunner, R. D. 2002. "Problems of Governance." In *Finding Common Ground: Governance and Natural Resources in the American West*, edited by R. D. Brunner, C. H. Colburn, C. M. Cromley, R. A. Klein, and E. A. Olson, 1–47. New Haven, CT: Yale University Press.

Brunner, R. D., C. H. Colburn, C. M. Cromley, R. A. Klein, and E. A. Olson, eds. 2002. *Finding Common Ground: Governance and Natural Resources in the American West.* New Haven, CT: Yale University Press.

Brunner, R. D., and A. H. Lynch. 2010. *Adaptive Governance and Climate Change.* Boston: American Meteorological Society.

Brunner, R. D., T. A. Steelman, L. Coe-Juell, C. M. Cromley, C. M. Edwards, and D. W. Tucker, eds. 2005. *Adaptive Governance: Integrating Science, Policy, and Decision-Making.* New York: Columbia University Press.

Casey, D., and T. W. Clark. 1996. *Tales of the Wolf: Fifty-One Stories of Wolf Encounters in the Wild.* Moose, WY: Homestead Publishing.

Chen, L. 1989. *An Introduction to Contemporary International Law: A Policy-Oriented Perspective.* New Haven, CT: Yale University Press.

Clark, D., D. S. Lee, M. M. R. Freeman, and S. G. Clark. 2008. "Polar Bear Conservation in Canada: Defining the Policy Problems." *Arctic* 61:347–60.

Clark, S. G. 2002. *The Policy Process: A Practical Guide for Natural Resource Professionals.* New Haven, CT: Yale University Press.

Clark, S. G., M. B. Rutherford, M. R. Auer, D. N. Cherney, R. L. Wallace, D. J. Mattson, D. A. Clark, et al. 2011. "College and University Environmental Programs as a

Policy Problem (Part 1): Integrating Knowledge, Education, and Action for a Better World?" *Environmental Management* 47:701–15.

Clark, T. W., and D. Casey. 1993. *Tales of the Grizzly*. Moose, WY: Homestead Publishing.

Cromley, C. M. 2000. "The Killing of Grizzly Bear 209: Identifying Norms for Grizzly Bear Management." In *Foundations of Natural Resources Policy and Management*, edited by S. G. Clark, A. R. Willard, and C. M. Cromley, 173–220. New Haven, CT: Yale University Press.

Dahl, R. A. 1989. *Democracy and Its Critics*. New Haven, CT: Yale University Press.

———. 1998. *On Democracy*. New Haven, CT: Yale University Press.

Graves, W. N. 2007. *Wolves in Russia: Anxiety through the Ages*. Calgary: Detselig Enterprises.

Grumbine, R. E. 1997. "Reflections on 'What Is Ecosystem Management?'" *Conservation Biology* 11:41–47.

Howlett, M. 2012. "The Lessons of Failure: Learning and Blame Avoidance in Public Policy-Making." *International Political Science Review* 33:539–55.

Kellert, S. R. 1996. *The Value of Life: Biological Diversity and Human Society*. Washington, DC: Island Press.

Kellert, S. R., M. Black, C. R. Rush, and A. J. Bath. 1996. "Human Culture and Large Carnivore Conservation in North America." *Conservation Biology* 10:977–90.

Laliberte, A. S., and W. J. Ripple. 2004. "Range Contractions of North American Carnivores and Ungulates." *BioScience* 54:123–38.

Lasswell, H. D. 1936. *Politics: Who Gets What, When, How.* New York: McGraw-Hill.

———. 1971. *A Pre-View of Policy Sciences*. New York: American Elsevier.

Lasswell, H. D., and M. S. McDougal. 1992. *Jurisprudence for a Free Society: Studies in Law, Science, and Policy*. New Haven, CT: New Haven Press.

Lee, K. N. 1993. *Compass and Gyroscope: Integrating Science and Politics for the Environment*. Washington, DC: Island Press.

Manfredo, M. J., and T. L. Teel. 2008. "Integrating Concepts: Demonstration of a Multilevel Model for Exploring the Rise of Mutualism Value Orientations in Post-Industrial Society." In *Who Cares About Wildlife? Social Science Concepts for Exploring Human–Wildlife Relationships and Conservation Issues*, edited by M. J. Manfredo, 191–217. New York: Springer.

Mattson, D. J. 2004. "Living with Fierce Creatures? An Overview and Models of Mammalian Carnivore Conservation." In *People and Predators: From Conflict to Coexistence*, edited by N. Fascione, A. Delach, and M. E. Smith, 151–76. Washington, DC: Island Press.

Mattson, D. J., K. L. Byrd, M. B. Rutherford, S. R. Brown, and T. W. Clark. 2006. "Finding Common Ground in Large Carnivore Conservation: Mapping Contending Perspectives." *Environmental Science and Policy* 9:392–405.

Mattson, D. J., and E. J. Ruther. 2012. "Explaining Puma-Related Behaviors and Behavioral Intentions among Northern Arizona Residents." *Human Dimensions of Wildlife* 17:91–111.

Montana Fish, Wildlife and Parks. 2011. *2011 Montana Wolf Hunting Season Report.* Available from http://fwp.mt.gov/fishAndWildlife/management/wolf/, accessed February 12, 2013.

Nie, M. A. 2003. *Beyond Wolves: The Politics of Wolf Recovery and Management.* Minneapolis: University of Minnesota Press.

Ostrom, E. 2009. "A General Framework for Analyzing Sustainability of Social-Ecological Systems." *Science* 325:419–22.

———. 2011. "Background on the Institutional Analysis and Development Framework." *Policy Studies Journal* 39:7–27.

Peacock, E., A. E. Derocher, G. E. Thierman, and I. Stirling. 2011. "Conservation and Management of Canada's Polar Bears (*Ursus maritimus*) in a Changing Arctic." *Canadian Journal of Zoology* 89:371–85.

Peters, G. B. 2012. "Governance as Political Theory." In *The Oxford Handbook of Governance*, edited by D. Levi-Faur, 19–32. New York: Oxford University Press.

Pielke, R. A., Jr. 2007. *The Honest Broker: Making Sense of Science in Policy and Politics.* Cambridge: Cambridge University Press.

Reisman, W. N. 1995–1996. "A Jurisprudence from the Perspective of the 'Political Superior.'" *Kentucky Law Review* 23:604–28.

Rittel, H. W. J., and M. M. Webber. 1973. "Dilemmas in a General Theory of Planning." *Policy Sciences* 4:155–69.

Rutherford, M. B., and T. W. Clark. 2005. "Coexisting with Large Carnivores: Lessons from Greater Yellowstone." In *Coexisting with Large Carnivores: Lessons from Greater Yellowstone*, edited by T. W. Clark, M. B. Rutherford, and D. Casey, 254–70. Washington, DC: Island Press.

Rutherford, M. B., M. L. Gibeau, S. G. Clark, and E. C. Chamberlain. 2009. "Interdisciplinary Problem Solving Workshops for Grizzly Bear Conservation in Banff National Park, Canada." *Policy Sciences* 42:163–88.

Steelman, T. A., and M. E. DuMond. 2009. "Serving the Common Interest in US Forest Policy: A Case Study of the Healthy Forests Restoration Act." *Environmental Management* 43:396–410.

Walters, C. 1986. *Adaptive Management of Renewable Resources.* New York: MacMillan.

Young, O. R. 2013. *On Environmental Governance: Sustainability, Efficiency, and Equity.* Boulder, CO: Paradigm Press.

Xiang, W. 2013. "Working with Wicked Problems in Socio-Ecological Systems: Awareness, Acceptance, and Adaptation." *Landscape and Urban Planning* 110:1–4.

2

State-Level Management of a Common Charismatic Predator Mountain Lions in the West

DAVID J. MATTSON

Introduction

This chapter on mountain lion (*Puma concolor*) management in the western United States focuses on a central theme of this book: the capacity of participants and institutions to foster respectful, dignified, and civil decision-making processes focused on common interest outcomes. Unlike other case studies in this volume that feature on-the-ground innovations, mountain lion management highlights the extent to which the models of governance employed by participants affect civility. This case also brings into sharp relief the symbolic rather than material stakes of most of the people involved. Finally, mountain lion management is yet another example where rapid change and diversification in society's demands have outstripped adaptive responses in traditional management institutions, leading to high levels of anxiety and conflict.

I draw on trends and conditions associated with mountain lion management in the West, featuring Arizona, to explain the conflict, especially between animal welfare advocates and hunters. I focus on the roles played by current power and wealth arrangements and on the governance models consciously or otherwise adopted by traditional management agencies. State-level wildlife agencies are among the remaining bastions of bureaucratic, customer-oriented, scientific management (Mattson and Clark 2010b). To some, this may seem like a good thing. However, I examine here how this model of management works together with current agency cultures against the cultivation of common ground among people with diverse conflicted interests (see chapters 8, 9, and 10). In particular, I examine how current modes of governance have impeded otherwise laudable efforts to engage previously marginalized stakeholders in building durable policies for managing mountain lions. In the first half of this chapter I review the history of mountain lion management, and in the second half I attempt to explain problematic dynamics and draw lessons that could be used to improve decision-making and policy-making processes.

The method for my analysis was framed by concepts of the policy sciences (Lasswell 1971) and informed by other sociological and psychological theories, personal experience, and a review of all materials that I could obtain germane to the management of mountain lions in the West. My analysis presupposes that the aspiration of virtually all humans is a life of dignity (Mattson and Clark 2011) and that within Western cultures such as the United States the best means of fostering widespread dignity is through the sustenance of civil and equitable governance (Shils 1997; Dahl 2006). These are the goal and the core standards by which I judged outcomes and effects of mountain lion management. The basic structure of my analysis is problem orientation, which entails an examination of trends, current conditions, and projections in the realms of biophysical, social, and decision-making phenomena, leading to the formulation of alternatives (Clark 2002, 87). I also mapped social and decision processes, with explicit reference to standards of decision process (Clark 2002, 60). Social process encompasses participants, their perspectives, their strategies, and the arenas and situations within which they characteristically operate (Clark 2002, 33). My view of participant perspectives was heavily influenced by a schematic of nature-views by Stephen Kellert (1996) and by precepts of existential psychology (Yalom 1980). The supporting analysis for this chapter was reported earlier by Mattson and Clark (2010b).

My perspective has been shaped by my life experience as well as theory and research germane to human psychological and social dynamics. I was formally trained as an ecologist, and my research has focused primarily on large carnivores—grizzly bears (*Ursus arctos horribilis*) in the Rocky Mountains and mountain lions on the Colorado Plateau. But this experience led me to the interface of science and policy and confronted me with troubling conflict and the politicization of science (Mattson and Craighead 1994; Wilkinson 1998, 65). My efforts to understand people led me to new domains of inquiry and thought. Here I offer a provisional understanding of human-centered dynamics in mountain lion management, acknowledging that there are numerous views of this business and that I, like everyone else, am captive to my subjectivity.

Context and Problem Definition

I start by briefly recapitulating trends in mountain lion range and populations, interactions of lions with people, and key features of historical mountain lion management. This sets the stage for an overview of changes in social process and management that occurred during the 1990s to 2000s, featuring events in Arizona. Taken together, this information provides a context for understanding the problems of governance that typify mountain lion management in the American West.

BRIEF HISTORY OF MANAGEMENT

Mountain lions are one of the most widely distributed mammals in the world, once ranging through nearly all of South America and North America south of the boreal forests (Hornocker and Negri 2010, vii). With the exception of southern Florida, North American mountain lions were extirpated from the eastern half of their range by the 1950s as a result of concerted eradication efforts. Mountain lions killed livestock, but perhaps more importantly, they symbolized untamed, savage, and threatening nature, thus provoking eradication as much out of principle as pragmatism (Gill 2010). Given the diligence with which lions were persecuted, it is amazing that they survived in as many places as they did, at the same time that grizzly bears and wolves (*Canis lupus*) were virtually eradicated from the United States (Laliberte and Ripple 2004). In Arizona and neighboring states, mountain lions probably occupy much the same range as they did prior to the arrival of Europeans. This resilience to persecution is a predictable consequence of the secretive nature of mountain lions, which tend to be active at night and favor rugged brushy or forested terrain (Murphy and Ruth 2010). Mountain lions are also somewhat unique among large predators in having made what appears to be a comeback in recent decades. Within core western range, circumstantial evidence, based largely on sightings, harvest, and depredation records, suggests that their populations increased between the 1970s and 1990s (Anderson et al. 2010). More dramatically, there are numerous records of mountain lions in the eastern United States, some reliable and some not, in places where they have not been seen for decades or even centuries, which suggests a wave of recolonization since the 1980s (Anderson et al. 2010; Beier 2010). Mountain lion range has also expanded northward from British Columbia, even into the southern Yukon Territory. Much of this comeback can be attributed to the classification and management of mountain lions in most states as game animals rather than varmints, together with the termination of intensive eradication efforts (Anderson et al. 2010; Gill 2010).

These apparent increases in populations and ranges have been accompanied by increases in problematic encounters between lions and people, especially since the 1970s, including some human fatalities. Attacks on people increased from an average of 0.5 per year during the 1960s to around 4.0 to 7.0 per year during the 1990s and 2000s (Mattson, Sweanor, and Logan 2011). Even though no more than twenty-nine people have been killed in all of Canada and the United States since the 1890s, the recent increase in attacks has gotten the attention of wildlife management agencies and the general public (e.g., Baron 2004). Mountain lion attacks are often sensational and well covered by the news media. At the same time, the public has increasingly litigated any harm attributable to wildlife, and because the courts have established that states own most wildlife, the state wildlife management agencies have typically been the target of litigation (Mangus 1991; Parker 1995). Given the twin threats of litigation and media coverage to the resources and legitimacy of wildlife management agencies, the easiest way for agencies to deal with lion threats to human safety has been to kill threatening animals (Perry and DeVos 2005; Gill 2010). From an agency perspective, delaying action clearly entailed greater potential costs compared to acting proactively.

The tendency to resolve human-lion conflicts by killing lions was consistent with a marked ambivalence about mountain lions among most wildlife management agencies in the West (Mattson and Clark 2010b). At the same time that many mountain lion populations apparently increased and spread, mule deer (*Odocoileus hemionus*) populations throughout most of the West were declining (Gill 1999; Mackie et al. 2003). Deer, the preferred prey of mountain lions (Murphy and Ruth 2010), are, through the sale of hunting licenses, a major source of revenue for most wildlife management agencies. There is a certain straightforward logic to multiplying the number of deer killed per year by the estimated number of mountain lions and, from that, deducing the number of deer that are thus not available for hunter harvest. Even though the best synopses of relevant research suggest that mountain lions only rarely limit mule deer populations, especially compared to weather and habitat (Ruth and Murphy 2010), simple math seems to have convinced many deer hunters and wildlife management agency commissioners that killing lions would lead to more harvestable deer (Brown 1984; Shaw 1994).

The plight of bighorn sheep (*Ovis canadensis*) has likely compounded managers' ambivalence about mountain lions. Bighorn sheep are typically another money maker for wildlife management agencies[1] and symbolically potent for many hunters (Mattson and Chambers 2009). Several vulnerable bighorn sheep populations in the desert Southwest went extinct at the same time that mountain lion populations apparently increased (Berger 1990). Agencies were also trying with varied success to restore sheep to areas where they had previously been extirpated (Rominger et al. 2004; McKinney et al. 2006). Mountain lions entered into this picture as known sheep killers and as the implicated agents of decline for several sheep populations in or near the Mohave Desert (Hayes et al. 2000; Holl, Bleich, and Torres 2004). Concerns about bighorn sheep plausibly further diminished sympathy for lions among managers, who strongly identified with the rewards (both material and symbolic) of producing large numbers of harvestable deer and bighorn sheep (Brown 1984; Shaw 1994; Baron 2004; Gill 2010).

Given this context, the historic approaches to managing mountain lions are relatively easy to understand and explain. Because the lion harvest in most states has amounted to no more than a few hundred animals (Hornocker and Negri 2010, 252), management agencies typically have had little monetary stake in the harvest and little incentive to invest the substantial sums needed to obtain reliable estimates of population sizes and trends, given the difficulties of monitoring this cryptic, low-density species (Anderson et al. 2010). For the purposes of agencies that are very likely more concerned about harvestable surpluses of deer, the plight of bighorn sheep populations, and threats of litigation, imprecise indices of population trend and size have apparently sufficed for managing mountain lions. Similarly, there was likely little or no incentive to tolerate lions perceived to pose a threat of any kind, whether to livestock or people. Quick resolution of conflict situations by killing lions has been the evident norm of lion management during much of the twentieth century in places like Arizona (Gill 2010).

CHANGE DURING THE 1990S AND 2000S

Beginning in the 1980s and accelerating during the 1990s, the context of mountain lion management changed in the West (Mattson and Clark 2010b). Previously unrecognized perspectives were voiced and empowered, particularly those of the animal welfare and environmental movements. This sea change was dramatically signaled by a successful referendum in California banning sport hunting of mountain lions, succeeded by other ballot initiatives banning the use of hounds (e.g., in Washington State, Negri and Quigley 2010). Banning hounds can reduce overall lion harvests because hounds are often important to successfully tracking these otherwise elusive animals (Zornes, Barber, and Wakeling 2006). Increased public activism took place in the context of an overall greater scrutiny of wildlife management policy by the public, accompanied by greater reliance of special interest groups on instruments such as ballot initiatives to intervene in wildlife-related policy processes. The success of many ballot initiatives created anxiety among wildlife management professionals, who self-evidently experienced diminished control and prestige and probably feared more of the same (Beck 1998; DeVos, Shroufe, and Supplee 1998). Hunters, the traditional clients of wildlife management, evinced similar angst about the increased power of groups with other interests (Mattson and Clark 2010b).

In Arizona several incidents substantially perturbed the mountain lion management arena. During the mid-1990s animal rights groups such as the Fund for Animals and the Animal Defense League began critiquing mountain lion management policies set by the Arizona Game and Fish Department (AGFD; Schubert 2002). Until 2004 Arizona had some of the most liberal regulations for "taking" lions, second only to Texas, which still managed lions as varmints. Lion hunting in Arizona was allowed year round, with no limits on age, gender, or reproductive status of the lion, with each licensed hunter annually allowed one lion kill, and with no area-specific limits on kills (e.g., AGFD 2002). In some game management units hunters were given incentives to kill additional lions (e.g., AGFD 2006). The lengthy written critiques submitted to AGFD by the Fund for Animals were well informed by both the scientific literature and AGFD's own data (Schubert 2004), which would have made these documents hard to discount. Although these animal welfare organizations "opposed sport hunting of any kind," they were also more pragmatically advocating, among other measures, establishment of a limited hunting season, physical inspection of lion kills by AGFD personnel (rather than by hunters), prohibition of hounds as a hunting aid, and increased public involvement in setting policies (Schubert 2002).

Between 2001 and 2004 issues related to mountain lion management in Arizona came to a head, catalyzed by responses of AGFD to several incidents. During 2000 and 2001 on Mt. Elden (near Flagstaff) and during 2004 in Sabino Canyon (near Tucson), mountain lions were seen exhibiting what was considered to be threatening behavior toward hikers in popular recreation areas. In both instances AGFD (the responsible wildlife management agency), in coordination with the US Forest Service (the responsible land management agency), set out to track down and remove the threatening animals (Perry and DeVos 2005). To the apparent surprise of both agencies, these actions unleashed a maelstrom of media coverage and heated public exchanges between those who opposed killing or removing the lions and those who supported the agencies' measures. A review of newspaper articles from the Flagstaff *Daily Sun* and the Tucson *Daily Star* showed a huge spike in references to mountain lions during these episodes (Mattson and Clark 2012). Perhaps more importantly for policy, the topical focus of the articles shifted during these spikes from a normal preoccupation with biology and routine reporting of more mundane encounters to a pointed critique of AGFD policies and advocacy of nonlethal approaches to resolving human-mountain lion conflicts (Mattson and Clark 2012). Among the more important voices in this discourse was Governor Janet Napolitano, who publicly expressed dissatisfaction with the performance of AGFD and even went so far as to publicly entertain the idea of eliminating the commission system in order to make the agency more responsive to elected officials and the public (Perry and DeVos 2005). Although AGFD had its defenders, the agency was squarely in the bull's-eye of public discontent, articulated as dissatisfaction with current policies as well as questioning of agency competence and bias.

THE AGFD RESPONSE AND OTHER OUTCOMES

Mt. Elden and Sabino Canyon, along with other less publicized incidents, placed AGFD in a crucible. Historical norms were apparently not working well as far as agency interests were concerned. The agency's professionalism had been called into question in highly publicized ways. The scientific basis for its management had been roundly criticized, and its policies were shown to be outside the norms of other western states. Moreover, its policies allowed for hunting practices (i.e., killing kittens and dams with kittens) that overtly offended the sensibilities of many Arizona citizens. Several interest groups, most notably animal welfare activists, increasingly expressed their frustration about being marginalized in the decision-making process and about the deference given to the special interests of hunters. Perhaps most attention getting was the dissatisfaction of the governor and some state legislators. There were good reasons for AGFD to look at options to restore the agency's credibility with key stakeholders and pacify its critics.

AGFD undertook several tasks to ameliorate this crisis. First and perhaps most important, the agency held three public workshops to generate input for a protocol to respond to situations where mountain lions were thought to pose a threat to human safety (Perry and DeVos 2005). These workshops were independently facilitated and entailed substantive input from a spectrum of interests. The outcomes were two reports (AGFD 2004a, 2004b) and an action plan (AGFD 2005) that responded explicitly to verbal and written input by the public. The intent was to design a protocol that stabilized expectations and garnered public support. Second, the agency revised its policies for hunting mountain lions. Kittens and dams with kittens were protected, the general season was limited to September 1–May 31, "bag" limits were set for some game management units, and more stringent reporting requirements were established (e.g., AGFD 2010). However, in units where AGFD was either trying to establish or reverse declines in bighorn sheep populations the season was still year round, the harvest objectives were to reduce lion numbers substantially, and bag limits were one lion per hunter per day until annual harvest objectives were reached (e.g., AGFD 2000, 2006). AGFD also contracted with a public relations firm to educate the public about safety around mountain lions in targeted high-risk areas (AGFD 2005). These changes served to bring Arizona's policies within the norms of other states (AGFD 2010), which was also important to AGFD interests.

Using the standards of sustainable and ameliorative decision processes, there were several aspects of the agency's response that were problematic, despite a positive overall direction. Each of the measures taken built on accepted norms of state-level wildlife management agency culture and practice. The move to educate fit a well-developed tradition of informing the public about wildlife and the rationale for current management practices (e.g., Shroufe 1988; Perry and DeVos 2005; AGFD 2005). The onetime public engagement to solicit input for a protocol to deal with threatening lions was consistent with a tradition of soliciting feedback about prospective policies and hunting regulations through annual public meetings and open houses. The tightening of lion hunting regulations also was comparatively natural given that legitimacy is often reckoned in terms of broader management norms. However, it could be argued that these measures, which did not stray far from traditional norms, did not go far enough.

Still lacking was an ongoing means of substantively engaging the full spectrum of interests in the development of policy. Observers of natural resource management have remarked that important differences exist between superficial engagement with stakeholders through education, onetime workshops, and formal meetings and comment periods, and more in-depth engagement through ongoing consultation and collaboration (Pimbert and Pretty 1995; Decker et al. 1996; Decker and Chase 1997). The report on the protocol workshop held in Tucson (AGFD 2004b) contained some anomalies that pointed to a deeper-rooted and persistent problem, primarily in understandings of governance. For example, the comment that "many participants stressed that AGFD needed to expand its constituency from a primary focus on hunters and anglers to the larger population that has an interest in wildlife and the outdoors" was placed in a section pertaining to public education. Similarly, a paragraph about the legitimacy of the commission system, including the comment that "the non-consumptive community feels like they have no voice on the Commission" was placed in a section devoted primarily to the need for legislation to provide the agency with immunity from litigation, protect wildlife habitat, and prohibit the feeding of wildlife. There was no section on governance or decision process as such, which would have been a logical place for these comments. Lack of such recognition signaled that AGFD may not have considered the very nature of decision making to be a vital topic of discussion.

Turning to revision of mountain lion hunting regulations, the engagement with non-hunter interests had the appearance of being more symbolic than substantive. One can array the demands of environmentalists and animal welfare activists on the basis of which ones are most likely to affect total lion harvest to those least likely to have an impact: (1) stop all lion hunting, (2) stop hunting females, (3) stop using hounds, (4) limit the hunting season, (5) promptly close game management units after reaching a kill quota, (6) revise and justify estimates of lion densities, (7) require agency inspection of hunter kills, and (8) prepare and implement a comprehensive management plan. Demands (1) through (3) are the only ones likely to effect a substantial reduction in harvest, especially if a limited hunting season were to encompass times of year when most kills are made. As it turned out, AGFD only adopted measures that would have little or no impact on current harvest levels, including a general season that encompassed the period when 93 percent of lion kills were made during 1995 to 1999 (Schubert 2004). By conforming to broader norms, systematizing its harvest, and maintaining an ability to kill as many lions as in the past, AGFD did what appeared to be a good job of maintaining its interests, bolstering its image, reducing vulnerability to obvious criticisms, and preserving cultural priorities. However, this begs the question of how well the agency addressed common interests of the general public and how well it cultivated its capacity to find common ground among those with conflicting interests.

The importance of fostering common ground and focusing on common rather than special interests is highlighted by the current stridency of the discourse on managing mountain lions in the West (Mattson and Clark 2010b). Hunters have publicly called environmentalists and animal rights activists "nuts" who threaten "scientific management," who "only want to push their beliefs on others," who have "no vision of conservation," who are "hysterical," and who inflame issues only to raise money, and more (Howard 1991; Einwohner 1999; AGFD 2004b). Conversely, hunters are represented by animal welfare advocates as callow and uncaring exploiters who are "killing the Earth." Although the wildlife management agencies are generally praised by hunters, the agencies are also subject to criticisms from across the spectrum of interests. Perhaps most pointed from an agency perspective are charges that call professional skill and ethics into question, for example, from a hunter who charged that "with the aid and protection of educated wildlife management experts—cougars are wiping out our western deer herds" (Zumbo 2002, 24), and from an environmentalist that "AGFD has inexplicably and, perhaps, purposefully altered . . . estimates of lion habitat . . . thereby manufacturing 'paper lions'" (Schubert 2002, 1–2). Some even suggest, for example, "a conspiracy to shut down local guides and to capitalize on the mountain lion" (Lermayer 2006, 8). This kind of uncivil discourse is corrosive to liberal democracy (Shils 1997).

Mountain Lion Management Decision Process

Here I provide an explanation for the history and dynamics described earlier. I start by describing different worldviews at play and how they engender diverse and often conflicted demands regarding the outcomes of mountain lion management. I then describe some key exacerbating dynamics organized around the rubric of in- and out-groups, emphasizing allocations of power and wealth. This leads to an appraisal of the governance models employed by AGFD in terms of fostering common ground. I conclude with lessons that are germane to changing the dynamics of mountain lion management and policy making in order to increase civility and dignity and focus on common rather than special interests.

WORLDVIEWS AND SOCIAL CAPITAL

A number of schemes have been developed to represent how people view wildlife and nature. Typically these are bipolar, including anthropocentric to bio- or ecocentric, utilitarian to protectionist, and extrinsic to intrinsic valuation (Dietz, Fitzgerald, and Shwom 2005). Stephen Kellert (1996) developed a classification for describing worldviews that had eight to ten categories, but that can be consolidated into five: fearful (negativistic), aspiring to domination and emphasizing utility (utilitarian/dominionistic), valuing ecological connection or wholeness and the opportunity to learn (ecologistic/scientistic), valuing beauty and naturalness (aesthetic/naturalistic), and viewing animals as human-like and experiencing moral obligation for their well-being (humanistic/moralistic). These various worldviews can engender quite different and potentially conflicting demands regarding the state of the world and related outcomes of wildlife management (Mattson and Clark

2010b; Mattson and Ruther 2012). They are also typically intimately intertwined with people's identities and self-stories (Kellert 1996).

Most state wildlife management agencies and their related structures for funding and governance were established during the late 1800s and early 1900s at a time when the utilitarian/dominionistic worldview was pervasive (Reiger 2001). Although hunters and the agencies that were constituted to serve their interests are the undisputed harbingers of wildlife conservation (Reiger 2001), they were also deeply imbued with the utilitarian worldview, including notions that hunting was virtuous and that "wildlife" consisted of huntable species (Shaw 1994; Reiger 2001; Dizard 2003). During the next hundred-plus years Western culture changed dramatically, including how people viewed wildlife and nature (Kellert 1996). The comparatively novel ecologistic/scientistic worldview emerged. For example, a survey of the Arizona public found that the primary reason people thought protection of mountain lions was important was because "if top predators are lost, the entire ecosystem and balance of nature are put at risk" (Decision Research 2004, 4). This notion is quite new. The humanistic/moralistic and aesthetic/naturalistic worldviews also became prevalent, politically empowered, and identified with the animal welfare and environmentalist movements (Kellert 1996). Overall, the utilitarian/dominionistic worldview declined with increased urbanization and social mobility and with the advent of postmaterialist values (Kellert 1996). Perhaps most relevant to this discussion is the fact that views of wildlife and nature have considerably diversified during the last forty years, and along with this diversification has come greater diversity of potentially incompatible public demands on wildlife management agencies (Mattson and Clark 2012; Mattson and Ruther 2012).

Compounding this increase in conflicting demands has been a decrease in social capital, trust, and civility throughout the United States. Robert Putnam (2000) has documented this decline and mapped its geographic extent by state. Civility is the means and social capital the reservoir available for citizens in a liberal democracy to reach peaceful resolution of their differences (Shils 1997; Putnam 2000). Without civility and social capital, violence or the threat of violence—overt or tacit—is a likely consequence. With the future of civil society at stake, such a potential outcome ought to be a nontrivial consideration for those involved in arenas, such as wildlife management, that are increasingly prone to conflict. Although Arizona and the Intermountain West are not at the bottom in terms of social capital, they are not as well endowed as the Northeast and upper Midwest (Putnam 2000). With intrinsically limited and probably declining capacity for civil settlement of differences, the stakeholders in mountain lion management clearly need ameliorative decision processes that are exceptionally good at fostering common ground (Clark and Munno 2005; Mattson and Clark 2010b).

One could argue that the greatest challenge facing wildlife management is helping participants find common ground on which to craft durable common interest policies (Decker et al. 1996; Nie 2004a; Clark and Rutherford 2005; Jacobson and Decker 2006). Yet this is clearly not how most participants, including wildlife managers and policy makers, see it. Most have remained focused on pursuit of their special interests, including those of their agency, organization, or group (Mattson and Clark 2010b). The quality of governance has not been a focus of attention. Instead, struggles over symbolic stakes and physical outcomes have dominated (Mattson and Clark 2010b). To understand this state of affairs, it is helpful to examine participant identities and demands, the ways in which participants interact and affect each other, how they define problems, and the models of governance they employ. As a useful starting point, one can define the in-group as those who have primary access to power over decision making under current arrangements, and the out-group as those who don't (Clark and Rutherford 2005). The in-group of mountain lion management is organized around what I call the utilitarian/dominionistic subsystem of social process.

THE POWER IN-GROUP

The in-group consists primarily of hunters and agency personnel and, to a lesser extent, those involved in agriculture (Mattson and Chambers 2009; Mattson and Clark 2010b; Mattson and Clark 2012). There is a strong tendency for members of this in-group to hold a utilitarian worldview and to be invested in domination and power (Mattson and Clark 2010a). Hunting is seen as a virtuous and threatened activity, and wildlife is valued primarily for its material qualities. This is not to say that members of the in-group hold these perspectives exclusively, but rather that these perspectives constitute strong modalities. Virtually all wildlife commissioners in the West, historically including Arizona, are self-identified hunters and self-identified with groups that promote hunting and other instrumental valuations of wildlife (Hagood 1997; Mattson and Clark 2010b). Employees of wildlife management agencies tend to have a similar profile (Mattson and Clark 2010a). As a consequence, hunters, agency managers, and agency commissioners naturally find common ground in shared worldviews and shared pursuits. These commonalities predictably engender receptiveness to persuasion among those who belong to the in-group, based on shared assumptions, especially about the priority of hunting and the hunting experience (Mattson and Ruther 2012; Mattson and Clark 2012). Most funding for wildlife management agencies comes directly or indirectly from hunters and gun owners (from license fees or federal grants derived from taxes on arms and ammunition), which reinforces a focus on hunting (Hagood 1997; Mattson and Clark 2010b). In Arizona these sources have accounted for roughly 80 percent of all funding during the last decade (http://www.azgfd.gov/inside_azgfd/annual_report .shtml). Most of the remainder came from funds generated by a state lottery. Virtually none of AGFD's funding comes from the general fund or is directly controlled by the legislature. Moreover, the commissioners, who set wildlife management policy, are appointed by the governor for five-year terms and are not subject to direct oversight by elected officials (http://www.azgfd.gov /inside_azgfd/commission.shtml).

These arrangements have several important consequences. First and foremost, wealth and legitimacy are generated almost entirely internal to the utilitarian/dominionistic subsystem, identified with the power in-group. There are few apparent rewards for responding to interests external to this subsystem and ample rewards for focusing almost entirely on those who are part of the in-group. This bounding of attention, especially within agencies, is predictably reinforced by traditional culture and the related emphasis on hunting as an activity and management tool. As a result, there is a strong tendency within wildlife management agencies to serve the perceived interests of hunters directly or indirectly (Decker et al. 1996; Gill 1996; Rutberg 2001; Nie 2004a, 2004b; Clark and Munno 2005; Jacobson and Decker 2006). Although nongame management has been increasingly legitimized in recent decades, in almost all cases it receives relatively little funding (http://www .azgfd.gov/inside_azgfd/annual_report.shtml). As an interesting derivative of this emphasis on hunting, AGFD has been at odds not only with animal welfare advocates, but also with environmental groups. For example, AGFD advocates maintaining or even expanding existing wildland road networks in order to allow hunter access and to maintain structures for artificially provisioning wildlife with water (Mattson and Chambers 2009). This puts the agency in opposition to environmentalists who are fighting for wilderness, legally designated or otherwise. As a bottom line, this configuration of wealth, authority, and worldviews predictably generates an intrinsic bias toward the special interests of the agency and hunters. As I discuss later, servicing special interests is often justified by scientistic narratives built around notions such as "the good of the resource" (Clark and Rutherford 2005; Mattson and Chambers 2009; chapter 8, this volume).

The utilitarian/dominionistic subsystem has come under considerable apparent stress in recent years. The agency's neglect of increasingly powerful interest groups, such as animal welfare advocates and environmentalists, has taken a predictable toll. Overall, the demands and expectations of the external world have dramatically changed, especially with the skyrocketing urbanization of Arizona (Albrecht 2008), making the internal world of the utilitarian/dominionistic subsystem increasingly out of sync. Hunters declined as a fraction of the population in Arizona between 1996 and 2006 from 15 to 9 percent of residents more than fifteen years old (US Fish and Wildlife Service and US Census Bureau 1996, 2006). This decline threatened not only revenues, but also the political clout of AGFD. It is not surprising that AGFD and other wildlife management agencies have been preoccupied with understanding and reversing declines in hunting participation (e.g., Enck, Decker, and Brown 2000; AGFD 2001), to the point, for example, of having made recruitment of more hunters a priority in AGFD's 2001–2006 Strategic Plan (AGFD 2001, 16). Tensions have also plausibly arisen because younger cohorts and the inclusion of nongame management have brought a less homogeneous outlook to agencies, in the process creating sometimes conflicting subcultures (Organ and Fritzell 2000). Perhaps out of pragmatism, there were numerous statements in the preamble to Arizona's 2001–2006 strategic wildlife management plan that affirmed the value of collaboration, cooperation, and nonconsumption (e.g., AGFD 2001, 2, 7). But when management objectives for species such as mountain lions are examined, they were, and continue to be, indisputably about providing a "quality" hunt along with as many hunting opportunities as possible (AGFD 2001, 37).

There is good reason to expect that structural tensions, including perceptions that traditional values and identities are under assault, lead to heightened anxiety among members of the in-group. This probably explains the harshness of some statements emanating in both public and private from hunters and even agency employees, especially regarding animal welfare advocates (which I address in more detail later). There is often a tendency to morally exclude those who disagree with one's positions, especially if they are not part of one's in-group (Opotow and Weiss 2000; Skogan and Krange 2003). But the hostility of some statements is disturbing to anyone who cares about civility. A large body of psychological research suggests that anger is often indicative of an existential crisis being resolved by demonizing those who pose a threat (Cooper 2003; Alon and Omer 2006). Such personal turmoil and resulting venting are evidence of why amelioration and conciliation are needed, not only for management of mountain lions (Mattson and Clark

2010b), but also for management of wildlife in general (Clark and Ruther-ford 2005).

THE POWER OUT-GROUP

Animal welfare advocates and non-hunting environmentalists constitute what could be considered the power out-group of wildlife management in Arizona and elsewhere in the West. This is not to say that the remaining large majority of Arizonans has any greater access to the decision-making process, but rather that animal welfare advocates and, to a lesser extent, environmentalists are the most consistently engaged with wildlife issues, especially those related to mountain lions (Mattson and Clark 2010a). They are largely unempowered because they have virtually no access to agency budgets (for example, through fees or the legislature) or to commissions or managers (through shared worldviews or values). Animal welfare activists hold different worldviews, largely humanistic/moralistic, often expressed as intrinsic valuation of wildlife and demands for protection, as in the case of mountain lions (Mattson and Clark 2010a).

I focus here on animal welfare activists because they are the most consistently engaged with mountain lion issues, presumably because of the charismatic and symbolic nature of lions (Mattson and Clark 2010a). In fact, one of the problems facing animal welfare activists is that of engaging other participants. It is difficult enough to engage environmentalists, who tend to be focused on land management issues, much less the general public. For the most part, the general public in the West, including Arizona, is unaware of and largely ambivalent about wildlife management (Mattson and Clark 2010a). This is evidenced by the fact that in recent years the majority of Arizona residents could not even name the state agency responsible for wildlife management (Responsive Management 2004). Confounding this for the purposes of animal welfare activists is the fact that most Americans also support hunting, especially for meat and by Native Americans (Heberlein and Willebrand 1998).

Opportunities for animal welfare activists to access wildlife management decision processes have occurred after incidents that highlighted increasingly unacceptable management practices or by the strategic disclosure of practices that were patently offensive to most of the public. In this regard, the Mt. Elden and Sabino Canyon incidents were especially important because they mobilized a broad and temporarily vocal constituency for change, organized around both humanistic/moralistic and ecologistic/scientific worldviews (Perry and DeVos 2005; Mattson and Clark 2012). This broader constituency for nonlethal solutions was then able to engage the governor, who had sympathetic tendencies. Similarly, the hunting of kittens and dams with kittens created opportunities, primarily because this hunting practice was offensive to many people, not for reasons related to conservation of lion populations but rather for reasons largely symbolic and emotive (Mattson and Clark 2010a). Even though surveys show that the general public supports hunting, there is little support if it is to obtain trophies or is somehow considered to be unethical (Loker and Decker 1995; Heberlein and Willebrand 1998). The sometimes inflammatory strategies employed by animal welfare activists can be problematic for a democratic society. However, these strategies are a predictable outgrowth of decision processes that are closed, serve special interests, and neglect the common good—which tends to be the case with state-level mountain lion management (Clark and Munno 2005; Mattson and Clark 2010b).

Animal welfare activists and hunters have characteristic identities that, not surprisingly, inflame the latent potential for conflict (Mattson and Clark 2010b). For one, these two groups and their allies have strikingly different modal demographic profiles (Mattson and Clark 2010a). Animal welfare activists and others who espouse the humanistic/moralistic worldview tend to be disproportionately urban-dwelling, highly educated, professional women. The proportion with postgraduate and advanced professional training is remarkable. By contrast, hunters and those who hold a utilitarian/dominionistic worldview are disproportionately male and rural. They also tend to be less well educated and not as affluent. Perhaps most significant, hunters tend to be interested in the expression of power and domination in their lives, in part through hunting (Mattson and Clark 2010a). Some researchers have observed a problematic tendency for this primarily male-expressed interest in domination to blur with gender relations in the physical act of hunting (Einwohner 1999; Kalof, Fitzgerald, and Baralt 2004). There is good reason to suspect that well educated and self-affirmed females, especially those who espouse a different worldview, constitute an existential threat to the average hunter, and vice versa. Compounding this is the fact that animal welfare activists tend to pursue their interests with what has been described as religious fervor (Galvin and Herzog 1992; Shaw 1994; Jamison, Wenk, and Parker 2000). Edward Shils (1997) has noted the corrosive effect of such ideological vehemence on civility. Put together, these aspects of identity are a veritable recipe for harsh conflict. With such strongly differentiated worldviews organized around such distinct demographic communities, the sharp delineation of in-group boundaries is perhaps inescapable.

UNHELPFUL MODELS OF GOVERNANCE

The identities of key participants coupled with configurations of power create the conditions for conflict within the community of stakeholders pursuing their interests in mountain lion management. To overcome this built-in tendency for negative interactions, it is crucial to institutionalize means of engagement that are ameliorative rather than inflammatory (Brunner 2002; Brunner and Steelman 2005). Liberal democracy depends on the existence of cultural and institutional mechanisms that curb unbridled conflict and elevate and empower the common good in spite of citizens pursuing their special interests (Dahl 1982; Lasswell and McDougal 1992; Shils 1997). As I pointed out earlier, reservoirs of social capital and trust that facilitate a focus on common rather than special interests have diminished in the United States (Putnam 2000). The burden of fostering civility has fallen increasingly on institutionalized processes for developing and implementing public policy. The models of governance employed by public agencies such as AGFD largely determine whether or not policy processes degenerate into unalloyed and divisive pursuit of special interests. The focus here is legitimately on wildlife management agencies, which are trustees of the public or common interest and, through their authority, largely determine the nature of decision making.

Most wildlife management agencies, including AGFD, tacitly or otherwise use a mix of three models of governance: bureaucratic, business, and scientistic (Nie 2004a, 2004b; Clark and Rutherford 2005; chapter 8, this volume). The typical bureaucratic model holds that agencies are entrusted with authority and control for managing wildlife and are therefore wholly responsible and accountable. In the case of agencies governed by a commission, policies are not only implemented, but also formulated, prescribed, and evaluated by the agency. Accountability is not directly democratic, but rather is indirectly enforced through commission appointments, "customer" agitation, litigation, and ballot initiatives. The business model prioritizes service to customers and has been embraced under the rubric of Total Quality Management, or TQM (Hunt 1993). This approach originated in the business sector, but has subsequently been picked up by a number of government agencies. The scientistic model, which has a legacy dating back to Gifford Pinchot in the early 1900s, prioritizes the rational management of resources by experts in accordance with scientific principles and information (chapter 8, this volume). These three models amalgamate as an approach that prioritizes providing customers with high-quality service and products, efficiently developed or produced by experts according to scientific principles, but with ultimate authority and control residing with the agency and its commission. In most cases the customer is implicitly or explicitly hunters, although there is frequent verbal and written reference to serving the broader public (e.g., AGFD 2001).

At this point it is probably worth highlighting again the central business and challenge of a liberal democracy: that of the individual discerning and pursuing his or her interests, but in a civil and sustainable way (Dahl 1982, 2006; Lasswell and McDougal 1992; Shils 1997). My own perspective on the policy process encompasses the full spectrum of activities associated with individuals and groups in a society developing, implementing, appraising, and terminating policies that are ideally, in a liberal democratic society, an embodiment of long-term common interests. Competition is natural and integral, but so is respect for and accommodation of the interests of others, especially when codified in processes that facilitate peaceful resolution of competition. There is increasing appreciation of the fact that all facets of policy making play out in the bureaucratic arena of government agencies, every bit as much as in legislative and other executive arenas (Peters 2001). It is also increasingly evident that any claim to an impartial and mechanistic implementation of policy by government agencies is at best illusory and at worst deceptive (Wilson 1989; Box 2004). A responsibility for reconciling and otherwise arbitrating diverse societal interests in the form of common interest policies inescapably rests with agencies such as those charged with state wildlife management, especially given the insular commission structure.

The hybrid approach to governance employed by AGFD and most other wildlife management agencies is highly problematic if the goal is a civil liberal democracy. The bureaucratic impulse often engenders unwillingness to share or otherwise delegate power in the development and implementation of policy (Wilson 1989; Peters 2001). This is expressed in the form of statements such as "we cannot turn over decision-making to stakeholders because we have been given the authority for management, and bear financial liability for our decisions" (i.e., accountability through litigation), or "if our decisions led to some kid being killed by a mountain lion, we would have to live with the responsibility, regardless of whether or not the process was based on consultation with the public" (J. DeVos, R. Miller, AGFD, personal communication).

These kinds of comments invoke an implicit contract with society, the agency's financial liability, and personal guilt. However, these justifications do not hold up to critical scrutiny. Guilt rooted in personal sensibilities is typically not a good basis for public policy (e.g., Kaplan 1958; Jonsen and

Butler 1975; Willbern 1984; Etheridge 2005). Agency financial liability is on its face a special (or agency) rather than a common societal interest. The societal contract has more merit, but on closer examination may not be warranted in the case of wildlife management. A key part of the societal contract with government bureaucracies is that agencies will be fair, impartial, and devoted to the public interest (Box 2004). As I have suggested earlier, configurations of agency power, wealth, and culture lead to considerable partiality for a relatively narrow set of special interests—those of hunters and fishermen. Accountability is also not directly democratic. As a bottom line, the bureaucratic model employed by most wildlife management agencies does not recognize, in practice, that responsibility and accountability extend to the entire public and to providing venues that are fair, balanced, and otherwise impartial (Decker et al. 1996; Gill 1996, 2001; Nie 2004a, 2004b; Clark and Rutherford 2005; Jacobson and Decker 2006). This last proviso is especially germane given that wildlife management policy is set, not by the elected legislature or executive, but by the appointed commission in consultation with agency experts.

Numerous public policy experts have described the axiomatic differences between missions of for-profit businesses and government bureaucracies. Unlike businesses, government agencies are tasked with securing the common or public interest (Steelman and DuMond 2009), and the primary criteria for performance are not financial profit (Lasswell 1971; Box 2004). For most businesses, the customer is closely linked to the product, which is almost wholly a private matter of the company. For public agencies the legitimate customer is the entire public (Terry 2003; Box 2004), which, as I have shown, holds widely divergent notions of what the product of wildlife management is or should be (Mattson and Clark 2010b). In this public policy context, the notion of customer has little relevance and in fact can potentially divert attention from the task of amelioration. In practice, the notions of TQM and "customers" seem to combine with the cultural biases of wildlife management agencies to legitimize the special standing of hunters and huntable game in contrast to all other interest groups and wildlife management products. In short, although the business model of government potentially fosters responsiveness to customer input, this model does not provide any explicit help in identifying who the customer or related product is or should be (Mattson and Clark 2010b). For that reason, the TQM movement potentially distracts from a focus on all that is needed to create common interest policies for managing animals such as mountain lions, while providing a rationale for continued service of special interests.

As used by state wildlife agencies, scientistic management provides an integral rationale for the bureaucratic and business models of governance (Clark and Rutherford 2005; Mattson and Chambers 2009; chapter 8, this volume). I use "scientistic" rather than "scientific" here to emphasize a paradigm or even philosophy rather than scientific practice or the use of scientific information. Of greatest relevance to this discussion, scientistic wildlife management is based on the premise that problems are objective and biophysical in nature and therefore solvable by experts (wildlife managers and biologists) applying their biophysical knowledge (about wildlife and habitats, chapter 8, this volume). In this model, managers determine what the problems of wildlife management are and how to solve them, using principles such as "the good of the resource," "carrying capacity," or "sustainable harvest." This paradigm is a key bastion of the bureaucratic approach in that it places sole authority and control over policy in the hands of wildlife management agencies because these agencies are central repositories of expertise.[2] Scientistic management is also consistent with the business model of governance, although the match is not as good. Wildlife managers apply their expert knowledge to provide a better product for the customer, with huntable wildlife typically the primary product. A mismatch occurs insofar as the customer, rather than solely the expert, provides substantive input on the nature of the problem, typically defined in terms of amounts of game habitat or harvestable wildlife.

There are numerous critiques of scientistic (i.e., "scientific") management as applied to natural resources (e.g., Clark 1993; Brunner 2002; Sarewitz 2004; Pielke 2007; Ascher, Steelman, and Healy 2010). Its biggest shortcomings are in failing to recognize that problem definition is at the heart of any policy process (Weiss 1989; Dery 2000) and in giving unwarranted deference to experts. A problem can be understood as any discrepancy between the way the world is and the way we desire it (Dery 1984; Clark 2002, 100). Given the many worldviews engaged with mountain lion management, there are numerous different ways that those involved want the world to be (Mattson and Clark 2010a). For some there would be more lions to hunt, for some there would be fewer lions preying on deer or bighorn sheep, for some there would be enough lions to fulfill an ecological role, for some there would be no lion hunting, for some there would be ample opportunities simply to see lions in the wild, and for some there would be no lions at all to pose a threat to their safety. Each of these demands engenders a different understanding of "the problem" of mountain lion management. In short, there is no one objective problem of mountain lion management rooted in biological conditions, to be divined by an expert. There are multiple problems held by multiple interest groups in need of reconciliation. To presume otherwise is tacitly to impose the typically hidden values of agency experts on all others involved and implicitly to contravene the social contract between public and agency. Scientistic management, combined with bureaucratic and business models of governance, predictably complicates the cultivation of common ground and the clarification of common interests among conflicted stakeholders (Brunner 2002; Brunner and Steelman 2005; Ascher, Steelman, and Healy 2010; chapter 8, this volume).

The conflict and acrimony surrounding mountain lion management are not difficult to understand, given the identities of those who are most passionately engaged, the configuration of power and wealth, and the models of governance at play. It is not surprising that public input pertaining to governance had no logical place in the AGFD report on the public mountain lion management workshop held in Tucson. Simply put, the scientistic, bureaucratic, and business models do not provide a means of framing issues in terms of governance, common interests, and civility. Overall, the configuration of models, identities, and resources also explains why AGFD did not undertake sustained and meaningful consultation or collaboration with the full spectrum of stakeholders who were substantively engaged with mountain lion management and instead opted for a onetime process coupled with traditional formal public meetings geared toward hunting regulations. It is also clear why AGFD implemented changes in lion hunting regulations that seemed to be designed more to advance the strategic interests of the agency and hunters rather than seriously address the concerns of animal welfare activists and others holding the humanistic/moralistic worldview. When it comes to mountain lion management in Arizona and elsewhere in the West, the need for amelioration is great and the currently employed tools insufficient to the task.

Moving Forward: Lessons and Recommendations

Mountain lion management will continue to serve a narrow set of special interests organized around hunting as long as revenues are primarily hunting-related, commissioners are deeply imbued with the ethos of hunting, and management agencies are dominated by a hunting culture (see chapter 9, this volume; appendix for this chapter). This is not to say that hunting is intrinsically bad, but rather that any policy process that patently serves narrow special interests while marginalizing all others is fundamentally incompatible with a liberal democracy grounded in civil discourse (Dahl 1982, 2006;

Lasswell and McDougal 1992; Shils 1997; Brunner 2002; Clark 2002; see chapter 8, this volume). Clearly, the stakeholders in mountain lion management value lions in a wide variety of ways, more than simply as game. Additionally, wildlife management is also more than just a means of providing hunting opportunities, and the tools of management clearly consist of more than hunters with firearms. Moving beyond the current paradigm will likely require diversifying revenues so that no one interest group has a lock on agency financial well-being, diversifying commission membership to represent the full spectrum of ways that people value animals such as mountain lions, and diversifying the cultures of management agencies and the academic institutions that train prospective employees (Decker et al. 1996; Gill 1996; Hagood 1997; Beck 1998; Pacelle 1998; Rutberg 2001; Nie 2004b; Jacobson and Decker 2006). These trends are already afoot, driven by broad societal forces. Change of this nature will likely continue to happen. The question is the extent to which those with power over the institutions of wildlife management will be willing participants.

Aside from addressing these broader conditioning factors, the question remains whether other means exist for changing current mountain lion management. Perhaps the best prospect is to create respectful venues where people who represent the full spectrum of interests can engage face to face over pragmatic rather than symbolic issues (Gill 2001; Nie 2002, 2004b; Durant, Fiorino, and O'Leary 2004; Clark and Munno 2005; McLaughlin, Primm, and Rutherford 2005; Mattson et al. 2006). Considerable research and on-the-ground experience demonstrate that when people are given the opportunity to interact with others about substantive matters rather than symbolic constructs and do so according to ground rules that enforce a measure of civility, a surprising capacity for empathy often emerges (Koger and Winter 2010; Wondolleck and Yaffee 2000). This process is typically facilitated when an otherwise conflicted group is given a specific pragmatic task, such as developing a plan to increase human safety in a popular recreation area used by lions, and diverted from issues that are symbolically loaded, such as whether all lion hunting in the state of Arizona should be banned or not (McLaughlin, Primm, and Rutherford 2005).

However, such groups will not emerge unless there are substantive rewards for stakeholders to participate, for example, by being able to contribute authoritatively to setting management policies (Durant, Fiorino, and O'Leary 2004; Sabatier et al. 2005; Armitage, Berkes, and Doubleday 2007). These conditions largely depend on divestiture of some power by wildlife

Box 2.1

Lessons for Managers from the Case of Mountain Lions in the West

- Large carnivore management has become much more complicated in recent decades because of the proliferation of different views regarding proper relations between people and wildlife, rooted in phenomena such as urbanization and higher education. This diversification of views has led to increasingly conflicted expectations and demands regarding the outcomes of large carnivore management.
- Current institutions of state-level wildlife management are intrinsically corrosive to civil society. Current structures lead to tendencies to serve the special interests of hunters, anglers, and agricultural producers and to disregard the interests of all others. Reforms are needed in terms of funding, culture, and representation if state wildlife management is to serve the common interest.
- Scientized and business models are intrinsically problematic for large carnivore management. Scientized management presupposes that problems are objective phenomena to be discerned and solved by technical experts, who therefore logically hold power. Business models of management presuppose that government agents serve customers with products, without defining who the customers are or what the products should be. Neither model fosters attention to quality of governance or civil negotiation among those with different interests.
- Symbolic projections by participants, whether of their identities or worldviews, often have a strong inflammatory effect on conflict in management of large carnivores. Gains in the common interest are likely to be made by refocusing participants on solving practical problems that are of limited scope and scale.

management agencies. Here we come back full circle, because few agencies or their governing commissions are apparently inclined to divest power, probably for various psychological reasons, but more certainly because of the management models and related notions of governance that they embrace (Nie 2004b; Brunner and Steelman 2005; Clark 2008). Under such circumstances, the opportunities for constituting alternative structures (organized around civil interactions among stakeholders) often reside in the outcome of crises that make status quo arrangements temporarily untenable for the management agencies. In this regard, crises such as those of Mt. Elden and Sabino Canyon, although often uncivil and uncomfortable for participants, also constituted crucibles of potentially constructive change.

Conclusion

Mountain lion management in the western United States exemplifies natural resource cases where the policy process and related discourse are organized largely around symbolic rather than material stakes. Few individuals make a living off mountain lion hunting (Mattson and Clark 2010a). Wildlife management agencies, AGFD in particular, typically make little money from selling permits to hunt lions (in 2010, $14.50 per tag). Comparatively few agricultural producers are substantively affected by lion predation on livestock (Mattson and Clark 2010a). Fewer people yet actually see these elusive, largely nocturnal predators. People engaged in mountain lion management seem to be primarily motivated by the idea of mountain lions, whether for good or bad, depending on the individual's view of the proper relations between people and wildlife (Mattson and Clark 2010b). Participants are also seemingly caught up in the symbolic identities of those who threaten them and their core beliefs (Charon 2007). In this respect, these symbolic dynamics predictably touch on deep existential issues (Yalom 1980; Cooper 2003). Under such symbolically and existentially potent circumstances, people typically have little access to rationality and as often as not end up driven by anxieties translated into fear and anger. Moreover, these emotions are often organized around group identities and the demonization of threatening others who are outside the group (Alon and Omer 2006). Simply appealing to logic or rationality has little prospect of moving participants toward greater civility (Koger and Winter 2010). Hope for change resides primarily in changing incentive structures and providing venues that both deflate the symbolic issues and foster intrinsic capacities for empathy (Wondolleck and Yaffee 2000; Armitage, Berkes, and Doubleday 2007; Mattson and Clark 2010b).

As a larger issue, mountain lion management highlights some problematic aspects of American culture that drive these dynamics perhaps every bit as much as does the nature of the wildlife management institutions. The success of a liberal democracy hangs in the balance of people ardently pursuing their own interests, yet honoring and respecting the interests of others, especially as expressed in decision processes that are designed to arrive at common interest policies (McDougal, Lasswell, and Chen 1980; Dahl 1982).

This notion of honoring and respecting is encapsulated in the notion of civility (Shils 1997). Civility is especially vulnerable to not only hedonistic and narcissistic impulses, but also to paranoia and strident ideology. In the case of mountain lion management, ideology and fear appear to characterize both the identities and interactions of many participants. The United States has long had social and cultural institutions that foster civility, already well developed at the time of Alexis de Tocqueville's writings (Tocqueville 1839). However, these founts of civility and social capital are in decline (Putnam 2000). Mountain lion management is perhaps a microcosm of our diminished capacity for respectful engagement on common ground to seek common interest solutions, regardless of the institutional context. Reform of mountain lion management is contingent on stakeholders' democratic character, but ideally lion management itself would contribute, in turn, to building civility and trust.

Appendix
Decision Activities, Standards, and Effectiveness in Mountain Lion Conservation in the West

INTELLIGENCE
Recognizing the problem and gathering information.
Dependable
Comprehensive
Selective
Creative
Open

Intelligence in this case has been moderately comprehensive but nonetheless strongly biased toward ecological factors. Since 2000 three different books have synopsized the state of knowledge regarding mountain lion ecology and management in terms understandable by the educated public. Even so, routine management of lions is based on indices of ecological and human factors that are of limited reliability. The intelligence function in mountain lion management has often been highly politicized, very likely because the arena has lacked open fact-finding approaches that jointly involve stakeholders and help build shared and accepted understandings of how the world might work as a basis for creating mutual gains policies and solving concrete problems.

PROMOTION

Open debate, in which various groups advocate for their interests or preferred policy.

Rational

Integrative

Comprehensive

Promotional activities in mountain lion management have often been polarizing and divisive and, in authoritative venues, typically structured to serve the interests of those who are part of the power in-group. Nonconsumptive stakeholders have at times resorted to inflammatory rhetoric and opportunities provided by news media to promote their preferred policies, primarily because routine access to authoritative decision making has been lacking. Promotion also often takes place under auspices of ballot initiatives designed to intervene authoritatively in routine decision making by agency personnel and commissioners. Politicization of promotion has often precluded rational, integrative, and comprehensive debate.

PRESCRIPTION

Setting the policy, rules, or guidelines.

Effective (stable expectations)

Rational

Comprehensive

Authoritative prescriptions have typically served the interests of hunters, under the premise that hunting is both necessary and virtuous and the often tacit assumption that mountain lions are competitors for deer, elk, and bighorn sheep. Avoidance of litigation has also often been a driver of sport-hunting prescriptions putatively designed to protect human safety. Many prescriptions end up not being particularly rational or effective because there is little or no science supporting the notion that sport hunting of mountain lions yields more deer or makes the world safer for people. Prescriptions have also often been bounded in scope because participation by nonconsumptive stakeholders has been institutionally limited.

INVOCATION

Implementation.

Timely

Dependable

Rational

Nonprovocative

Policy prescription and invocation often go hand-in-hand in management of mountain lion sport hunting. The circumstances under which prescriptions are to be implemented are often clearly specified during policy development. Under these circumstances, invocation is typically timely, rational, and dependable. However, invocation of policies designed to protect human safety or limit depredation of livestock or bighorn sheep have sometimes been highly provocative, largely because of the de facto exclusion of certain stakeholders from earlier authoritative promotion and prescription activities. Invocation then becomes a time when those who have been marginalized have the opportunity to revisit promotion and prescription either through highly politicized public discourses or through litigation.

APPLICATION

Dispute resolution and enforcement of the prescription.

Rational and realistic

Uniform

Resolution of disputes in mountain lion management has typically been uncivil, divisive, and inflammatory. Resolution has often been the outcome of litigation, ballot initiatives, and media-framed incidents, none of which are intrinsically ameliorative. In this regard, application has not often been realistic, at least relative to the goal of sustaining civil society and serving the public trust. Nor has resolution often been particularly rational within or consistent across cases, largely because of the often politicized way in which information has been handled by everyone involved, as well as the inflamed emotions evident during this phase of the policy process. Perhaps the most rational and consistent policy applications, at least framed in terms of policy fulfillment, have occurred as an outcome of deliberations entrained by litigation, primarily because legal deliberations explicitly mandate some measure of rationality.

APPRAISAL

Review and evaluation of the activities so far.

Dependable

Rational

Comprehensive

Selective
Continuing
Independent

Appraisal has been consistently weak to nonexistent in mountain lion management, especially related to performance of decision-making processes. Even in assessing ecological factors such as effects of sport hunting regimes on lion populations, appraisal has often appeared to be more a defense of past policies than a dependable and rational assessment. Authoritative appraisals have virtually never been independent. Even appraisals by nonconsumptive stakeholders have not been truly independent and seem designed more to promote a partisan cause than reliably evaluate management practices. Most appraisals of decision making in mountain lion management have been done by academics, who have probably come closest to reaching the standards of a good evaluation, but with little prospect of influencing those in power.

TERMINATION/SUCCESSION

Ending or moving on.
Timely
Comprehensive
Dependable
Balanced
Ameliorative

The ending of policies has been perhaps one of the most conflicted and politicized aspects of the mountain lion policy process. Much conflict has been organized around nonconsumptive stakeholders attempting to end policies centered on killing lions, whether for sport, human safety, depredation control, or conservation of deer, bighorn sheep, and elk. Conflict and debate have often dragged on for years and have involved highly partisan uses of information. The result has been termination processes that are untimely, unbalanced, undependable, highly selective, and certainly not ameliorative.

OVERALL STANDARDS

Honest
Economical
Technically efficient
Loyal and skilled personnel

Complementary and effective impacts
Differentiated structures
Flexible and realistic in adjusting to change
Deliberate
Responsible

Overall, mountain lion management in the West has been corrosive to civil society. Authoritative policy processes have been organized around serving the special interests of hunters and agricultural producers, largely to the exclusion of all others. As a result, marginalized stakeholders have resorted to inflammatory methods to gain access to authoritative decision making. Those in power have responded largely by defense of the status quo. Throughout, use of information, scientific or otherwise, has been highly politicized. These dynamics have led to chronic disservice of the public interest, inefficiencies in use of all resources, institutional rigidity, and lack of adaptive responses to changes in culture and other external circumstances. Virtually everyone involved has apparently felt disrespected and demoralized to some extent.

NOTES

I thank many people, too numerous to mention, for sharing with me their widely varied perspectives on mountain lions and mountain lion management. Regardless of their views, all cared about lions and the health of our American society. I am especially grateful to Denise Casey, Susan Clark, Murray Rutherford, and other reviewers for their helpful comments and editing.

The perspectives here do not represent the official views of the US Geological Survey or the US Government, with which D. Mattson was affiliated at the time he wrote this chapter.

1. I accessed information on tag fees charged during 2010 for big game species in eleven western US states (Washington, Idaho, Montana, Wyoming, Oregon, California, Nevada, Utah, Colorado, New Mexico, and Arizona) using online access provided by each state's wildlife management agency. Fees charged for bighorn sheep tags averaged 7.8 times more than fees charged for cougars for state residents ($222 versus $28), and 6.5 times more for nonresidents ($1,438 versus $222). Similarly, fees for bighorn sheep tags were 7.1 and 5.2 times greater than fees for deer tags for residents and nonresidents, respectively.

2. The document edited by Nobile and Duda (2008) constitutes an authoritative statement of the philosophic underpinnings of current state-level wildlife management, called the North American Model of Wildlife Conservation. The paramount role of scientists and technical experts is described. Mahoney et al. (2008) articulate the seven core principles of the North American Model and then identify problems, challenges, and oppor-

tunities for each. Even though explicit reference is made to the principles of public trust, democratic rule of law, opportunity for all, and legitimate use, which bespeak equitable goals, the content makes clear that the primary beneficiaries of wildlife management are hunters and that the defining management tool is hunter harvest. The restriction of equity principles through partial incorporation in symbolic rather than substantive form is remarkable ("restriction through partial incorporation," Brunner and Steelman 2005).

REFERENCES

Albrecht, D. E. 2008. *Population Brief: The State of Arizona*. Logan, UT: Western Rural Development Center.

Alon, N., and H. Omer. 2006. *The Psychology of Demonization: Promoting Acceptance and Reducing Conflict*. Mahwah, NJ: Lawrence Erlbaum.

Anderson, C. R., Jr., F. Lindzey, K. H. Knopff, M. G. Jalkotzy, and M. S. Boyce. 2010. "Cougar Management in North America." In *Cougar: Ecology and Conservation*, edited by M. Hornocker and S. Negri, 41–54. Chicago: University of Chicago Press.

Arizona Game and Fish Department. 2000. *Predator Management Policy*. Phoenix: Arizona Game and Fish Department. Available from http://www.azgfd.gov/w_c /predator_management.shtml. Accessed November 11, 2012.

———. 2001. *Wildlife 2006: The Arizona Game and Fish Department's Wildlife Management Program Strategic Plan for the Years 2001–2006*. Phoenix: Arizona Game and Fish Department.

———. 2002. *2002–2003 Arizona Hunting and Trapping Regulations*. Phoenix: Arizona Game and Fish Department.

———. 2004a. *Report of the Flagstaff and Phoenix Mountain Lion Workshops*. Phoenix: Arizona Game and Fish Department.

———. 2004b. *Report of the Mountain Lion Workshop, May 1, 2004, Tucson, Arizona. Final Report*. Phoenix: Arizona Game and Fish Department.

———. 2005. *Action Plan for Minimizing and Responding to Lion/Human Interactions*. Phoenix: Arizona Game and Fish Department.

———. 2006. *Black Mountain Predation Management Plan*. Phoenix: Arizona Game and Fish Department.

———. 2010. *2010–11 Arizona Hunting and Trapping Regulations*. Phoenix: Arizona Game and Fish Department.

Armitage, D., F. Berkes, and N. Doubleday. 2007. *Adaptive Co-Management: Collaboration, Learning, and Multi-Level Governance*. Vancouver: University of British Columbia Press.

Ascher, W., T. A. Steelman, and R. Healy. 2010. *Knowledge and Environmental Politics: Re-Imagining the Boundaries of Science and Politics*. Cambridge, MA: MIT Press.

Baron, D. 2004. *The Beast in the Garden: A Modern Parable of Man and Nature*. New York: Norton.

Beck, T. D. I. 1998. "Citizen Ballot Initiatives: A Failure of the Wildlife Management Profession." *Human Dimensions of Wildlife* 3:21–28.

Beier, P. 2010. "A Focal Species for Conservation Planning." In *Cougar: Ecology and Conservation*, edited by M. Hornocker and S. Negri, 177–89. Chicago: University of Chicago Press.

Berger, J. 1990. "Persistence of Different-Sized Populations: An Experimental Assessment of Rapid Extinctions in Bighorn Sheep." *Conservation Biology* 4:91–98.

Box, R. C. 2004. *Public Administration and Society: Critical Issues in American Governance*. Armonk, NY: M. E. Sharpe.

Brown, D. 1984. "A Lion for All Seasons." In *Proceedings of the Second Mountain Lion Workshop*, edited by J. Roberson and F. Lindzey, 13–22. Salt Lake City: Utah Division of Wildlife Resources.

Brunner, R. D. 2002. "Problems of Governance." In *Finding Common Ground: Governance and Natural Resources in the American West*, edited by R. D. Brunner, C. H. Colburn, C. M. Cromley, R. A. Klein, and E. A. Olson, 1–47. New Haven, CT: Yale University Press.

Brunner, R. D., and T. A. Steelman. 2005. "Toward Adaptive Governance." In *Adaptive Governance: Integrating Science, Policy, and Decision Making*, edited by R. D. Brunner, T. A. Steelman, L. Coe-Juell, C. M. Cromley, C. M. Edwards, and D. W. Tucker, 268–304. New York: Columbia University Press.

Charon, J. M. 2007. *Symbolic Interactionism: An Introduction, an Interpretation, an Integration*. Upper Saddle River, NJ: Prentice Hall.

Clark, S. G. 2002. *The Policy Process: A Practical Guide for Natural Resource Professionals*. New Haven, CT: Yale University Press.

———. 2008. *Ensuring Greater Yellowstone's Future: Choices for Leaders and Citizens*. New Haven, CT: Yale University Press.

Clark, T. W. 1993. "Creating and Using Knowledge for Species and Ecosystem Conservation: Science, Organization, and Policy." *Perspectives in Biology and Medicine* 36:497–525.

Clark, T. W., and L. Munno. 2005. "Mountain Lion Management: Resolving Public Conflict." In *Coexisting with Large Carnivores: Lessons from Greater Yellowstone*, edited by T. W. Clark, M. B. Rutherford, and D. Casey, 71–98. Washington, DC: Island Press.

Clark, T. W., and M. B. Rutherford. 2005. "The Institutional System of Wildlife Management: Making It More Effective." In *Coexisting with Large Carnivores: Lessons from Greater Yellowstone*, edited by T.W. Clark, M. B. Rutherford, and D. Casey, 211–53. Washington, DC: Island Press.

Cooper, M. 2003. *Existential Therapies*. Thousand Oaks, CA: Sage.

Dahl, R. A. 1982. *Dilemmas of Pluralistic Democracy: Autonomy vs. Control*. New Haven, CT: Yale University Press.

———. 2006. *On Political Equality*. New Haven, CT: Yale University Press.

Decision Research. 2004. *Arizona Statewide Polling Results*. San Diego: Decision Research.

Decker, D. J., and L. C. Chase. 1997. "Human Dimensions of Living with Wildlife— A Management Challenge for the 21st Century." *Wildlife Society Bulletin* 25:788–95.

Decker, D. J., C. C. Kruger, R. A. Baer Jr., B. A. Knuth, and M. E. Richmond. 1996.

"From Clients to Stakeholders: A Philosophical Shift for Fish and Wildlife Management." *Human Dimensions of Wildlife* 1:70–82.

Dery, D. 1984. *Problem Definition in Policy Analysis*. Lawrence: University of Kansas Press.

———. 2000. "Agenda Setting and Problem Definition." *Policy Studies* 21:37–47.

DeVos, J. C., Jr., D. L. Shroufe, and V. C. Supplee. 1998. "Managing Wildlife by Ballot Initiative: The Arizona Experience." *Human Dimensions of Wildlife* 3:60–66.

Dietz, T., A. Fitzgerald, and R. Shwom. 2005. "Environmental Values." *Annual Review of Environment and Resources* 30:335–72.

Dizard, J. E. 2003. *Mortal Stakes: Hunters and Hunting in Contemporary America*. Amherst: University of Massachusetts Press.

Durant, R. F., D. J. Fiorino, and R. O'Leary. 2004. *Environmental Governance Reconsidered: Challenges, Choices and Opportunities*. Cambridge, MA: MIT Press.

Einwohner, R. L. 1999. "Gender, Class, and Social Movement Outcomes: Identity and Effectiveness in Two Animal Rights Campaigns." *Gender and Society* 13:56–76.

Enck, J. W., D. J. Decker, and T. L. Brown. 2000. "Status of Hunter Recruitment and Retention in the United States." *Wildlife Society Bulletin* 28:817–24.

Etheridge, L. S. 2005. "Wisdom in Public Policy." In *Wisdom: Psychological Perspectives*, edited by R. Sternberg and J. Jordan, 297–328. New York: Cambridge University Press.

Galvin, S. L., and H. A. Herzog, Jr. 1992. "Ethical Ideology, Animal Rights Activism, and Attitudes Towards Treatment of Animals." *Ethics and Behavior* 2:141–49.

Gill, R. B. 1996. "The Wildlife Professional Subculture: The Case of the Crazy Aunt." *Human Dimensions of Wildlife* 1:60–69.

———. 1999. *Declining Mule Deer Populations in Colorado—Reasons and Responses: A Report to the Colorado Legislature*. Denver: Colorado Division of Wildlife.

———. 2001. "Professionalism, Advocacy, and Credibility: A Futile Cycle?" *Human Dimensions of Wildlife* 6:21–32.

———. 2010. "To Save a Mountain Lion: Evolving Philosophy of Nature and Cougars." In *Cougar: Ecology and Conservation*, edited by M. Hornocker and S. Negri, 5–16. Chicago: University of Chicago Press.

Hagood, S. 1997. *State Wildlife Management: The Pervasive Influence of Hunters, Hunting, Culture and Money*. Washington, DC: The Humane Society of the United States.

Hayes, C. L., E. S. Rubin, M. C. Jorgensen, R. A. Botta, and W. M. Boyce. 2000. "Mountain Lion Predation of Bighorn Sheep in the Peninsular Ranges, California." *Journal of Wildlife Management* 64:954–59.

Heberlein, T. A., and T. Willebrand. 1998. "Attitudes Toward Hunting across Time and Continents: The United States and Sweden." *Gibier Faune Sauvage* 15:1071–80.

Holl, S. A., V. C. Bleich, and S. G. Torres. 2004. "Population Dynamics of Bighorn Sheep in the San Gabriel Mountains, California, 1967–2002." *Wildlife Society Bulletin* 32:412–26.

Hornocker, M., and S. Negri, eds. 2010. *Cougar: Ecology and Conservation*. Chicago: University of Chicago Press.

Howard, W. E. 1991. "Mountain Lion and the Bambi Syndrome." In *Mountain Lion–Human Interaction Symposium and Workshop*, edited by C. S. Braun, 96–97. Denver: Colorado Division of Wildlife.

Hunt, V. D. 1993. "Quality Management for Government." In *A Government Manager's Guide to Quality Management*, 21–52. Milwaukee: ASQC Quality Press.

Jacobson, C. A., and D. J. Decker. 2006. "Ensuring the Future of State Wildlife Management: Understanding Challenges for Institutional Change." *Wildlife Society Bulletin* 34:531–36.

Jamison, W. V., C. Wenk, and J. V. Parker. 2000. "Every Sparrow That Falls: Understanding Animal Rights Activism as Functional Religion." *Society and Animals* 8:305–30.

Jonsen, A. R., and L. H. Butler. 1975. "Public Ethics and Policy Making." *The Hastings Center Report* 5:19–31.

Kalof, L., A. Fitzgerald, and L. Baralt. 2004. "Animals, Women, and Weapons: Blurred Sexual Boundaries in the Discourse of Sport Hunting." *Society and Animals* 12:237–51.

Kaplan, A. 1958. "American Ethics and Public Policy." *Daedalus* 87:48–77.

Kellert, S. R. 1996. *The Value of Life: Biological Diversity and Human Society*. Washington, DC: Island Press.

Koger, S. M., and D. D. Winter. 2010. *The Psychology of Environmental Problems*. New York: Psychology Press.

Laliberte, A. S., and W. J. Ripple. 2004. "Range Contractions of North American Carnivores and Ungulates." *BioScience* 54:123–38.

Lasswell, H. D. 1971. *A Pre-View of Policy Sciences*. New York: American Elsevier.

Lasswell, H. D., and M. S. McDougal. 1992. *Jurisprudence for a Free Society*. New Haven, CT: New Haven Press.

Lermayer, R. M. 2006. "Editor's Response: Mountain Lion Conspiracy and Mountain Lion Mismanagement." *Predator Xtreme*, February:6–8.

Loker, C. A., and D. J. Decker. 1995. "Colorado Black Bear Hunting Referendum: What Was Behind the Vote?" *Wildlife Society Bulletin* 23:370–76.

Mackie, R. J., J. G. Kie, D. F. Pac, and H. L. Hamlin. 2003. "Mule Deer." In *Wild Mammals of North America*, edited by G. Feldhamer, B. Thompson, and J. Chapman, 889–904. Baltimore, MD: Johns Hopkins University Press.

Mahoney, S. P., et al. 2008. "The North American Model of Wildlife Conservation: Enduring Achievement and Legacy." In *Strengthening America's Hunting Heritage and Wildlife Conservation in the 21st Century: Challenges and Opportunities*, edited by J. Nobile and M. D. Duda, 7–24. Harrisonburg, VA: Responsive Management.

Mangus, G. 1991. "Legal Aspects of Encounters on Federal Lands and in State Programs." In *Mountain Lion–Human Interaction Symposium and Workshop*, edited by C. S. Braun, 43–44. Denver: Colorado Division of Wildlife.

Mattson, D. J., and N. Chambers. 2009. "Human-Provided Waters for Desert Wildlife: What Is the Problem?" *Policy Sciences* 42:113–35.

Mattson, D. J., and S. G. Clark, 2010a. "Groups Participating in Cougar Management."

In *Cougar: Ecology and Conservation*, edited by M. Hornocker and S. Negri, 254–59. Chicago: University of Chicago Press.

———. 2010b. "People, Politics, and Cougar Management." In *Cougar: Ecology and Conservation*, edited by M. Hornocker and S. Negri, 206–20. Chicago: University of Chicago Press.

———. 2011. "Human Dignity in Concept and Practice." *Policy Sciences* 44:103–33.

———. 2012. "The Discourse of Incidents: Cougars on Mt. Elden and in Sabino Canyon, Arizona." *Policy Sciences* 45:315–43.

Mattson, D. J., and J. J. Craighead. 1994. "The Yellowstone Grizzly Bear Recovery Program: Uncertain Information, Uncertainty Policy." In *Endangered Species Recovery: Finding the Lessons, Improving the Process*, edited by T. W. Clark, R. P. Reading, and A. Clarke, 101–29. Washington, DC: Island Press.

Mattson, D. J., and E. J. Ruther. 2012. "An Explanation of Reported Puma-Related Behaviors and Behavioral Intentions among Northern Arizona Residents." *Human Dimensions of Widlife* 17:91–111.

Mattson, D. J., L. Sweanor, and K. Logan. 2011. "Factors Governing Cougar Attacks on Humans." *Human-Wildlife Interactions* 5:135–58.

Mattson, D. J., K. L. Byrd, M. B. Rutherford, S. R. Brown, and T. W. Clark. 2006. "Finding Common Ground in Large Carnivore Conservation: Mapping Contending Perspectives." *Environmental Science and Policy* 9:392–405.

McDougal, M. S., H. D. Lasswell, and L. Chen. 1980. *Human Rights and World Public Order: The Basic Policies of an International Law of Human Dignity*. New Haven, CT: Yale University Press.

McKinney, T., J. C. DeVos Jr., W. B. Ballard, and S. R. Boe. 2006. "Mountain Lion Predation of Translocated Desert Bighorn Sheep in Arizona." *Wildlife Society Bulletin* 34:1255–63.

McLaughlin, G. P., S. Primm, and M. B. Rutherford. 2005. "Participatory Projects for Coexistence: Rebuilding Civil Society." In *Coexisting with Large Carnivores: Lessons from Greater Yellowstone*, edited by T. W. Clark, M. B. Rutherford, and D. Casey, 177–210. Washington, DC: Island Press.

Murphy, K., and T. K. Ruth. 2010. "Diet and Prey Selection of a Perfect Predator." In *Cougar: Ecology and Conservation*, edited by M. Hornocker and S. Negri, 118–37. Chicago: University of Chicago Press.

Negri, S., and H. Quigley. 2010. "Cougar Conservation: The Growing Role of Citizens and Government." In *Cougar: Ecology and Conservation*, edited by M. Hornocker and S. Negri, 221–34. Chicago: University of Chicago Press.

Nie, M. 2002. "Wolf Recovery and Management as Value-Based Political Conflict." *Ethics, Place and Environment* 5:65–71.

———. 2004a. "State Wildlife Governance and Carnivore Conservation." In *People and Predators: From Conflict to Coexistence*, edited by N. Fascione, A. Delach, and M. E. Smith, 197–218. Washington, DC: Island Press.

———. 2004b. "State Wildlife Policy and Management: The Scope and Bias of Political Conflict." *Public Administration Review* 64:221–33.

Nobile, J., and M. D. Duda, eds. 2008. *Strengthening America's Hunting Heritage and Wildlife Conservation in the 21st Century: Challenges and Opportunities*. Harrisonburg, VA: Responsive Management.

Opotow, S., and L. Weiss. 2000. "Denial and the Process of Moral Exclusion in Environmental Conflict." *Journal of Social Issues* 56:475–90.

Organ, J. F., and E. K. Fritzell. 2000. "Trends in Consumptive Recreation and the Wildlife Profession." *Wildlife Society Bulletin* 28:780–87.

Pacelle, W. 1998. "Forging a New Wildlife Management Paradigm: Integrating Animal Protection Values." *Human Dimensions of Wildlife* 3:42–50.

Parker, V. 1995. "Natural Resources Management by Litigation." In *A New Century for Natural Resources Management*, edited by R. L. Knight and S. F. Bates, 209–20. Washington, DC: Island Press.

Perry, G. L., and J. C. DeVos Jr. 2005. "A Case Study of Mountain Lion-Human Interaction in Southeastern Arizona." *Mountain Lion Workshop* 8:104–13.

Peters, B. G. 2001. *The Politics of Bureaucracy*, 5th ed. New York: Routledge.

Pielke, R. A., Jr. 2007. *The Honest Broker: Making Sense of Science in Policy and Politics*. Cambridge: Cambridge University Press.

Pimbert, M. P., and J. N. Pretty. 1995. "Parks, People and Professionals: Putting 'Participation' into Protected Area Management." Discussion Paper No. 57. Geneva: United Nations Research Institute for Social Development.

Putnam, R. D. 2000. *Bowling Alone: The Collapse and Revival of American Community*. New York: Simon and Schuster.

Reiger, J. F. 2001. *American Sportsmen and the Origins of Conservation*. Corvallis: Oregon State University Press.

Responsive Management. 2004. *Arizona Residents' Opinions on the Arizona Game and Fish Department and Its Activities*. Harrisonburg, VA: Responsive Management.

Rominger, E. M., H. A. Whitlaw, D. L. Weybright, W. C. Dunn, and W. B. Ballard. 2004. "Influence of Mountain Lion Predation on Bighorn Sheep Translocations." *Journal of Wildlife Management* 68:993–99.

Rutberg, A. T. 2001. "Why Agencies Should Not Advocate Hunting or Trapping." *Human Dimensions of Wildlife* 6:33–37.

Ruth, T. K., and K. Murphy. 2010. "Cougar-Prey Relationships." In *Cougar: Ecology and Conservation*, edited by M. Hornocker and S. Negri, 138–62. Chicago: University of Chicago Press.

Sabatier, P. A., W. Focht, M. Lubell, Z. Trachtenberg, A. Vedlitz, and M. Matlock. 2005. *Swimming Upstream: Collaborative Approaches to Watershed Management*. Cambridge, MA: MIT Press.

Sarewitz, D. 2004. "How Science Makes Environmental Controversies Worse." *Environmental Science and Policy* 7:385–403.

Schubert, D. J. 2002. "Letter from The Fund for Animals (D. J. Schubert) to M. Golightly (Chairman) and D. Shroufe (Director), Arizona Game and Fish Department." The Fund for Animals. Available from http://www.fund.org/library/libraryuploads/huntcom.asp. Accessed September 5, 2002.

————. 2004. *Summary and Evaluation of Mountain Lion Management Practices and Procedures in Arizona*. Phoenix: The Fund for Animals.

Shaw, H. 1994. *Soul Among Lions: The Cougar as Peaceful Adversary*. Tucson: University of Arizona Press.

Shils, E. 1997. "The Virtue of Civility." In *The Virtue of Civility: Selected Essays on Liberalism, Tradition, and Civil Society by Edward Shils*, edited by S. Grosby, 320–55. Indianapolis: Liberty Fund.

Shroufe, D. L. 1988. "Welcoming Address." In *Proceedings of the Third Mountain Lion Workshop*, edited by R. H. Smith, 1. Phoenix: Arizona Game and Fish Department.

Skogan, K., and O. Krange. 2003. "A Wolf at the Gate: The Anti-Carnivore Alliance and the Symbolic Construction of Community." *Sociologia Ruralis* 43:309–25.

Steelman, T., and M. E. DuMond. 2009. "Serving the Common Interest in US Forest Policy: A Case Study of the Healthy Forest Restoration Act." *Environmental Management* 43:396–410.

Terry, L. D. 2003. *Leadership of Public Bureaucracies: The Administrator as Conservator*, 2nd ed. Armonk, NY: M. E. Sharpe.

Tocqueville, A. de. 1839. *Democracy in America*. 3rd ed. New York: George Adlakd.

US Fish and Wildlife Service and US Bureau of the Census. 1996. *1996 National Survey of Fishing, Hunting and Wildlife-Associated Recreation: Arizona*. US Fish and Wildlife Service FHW/96-AZ.

————. 2006. *2006 National Survey of Fishing, Hunting and Wildlife-Associated Recreation: Arizona*. US Fish and Wildlife Service FHW/06-AZ.

Weiss, J. A. 1989. "The Powers of Problem Definition: The Case of Government Paperwork." *Policy Sciences* 22:97–121.

Wilkinson, T. 1998. *Science under Siege: The Politician's War on Nature and Truth*. Boulder, CO: Johnson Books.

Willbern, Y. 1984. "Types and Levels of Public Morality." *Public Administration Review* 44:102–08.

Wilson, J. Q. 1989. *Bureaucracy: What Government Agencies Do and Why They Do It*. New York: Basic Books.

Wondolleck, J., and S. Yaffee. 2000. *Making Collaboration Work: Lessons from Innovation in Natural Resources Management*. Washington, DC: Island Press.

Yalom, I. D. 1980. *Existential Psychotherapy*. New York: Basic Books.

Zornes, M. L., S. P. Barber, and B. F. Wakeling. 2006. "Harvest Methods and Hunter Selectivity of Mountain Lions in Arizona." In *Managing Wildlife in the Southwest: New Challenges for the 21st Century*, edited by J. W. Cain III and P. Krausman, 85–89. Tucson: Southwest Section of the Wildlife Society.

Zumbo, J. 2002. "Predator Control: Trapping Is the Key to Saving Game Populations." *Outdoor Life* December: 24–25.

Wolves in Wyoming
The Quest for Common Ground

REBECCA WATTERS, AVERY C. ANDERSON, AND SUSAN G. CLARK

Introduction

Early in 1995, amid national press coverage and acclaim from environmental groups, fourteen grey wolves (*Canis lupus*) were released into Yellowstone National Park, the heart of the Greater Yellowstone Ecosystem. Their release represented the culmination of a convoluted epic in which the wolf had variously played the villain, the victim, and the hero. Prior to this release, the last known wolf in the West was shot in Wyoming in the 1940s, the end of an official government eradication program that began in 1914 but reflected a much longer cultural tradition of extirpating wolves wherever they were found (Loomis 1995; Casey and Clark 1996). From the role of villain, the wolf, with its disappearance from the landscape, over time assumed the role of victim in the wider American imagination. In 1987 a reintroduction plan for Yellowstone was written and by 1994 it had been approved. The wolves released in 1995, in the words of the Yellowstone Park superintendent at the time, "right[ed] a wrong" and restored ecological balance to the park (Sanders 2001). These wolves were widely perceived as heroes.

Widely—but not universally. In the ranching communities of Greater Yellowstone the roles of villain and victim were perceived differently: the federal government and wolves as villains, and ranchers and livestock as victims. The governments of Idaho, Montana, and particularly Wyoming were concerned about depredating wolves returning to livestock country. Over the next decade, wolves dispersing in the region occasionally killed livestock, leaving ranchers feeling helpless and infuriated. In response, federal managers lethally "controlled" wolves, leaving environmentalists equally angry.

As wolves reached the recovery numbers outlined in the reintroduction plan, ranchers and state governments looked forward to assuming management of the states' wolf populations from the federal government, while environmental groups sought new reasons to keep the animal under federal management. Between 2008 and 2011 the species was delisted, relisted, and delisted again in Idaho and Montana but not Wyoming, relisted again in all three states, and then delisted in Montana, Idaho, and parts of Utah, Washington, and Oregon pursuant to a new federal law passed as a rider to a Congressional budget bill (Chaney 2010; Volz 2010; Barringer and Broder 2011). Both sides of the debate (environmental advocacy groups and the state governments) resorted to suing the federal government whenever they disagreed with its decisions, resulting in gridlock that was only broken with unprecedented Congressional interference in the delisting process. This pattern of interaction suggests that the decision-making process for wolf management in the West is flawed. As the whiplash-inducing back-and-forth legal process monopolizes the spotlight, the people living on the ground with wolves continue to struggle to fit these animals into their lives, livelihoods, and narratives.

Two communities in Wyoming exemplify this struggle and the wider context in which intense debates about wolves occur. The Wind River Mountain Range in northwestern Wyoming runs over one hundred miles southeast of Yellowstone National Park, dividing the Green River watershed to the west from the Wind River watershed to the east. East of the mountains the Wind River Indian Reservation supports the Eastern Shoshone and the Northern Arapaho tribes, peoples with long-standing relationships to the land and wildlife. To the west, the ranching community of the Upper Green River basin maintains a different cultural tradition of interaction with landscape, resources, and wildlife. Both valleys provide habitat for elk, deer, bighorn sheep, and, in recent years, wolves. As the wolf population grows, these communities face tough choices about how to manage the controversial species.

This chapter examines the "problem" in these communities, ostensibly defined as wolf management. We begin by describing the communities in terms of players, context, and values. We discuss how different participants define the problem and how it might be redefined to reveal potential common ground. Then we analyze and evaluate the decision processes through which these communities have attempted to understand and resolve their problems. Finally, we offer recommendations to improve processes and outcomes for these communities and others facing similar issues.

Despite the differences in their cultures and governance systems, these two communities have experienced similar frustration in their quest to achieve durable outcomes around federal and state decision processes for wolf management. In both cases the decision process shows corrosive conflict in which respect and other values have been politicized. Common ground remains elusive. These similarities lend credence to our position that a deprivation of basic human values lies at the heart of the problem, rather than the scientific issues that are usually invoked to explain such failures in resource management.

METHODS AND STANDPOINT

Field research for this chapter was conducted by authors Watters and Anderson, who were master's students at the time under the guidance of third author Clark. After gathering general information on the decision processes in these communities, we identified the key players in wolf management and began talking with them about their values, expectations, demands, history, and present worldviews. We sought to understand the perspectives of people invested in the wolf debate, explore how and why values had become so politicized, and offer insights to improve social interactions and upgrade the decision process.

We interviewed nearly fifty people during the spring and summer of 2006, followed up with many of these contacts in 2007, and continued to track wolf issues in the region through 2011. Our interviews included ranchers and other landowners, cowboys, members of the Shoshone tribe from the Wind River Indian Reservation, leaders of environmental groups, consultants, scientists, business owners, academics, and employees from Wyoming Game and Fish Department (WYGF), the US Fish and Wildlife Service (USFWS), the US Forest Service, the National Park Service, Montana Fish, Wildlife and Parks, and a state-run elk feed ground. We also read official government reports, peer-reviewed and gray literature, and Wyoming newspapers. We used Clark and Brunner's (1996) methods for mapping social and decision processes to organize our observations and recommendations.

Recent arrivals from the East Coast, we had both been raised in a culture that prioritized wildlife over agricultural systems. We also had some preconceived notions about westerners' views on carnivores. In order to conduct useful research, we endeavored to put our own values and biases aside.

That said, we are pieces embedded in these processes as well, and hope to be understood from the standpoint described here. We are grateful to all of the people who took time to help us better understand these dynamics. We are self-conscious about our own roles and in every aspect of our work we sought to be respectful to all other parties.

Context and Problem Definition

Outgoing Secretary of the Interior Bruce Babbitt hailed the wolf reintroduction to Yellowstone as a success in 2001, and many environmental groups in the Greater Yellowstone region consider wolf reintroduction the pinnacle of conservation efforts over the past few decades (Northern Rockies Conservation Cooperative 2009). This narrative, however, ignores the ongoing social and political gridlock over wolf management, even after delisting. The current lawsuit-based approach to resolving the wolf conflict is intractable; it is easy to imagine that environmental groups and western states will be suing the federal government and each other long into the future. A new approach is called for, and in order to figure out what that new approach might be, we need to understand the perspectives of the players involved, the political and social contexts in which they operate, and the underlying foundation of their debate, that is, the way in which the "problem" of wolves is being defined.

THE PLAYERS

The conflict around wolf management in Wyoming involves a multitude of individuals and groups with differing worldviews, value demands, and cultural affiliations (Taylor and Clark 2005). Here we describe some of the main players in the wolf decision process in the Wind River and Upper Green River communities. Other people and organizations, especially hunters and those associated with energy or water development, also influence land, resource, and wildlife conservation in the region.

Native Americans on the Wind River Reservation

The reservation is home to the Eastern Shoshone and the Northern Arapaho tribes. Perspectives differ not only between the tribes, but also between traditionalists and those more integrated into the surrounding European-American economy. For example, we were told that some reservation ranchers were unenthusiastic at the prospect of the return of wolves, even though the animals have historic cultural significance. Others on the reservation see wolves as teachers and brothers and believe that they have a place and a right to exist in the natural world.

Ranchers in the Upper Green River Valley

Adhering to a strong work ethic and deep-rooted sense of land stewardship and conservation, ranchers in the Upper Green place value on the working landscape. Despite their grounded connection to the land, most generally resist being associated with environmentalists. Ranchers in some cases view the reintroduction of predators to the landscape as an attack on their livelihoods by the federal government. Because ranchers are invested in maintaining their way of life, losing livestock to predation constitutes not only a monetary loss, but also a loss to their sense of professionalism, self-reliance, and autonomy.

Environmentalists

Environmentalists in Greater Yellowstone tend to be liberal in outlook and nonlocal in origin. They often place a sense of rectitude (i.e., beliefs about what is morally right) above other values. Protection of endangered species and limitation of development are priorities for most environmentalists. Many see a working landscape as incompatible with their values, which are organized around the idea of wilderness. Despite their "outsider" status, environmentalists devote substantial resources and time to protecting the landscapes and wildlife that underpin Wyoming's booming tourist economy.

Wyoming Game and Fish Department

The regional offices of WYGF predominantly employ local people, and the agency's policies tend to align with both state policy and local opinion. In wolf management WYGF opposes the USFWS, advocating for state control of the species (Wyoming Game and Fish Press Release 2006). The WYGF depends on revenue from hunting and fishing licenses, and its policies are closely aligned with the interests of hunters (see chapter 9, this volume).

Federal Government Agencies (US Fish and Wildlife Service, US Forest Service, National Park Service)

Federal employees often take heat from both sides of the wolf debate: ranchers dislike them as representatives of the federal government, and some environmentalists object to federal agencies killing depredating wolves and pressure the federal government to keep wolves on the endangered species list. Many federal agency employees adhere to a belief in science-based management, though almost everyone we spoke to during our research recognized the need for a strong understanding of the social dimension for effective wolf management. The USFWS, tasked with managing the wolf population while it was listed, includes respected wildlife biologists who came to their professions out of a commitment to wildlife, but recognize the practical realities of managing carnivores in ranching communities.

TWO COMMUNITIES AND WOLF MANAGEMENT
Separated by the Wind River Range, but no more than fifty miles apart as the crow flies, the Native American communities on the Wind River Indian Reservation and the ranching community in the Upper Green River Valley share common biophysical resources, as well as a reliance on similar economic, cultural, and social resources and institutions. Both communities have seen increased natural gas drilling, but unlike the Upper Green River region, where the economy has experienced a rollercoaster of economic booms and busts, the socioeconomic situation on the reservation remains depressed.

In terms of cultural resources, both communities place authority in their elders or "old timers" and respect the traditional view of landscape and environment that these elders recall. Social institutions, such as family, schools, churches, and the Young Warrior's Society on the reservation, provide the backbone for these communities. Both communities feel their institutions are threatened by an influx of outsiders, increased trouble with methamphetamines, alcohol, and other drugs, and lack of interest from younger generations in maintaining traditional lifestyles. Both the Native American and the Upper Green River ranching communities express a strong feeling of stewardship toward the land and a sense that, despite their role on the landscape, their opinions about land and resource management have been largely ignored by government decision makers, the Wyoming government in the view of Wind River residents, and the federal government in the view of the Upper Green River ranching community. In both communities, this has led to a feeling of disrespect and loss of dignity.

Despite similarities between the two communities, important differences have also influenced wolf management, primarily with respect to the decision process or governance.

Wind River Reservation
The Wind River Reservation was delineated as the future home of the Eastern Shoshone by Chief Washakie in the 1868 Fort Bridger Treaty, which recognized Washakie's cooperation with the government during the settling of the frontier. Washakie chose the area because it was a relatively warm valley with mild winters that drew game animals down from the harsher surrounding highlands. He hoped that here his people would be able to survive with some degree of independence (O'Gara 2000).

Like most other reservations, this one was subsequently whittled down along with the rights and privileges promised to its inhabitants. In 1878 the federal government moved the defeated Northern Arapaho tribe onto the reservation even though the Arapaho and Shoshone were traditional enemies (Justia US Law 2011). During the early part of the twentieth century, reservation land was taken by white settlers and the Bureau of Reclamation for an agricultural and irrigation project that precipitated the major ongoing resource management issue for the tribe, namely, water rights to the Wind River, which have been consistently denied to Native Americans by the state of Wyoming (Meaney, Sullivan, and Clark 2010). This dispute has led to an enduring sense of injustice and anger at the state. Other environmental issues on the reservation include the remediation of uranium mining and creosote manufacturing sites, the presence of a potentially polluting sulfur plant, and the development of oil and gas resources.

As of 2007, 2,650 Shoshones and 7,400 Arapahos lived on the reservation, which today has an area of 2,268,008 acres (Eastern Shoshone Tribe Website 2007; Northern Arapaho Tribe Website 2007). Oil and gas revenues provide an important source of income in the community. Ranching also provides a livelihood for some tribal members, as well as for a number of white families with inholdings or property along reservation borders.

Each tribe maintains its own government, under the direction of a six-member tribal business council that works with the Bureau of Indian Affairs. Business council members are elected to two-year terms. A joint business council, consisting of the twelve members of the two tribal business councils, convenes to discuss issues relevant to both tribes. In addition, each tribe has a general council, consisting of all enrolled tribal members over the age of eighteen. The general council votes on decisions facing the tribe, with a majority vote carrying the decision. These tribal governments are recognized as autonomous by the federal government, and the Shoshones and Arapahos are considered sovereign nations. A fish and game department, under the direction of the joint business council, manages wildlife resources on the reservation, with technical assistance from the US Fish and Wildlife Service.

The last wolf on the reservation, in fact, in all of Wyoming, was shot in the 1940s by Leo Cottonear, a Shoshone who later spoke with regret of what he had done (*High Country News*, February 6, 1995). In 2005 wolves returned to the reservation in small numbers. Only portions of the reservation are considered suitable wolf habitat (Hnlicka, interview, July 2006), but wolves still have the potential to come into conflict with people.

Although the federal government consulted the Wind River Reservation before reintroducing wolves to Yellowstone, people on the reservation feel that they have not had adequate opportunity to participate in subsequent federal and state decision processes. A successful restoration program would eventually have direct effects on the reservation, and those effects would be tied to wider decisions about the wolf population, so the exclusion of the reservation from decision processes is problematic. When the wolves did return, the reservation had the opportunity to develop its own wolf management plan, but interviewees still expressed frustration that they had little voice in the larger discussion about statewide wolf management.

At the time of our interviews, the tribal wolf management plan was being written, but had not been implemented. The Tribal Game Code, which successfully rebuilt ungulate populations on the reservation, provided a precedent for the wolf management plan and was repeatedly referenced as a success story for tribal wildlife management. Developing a management plan for wolves, however, presented new challenges. Hunters have a shared interest in maintaining healthy populations of game animals, but large carnivores are less unifying. Traditionalists may see them as playing an important spiritual, cultural, and ecological role on the reservation, but ranchers and some hunters are less likely to be supportive. The decision-making process on the reservation, however, by which the joint business council must approve the plan and all enrolled members are entitled to vote on adopting it, has helped to ensure that the wolf management plan for the reservation is balanced, rather than drastically for or against carnivores.

The Wind River Wolf Management Plan was approved and adopted in April 2007. The plan was written by Tribal Fish and Game, in cooperation with USFWS officials, who made clear in interviews before the plan was finalized that it was unlikely to be radical. The plan states an intention to be neutral, sets no specific numerical goals for wolves or packs on the reservation, and allows lethal control for animals "in the act" of depredating or threatening domestic animals. On the other hand, the plan acknowledges traditional Shoshone and Arapaho views of wolves "as kin, as helpers, as strong, as deserving of respect, and as placed here by the Creator for a purpose" (Shoshone and Arapaho Tribal Fish and Game 2007, 2). In contrast to Wyoming's plan, the reservation's plan deems all reservation wolves trophy game even after delisting. Unless a wolf is caught harassing or harming livestock, no one will be able to kill a wolf without a trophy hunting tag (Shoshone and Arapaho Tribal Fish and Game 2007, 2).

The role that the Shoshone and Arapaho play in the broader political scene of Wyoming is limited, resulting in a sense of powerlessness and deprivation of respect. On the reservation, however, the process and the precedent exist for more democratic carnivore management.

Upper Green River Ranching Community

The Upper Green River Cattle Association (UGRCA) was founded in 1916 by rancher A. W. Smith, who wanted to organize and collaborate with other ranchers in the area and decided to lay claim to the eastern side of the Green River basin (Loizer and Platts 1982, 8). The UGRCA existed under a number of different names for several years prior to its official inception, but in 1916 it was formalized into the cattle association that exists today (Loizer and Platts 1982, 8). Nearly a hundred years later the association still persists in the southernmost portion of the Greater Yellowstone Ecosystem as one of the largest public grazing allotments in the national forest system, grazing 7,565 head of cattle and 27 horses on approximately 130,000 acres of the Bridger-Teton National Forest (Farquhar 2007; Sommers et al. 2010).

Historically, the big mountains, beautiful rivers, and seemingly endless sagebrush range drew people to the area who were interested in working the land. In the past twenty years, however, the rancher/homesteader influx has given way to new groups of settlers, many of them second home owners, typically not from Wyoming and typically connected to the landscape in terms of its use for recreation. The other significant group of new settlers to the area are workers on the Pinedale Anticline and Jonah Fields, two of the most productive natural gas sources in the continental United States. The second home owners and the gas field operators are generally disconnected from the history of the area as a ranching landscape.

In 2006 the UGRCA, better known as the "Drift," consisted of sixteen members, most of whom were local ranchers who owned land in the greater Pinedale/Daniel/Boulder area of the Upper Green (Farquhar 2007). The association provides its members with the opportunities afforded by a union, consolidated finances, shared cost of labor, and influential bargaining power. In addition, the Drift is made up of ranchers who play important leadership roles within the community. While the wealth generated by ranching in the region is far less than the revenue from natural resource extraction, the culture of ranching continues to prevail in the social identity of the community. Wyoming as a whole "produces less than 2 percent of the total value of America's cattle and calves" (Western 2002, 11), but small-scale cattle operations in the state generate other positive outputs beyond animal units. The Drift isn't supporting a nationwide cattle market but does have the opportunity to preserve critical open space, provide wildlife habitat, and supply other important ecosystem services. The role of ranchers as stewards of a landscape should not be obscured by other simple calculations of a rancher's worth (i.e., calves shipped at the end of the year). Much is at stake if the Drift ranchers go out of business and sell to development or resource extraction companies.

A long tradition of opposition to wolves is a hallmark of the ranching culture in this part of Wyoming. On April 22, 1915, a list of the more than one hundred members of the Wolf Bounty Association appeared in the *Jackson Hole Chronicle*, epitomizing the sentiment of the time. The association paid a high bounty on each wolf killed, and by 1928 WYGF reported that "only five wolves were known to remain in WY outside of Yellowstone National Park. Two of these ranged in Jackson Hole" (Platts 1989).

Hostility toward wolves is still common in the culture of the Upper Green River ranching community. The parents, grandparents, and great-grandparents of today's generation dedicated themselves to eliminating the very species that was then reintroduced to the area in 1995 by the federal government. "My grandfather killed every wolf on our place." "My father said that there were a few wolves around, but they never bothered the cattle" (anonymous interviews, July 2006). These comments help us to understand that many people in the ranching community of the Upper Green base their ideal vision for today's landscape on the picture painted for them by the old-timers in their community.

Even so, some ranchers expressed sympathy for the wolves, suggesting that federal culling, started so soon after reintroduction, was unfair to the animals themselves. Other ranchers expressed excitement as they recounted instances in which they had witnessed wolves in the area. Almost all the ranchers agreed that wolf populations should be allowed to persist in the national parks, but were adamant that livestock producers should have the unconditional right to protect their livestock. The UGRCA's first officially confirmed wolf depredation came in 2000 (Sommers et al. 2010). At the end of the summer of 2006, 50 of the 191 animals lost by UGRCA members were confirmed by federal and state wildlife agencies as predator kills (equal proportions by wolves and grizzly bears) (Farquhar 2007). For small livestock producers, this constitutes a significant loss. Offering up an apparently reasonable solution, one rancher commented: "We weren't proponents of wolf re-introduction from the start, but they are here now and we have no interest in eliminating them completely again. Instead, we just want the right to protect our stock. Ranchers are smart folks; we understand that if we take too many animals, the government will expand the protected areas, and re-list the species. If they just allowed me to take out the depredating animals, I'd leave the rest of them alone" (anonymous interview, July 2006).

Members of the ranching community expressed frustration with the federal government and the environmental groups who refuse to trust ranchers or at least recognize their decades of land stewardship. Ranchers, who consider themselves good stewards, are offended by rules that seem to demean their ability to make good decisions.

DEFINING THE PROBLEM OF WOLF MANAGEMENT

Wolf management poses a problem to people in Wyoming, but there are many different perceptions of what the problem is and how it should be addressed. The wolf issue illustrates the ways in which people define problems to reflect deeper concerns and the ways they participate in the management decision process to pursue their value demands. In this section we examine how different groups have defined the problem of wolf management in Wyoming. These different problem definitions illustrate the fact that the wolf is a catalyst through which people seek to advance their own beliefs, values, and demands. Many participants hold worldviews that exclude and discount the views of others, leading to disrespectful interactions, increasing conflict, and a hardening of attitudes. As a result, most participants are dissatisfied with the processes and outcomes.

The "Scientific Management" Problem Definition

The most prominent mantra for wolf management in the Greater Yellowstone Ecosystem is "appropriate scientific management." When wolves were first reintroduced into Yellowstone National Park, the problem was mainly seen as an issue of maps and numbers. How many wolves should be conserved, and where? How many wolves constituted a recovered population, eligible for removal from the endangered species list? How long would it take before they reached recovery? How quickly would they disperse beyond Yellowstone, and where would they go when they did? These questions were addressed and answers promoted in various ways by scientists, the government, and environmentalists in the belief that sound scientific information would ensure that reintroduction would be a biological, ecological, and social success.

Wolf management is still often defined as a problem of scientific management, particularly by environmentalists, federal resource managers, and the wolf-sympathetic public. In May 2011, when environmentalists launched their lawsuit challenging the latest wolf delisting, the executive director of the Alliance for the Wild Rockies stated, "We will not allow the fate of endangered species to be determined by politicians serving special interests. These decisions must be based on science, not politics" (Alliance for the Wild Rockies 2011). Proponents of scientific management tend to see science as a justification for the reintroduction, and they remain convinced that if everyone understood the science, then even those currently opposed would realize the necessity of reintroduction and continued protection (Mattson et al. 2006). The president of the Wolf Recovery Foundation affirmed this in a 2007 press release: "We've learned what an important role wolves play in our natural ecosystems. With this newly realized [scientific] information, there can be no reasonable justification for returning to the days of mindlessly killing wolves" (Maughan 2007). Many people, however, including some wildlife biologists, recognize that science alone will not solve the problems surrounding wolf reintroduction and advocate instead for a strong social and integrative approach to wolf management (Clark and Gillesberg 2001; Smith, personal interview, July 2006; Jimenez, personal interview, June 2007).

In another example of problem definition relying on scientific management, environmental advocacy groups responded to the 2009 delisting decision by forming a coalition and launching a website and ad campaign that urged the public to support "managing wolves with science, not politics" (www.westernwolves.org, 2009). The website and ad campaign gave a nod to the idea of social engagement with different stakeholders, but also included cultural references that could be interpreted as conflict-escalating rather than conciliatory. One ad stated:

"Wyoming's top predator isn't the wolf. It's the PICKUP TRUCK. . . . Wolves eat deer and elk. That's a fact of nature. Here are a few others. Vehicles kill many times the number of deer that wolves do. Elk herds are above Game and Fish target populations. And hunter success rates remain solid. Truth is, the first step toward sound wolf management is listening to science, not rhetoric, SUPPORT SCIENCE BASED WILDLIFE MANAGEMENT" (www .westernwolves.org, 2009). This ad attempts to address the concerns of hunters and reorient the debate toward scientifically verifiable information (the claims in the ad are technically correct), but the tone could be interpreted as condescending, and the reference to a pickup truck rather than simply a "car" or a "vehicle" seems like a backhanded jab at a particular cultural group. It is difficult to imagine that this ad would change a hunter's mind about wolves; it is easier to imagine the hunter becoming angry and more entrenched. Science cannot provide common ground for decisions unless there is a high degree of consensus on desired outcomes (Pielke 2007), which is not the case with wolves.

Problem Definitions Collide: The Symbolic Politics of Wolf Management

Just as Yellowstone is regarded by some as wilderness and by others as a highly managed landscape, wolves are the victims of competing mythologies. Environmentalists see wolves as majestic representations of wilderness itself; scientists strive for predator-prey balance in an ecosystem and consider wolves a necessary component; Native Americans see wolves as an emblem of tradition and, to a certain extent, as a rallying point for issues of sovereignty; and ranchers view wolves as a manifestation of a government that is determined to stand in their way. Each of these participants defines the wolf problem in accordance with his or her own political mythology. Wolves are manipulated as political symbols of wilderness, science, and encroachment by government, to fit different stories about how the world does and should work.

To environmentalists wolves represent the wild and the pristine, a pre-conquest wilderness ideal compatible with a mythology of the West as untouched nature. This attitude is reflected in the quote from the president of the Wolf Recovery Foundation in the preceding section. The ideal of an intact ecosystem is reinforced in Yellowstone itself, where ranger presentations speak of the way that humans, in removing wolves from Yellowstone, had "broken" an ecosystem that the reintroduction subsequently "fixed" (Canyon Campground presentation, July 2007). These recurring themes of hallowed wilderness and a "fully functional ecosystem" illustrate environmentalists' perception that the reintroduction of the wolf was the symbolic righting of past wrongs.

To ranchers who have been on the land for several generations, the myth of the West and the political symbolism of the wolf play out very differently. In broader American society the settling of the West was portrayed in the nineteenth and the first half of the twentieth centuries as the inevitable progress of civilization. Colonization of the West was seen as a vital stage in the birth of America, in which subduing the land and its native wildlife and indigenous people played an important part (Cronon 1994). In the minds of many Americans and particularly in ranching culture, the concept of the "cowboy frontier" lives on. The sense of land stewardship inherent in this mythology is strong, but the focus is on making the landscape productive, generating something of economic value. Before the advent of energy development, the only productive use of the semi-arid sagebrush country of

Wyoming was ranching, and the presence of large carnivores on the landscape was seen as directly undermining this use. Destroying wolves was considered an essential part of bringing the landscape into the European-American economic and social system, and for many people a wolf-free landscape remains a prerequisite to the perception of the land as productive. To ranchers and their narrative of self, place, and purpose, the return of wolves to this ecosystem is a step backwards and a repudiation by mainstream American culture of the central importance of ranching life and the legacy of the ranchers' ancestors.

The story that the ranchers tell about the settling of the West contrasts strongly with the story told by Native Americans. The native cultural narrative is one of recalled genocide, oppression, and military defeat, in which the fate of their cultures mirrors the historic fate of the wolf. The Shoshone and Arapaho tribes, like the adjacent ranchers, have a strong ethic of responsibility toward the environment, but their sense of stewardship embraces a different relationship with the land. For many Native Americans, particularly traditionalists, the wolf represents a relative and a teacher, and its return to the region has spiritual, cultural, and ecological significance, and represents a symbolic victory over the state of Wyoming.

Each of these symbolic definitions of the wolf management problem points to a different set of causes and a different set of potential solutions. Some observers argue that this is the real problem—the conflicting mythologies and political symbolism embraced and promoted by the participants in wolf management. However, defining the problem as simply symbolic politics is overly narrow and doesn't offer much hope for resolution. We suggest a new paradigm, including a new problem definition that has the potential to move us forward by recognizing the common ground that does exist.

A New Definition

The issues surrounding wolf management might be best understood as a problem of value deprivation. Harold Lasswell, a social scientist and a professor in the integrative sciences, recognized eight basic human values that all people seek while interacting with one another: power, wealth, respect, affection, rectitude, skill, enlightenment, and well-being (Lasswell 1971). In the case of wolves in Wyoming all values are at play, but in our conversations with community members across the two cultures it was evident that four—power, wealth, rectitude, and respect—are especially important. These values have been eroded in various ways in both communities, and this in turn has decreased participation, trust, and investment in the decision process. In the following we redefine the problem of wolf management for these two communities in value terms.

POLITICIZATION OF VALUES
Power

The power value and the pursuit of control are pivotal in the wolf case. All parties feel disempowered. Ranchers want to be granted the power to protect their lifestyle and livelihood, environmentalists want to gain political power to prevent wolf delisting under a Wyoming State plan that would allow unregulated killing of wolves, and Native Americans want the autonomy to make decisions as a sovereign state, a right granted to them by the federal government but often circumvented in resource management.

Several examples illustrate the links between the wolf debate and demands for power. During interviews both Native Americans and Upper Green River ranchers made quick shifts between the topics of wolves, water rights (on the reservation), and natural gas development (in the ranching community), suggesting the connection in people's minds between each example of natural resource management and their perceived lack of control over natural resources in general. In the Native American community one man emphasized his outrage that the Fremont County commission had passed a resolution banning wolves and bears from the county. The reservation is in Fremont County, but the county commission does not have authority on the reservation, and there is no Native American representation on the county commission, even though decisions made by the commission may have effects on the reservation. The carnivore ban was such a decision, made without consultation or inclusion of reservation voices. The interviewee then spoke about an American Civil Liberties Union lawsuit against the county commission, challenging the lack of tribal representation. He emphasized the continuing sense of disempowerment and disrespect experienced by Native Americans when local white government makes sweeping decisions with implications for the reservation, without legitimate authority (anonymous interview, July 2006).

In another example, even though we told each Upper Green River interviewee that we were interested in their thoughts on wolf management, all conversations included natural gas exploration. Ranchers spoke out particularly loudly. Mineral rights (distinct from surface rights) belong to the federal government, and ranchers and other landowners feel powerless about the allocation and development of those rights. Many people felt that it was only a matter of time before the Upper Green River Valley was transformed into a pockmarked industrial area with thousands of drilling rig pads.

Wealth

Some attempts to ameliorate conflicts arising from wolf reintroduction have addressed the wealth value; some even treat it as the most effective solution for promoting carnivore conservation (e.g., Sommers et al. 2010). The Bailey Wildlife Foundation Wolf Compensation Trust, developed by Defenders of Wildlife, attempted to resolve conflict by "eliminat[ing] a major factor in political opposition to wolf recovery and shift[ing] the economic burden of wolf recovery from livestock producers to those who support wolf reintroduction" (Defenders of Wildlife website, Bailey Wildlife Foundation Wolf Compensation Trust 2007). This was an innovative approach, but one that did not seem to be effective in the Upper Green River ranching community. The Defenders program only compensated ranchers for livestock found and confirmed as wolf kills, although UGRCA ranchers estimated that for every dead calf found there might be as many as seven that they didn't find. The Defenders program was also considered ineffectual because of the lag time between when a calf was killed and when the compensation check arrived in the mail. The compensation arrived too late to ease the financial hardship caused by a calf lost in the previous year.

Beyond procedural issues, members of the ranching community repeatedly told us that compensation alone is not a sufficient remedy. As one rancher put it, "What if you spent years and years building your dream house, and then one afternoon your next door neighbor came over and bulldozed it down, but handed you a check for its total monetary value. Would the fact that you had been compensated for your loss make it all OK?" (anonymous interview, July 2006). Addressing the wealth value while neglecting the other seven values does not solve the problem and may aggravate the issue. For ranchers the issue of wealth is a subset of the larger challenge of regional, national, and international economic change that threatens their way of life. Successfully raising livestock is an affirmation of the rancher's way of life, and monetary compensation for a lost animal does not relieve anxieties about the cultural validity and economic viability of ranching. Defenders of Wildlife ended its compensation program in 2010, but assisted the state of Montana in setting up its own compensation program (Defenders of Wildlife 2008; Urbigkit 2010).

Rectitude

Like power and wealth, rectitude also plays an important role in the demands of the groups involved in wolf management. Almost every debate about wolf management involves an assertion of the integrity and moral superiority of one group, one way of life, one worldview, over the others. Native Americans appeal strongly to a vision of autonomy and justice structured around traditional values and their history of oppression. Environmentalists evoke the moral order of nature. Ranchers speak of the land ethic and sense of community provided by the working landscape and the traditional Western lifestyle. Each group advocates for its position based on an idea of how the world *should* function. Rectitude is one of the more difficult values for which to find common ground because it is so closely tied to people's core identities. Whether these identities are rigid or flexible is open for study (see Rentfrow, Gosling, and Potter 2008).

Respect

In the ranching community, individuals demand respect by talking about the way that they as livestock producers feel they have been wronged, especially by the "feds" sitting behind desks in Washington. Implicit in this assertion is an appeal for validation of the rancher's way of life as important to the identity and character of the region and the nation, a demand that society respect the hard work and sacrifices that ranchers make to raise livestock. Reacting to an article on wolf management in the *Casper Star Tribune*, an anonymous commenter wrote: "I see that the unknowing tree-huggers from other areas of the country have chosen to tell us what is good for us. I think, for instance, that the cockroaches and rats in New York City should also be considered an endangered species, where the residents must feed and protect them. After all, we don't know anything about them or the problems that they cause" (Miller 2007, 1). This sentiment was common in Wyoming. People from the East Coast, and New York City in particular, were frequently referenced in our interviews as unknowledgeable and disrespectful of Western values and lifestyles.

The Native Americans of the Wind River demand respect by talking about the ways in which these animals were their spiritual ancestors and by using their wildlife management plans as a platform for claims to autonomy and sovereignty. The Shoshones and the Arapahos have struggled to gain control over natural resources on the reservation. Recognition of their right to manage large carnivores was an important step in showing a level of respect that has too often been withheld by the state and surrounding communities.

The problem of disrespect for Native Americans is not simply a matter of Native American perceptions. Several people from outside the reservation expressed the opinion that Native Americans were incapable of managing their wildlife, that the tribes should not have exclusive rights to game animals on the reservation, and even that the reservation should be taken away from the Native Americans because it hinders the integration of Native Americans into white culture (anonymous interview, July 2006). These remarks illustrate the depth of prejudice and the lack of contextual understanding on the part of some whites and are emblematic of the disrespect that Native Americans face. The fact that prejudiced whites use wildlife management as a way of pointing out perceived Native American incompetence and of advancing white interests suggests reasons for a staunch Native American defense of wildlife management as a platform for demanding respect.

Environmentalists demand respect by claiming that they are speaking on behalf of the animals themselves. In June 2007 President Bush's administration proposed a rule change to allow the killing of wolves because they were perceived to be "the major cause" of declines in elk or deer populations (Defenders of Wildlife 2007). In response, a regional representative of the Sierra Club stated: "The new rule would allow the killing of wolves as a first, rather than last, resort, and the government has no basis to do so. . . . Clearly, the wolves are not affecting hunting opportunities. This rule would promote the needless killing of wolves that eat elk and deer; the same animals that wolves have been preying upon for thousands of years" (Defenders of Wildlife 2007, 1). This speaker requests respect for the authority of evolution and natural order; animals have been killing their prey since time immemorial, and human authority pales in comparison.

In another example, an anonymous commenter wrote in response to a newspaper article on predators in the Upper Green River Valley:

If ranchers in the Upper Green would learn to live with predators as another rancher [mentioned in the same section] did, they would be a lot better off. Predators have a place on this earth even more so than livestock. Livestock are introduced animals, and if ranchers cannot learn to live with native species, they should not be in areas where they exist. They are a MINORITY in the public arena. The MAJORITY of the public cherishes predators along with other forms of wildlife. Wildlife of all kinds should be protected—not just those animals that are popular with hunters. I have no sympathy or patience with these ranchers who are unenlightened and greedy. (Farquhar 2007)

Like many other wolf advocates, this anonymous commenter demands respect for wolves and the "majority" who seek to protect them. In frustration, the commenter also takes a jab at the ranching community.

Lastly, many federal government employees demand respect by leaning on the platform of science and scientific management as the answer to all problems. In an article in the *Casper Star Tribune*, the director of the Mountain-Prairie Region of the US Fish and Wildlife Service at the time said, "It all zeros back to basing your decision on good science and not just anecdotal observation" (Miller 2007, 1). As discussed earlier, although good science is important for wildlife management, even the best science cannot resolve value-based disputes. A more realistic problem definition acknowledges the usefulness of science for understanding the way ecological systems function, but focuses primarily on wading through the human values and preferences that will ultimately drive management (see chapter 8, this volume). An anonymous response to this article read: "Wolves were flown and trucked in from another country and planted on that family homestead. They are not naturally occurring. In fact I can find no record of wolves that were here in the late 1800s or ever, for that matter, forming the huge packs that we see today" (Miller 2007, 1). This statement flies in the face of "good science," yet reflects a worldview held by many wolf opponents. Does the USFWS really believe that good science is the path to resolution of conflict in wolf management policy? Probably not, but agency representatives seek the respect, power, and authority that are associated with scientific expertise.

All these unsuccessful interactions reflect the fact that the wolf debate is fundamentally about values, with power, wealth, rectitude, and respect in the forefront. The carnivores themselves are just the medium through which value demands are being channeled. Any realistic problem definition must embody this reality. When wolf management is understood as a problem of value politicization and erosion, our attention is directed toward the real causes of the problem: decision processes that have failed to resolve value-based conflicts and advance the common interest.

Wolf Management Decision Process

Managing wolves in ways that serve common interests is an ongoing process of decision making through which values are generated and allocated. Most of these decisions are not about wolf biology, but about our own actions. For example, should we kill wolves, and if so when and how? When should we leave them alone? How should we set management goals? The wolf management process is about people and what we value, how we choose to interact, and especially how we talk about, set up, and carry out practices that affect wolves and the environment. Our decisions guide the ultimate management outcomes, and in order to improve these outcomes we must understand and carry out a high-quality decision process that emphasizes respect for all parties (Clark 2002).

OVERVIEW OF THE WOLF MANAGEMENT DECISION PROCESS IN THE GREATER YELLOWSTONE ECOSYSTEM

A decision process is a series of activities, typically portrayed as a cycle beginning with the gathering of intelligence, moving through phases of promotion, prescription, invocation, application, appraisal, and termination. For the Wind River Reservation and the Upper Green River ranching community, wolf reintroduction proceeded along a single path until the federal recovery goals were reached, at which point the tribes on the reservation initiated a separate decision process to create their own management plan, while remaining part of the Wyoming and federal decision processes. The following analysis takes into account the nested nature of these two decision processes.

Wolves were reintroduced to Yellowstone National Park in 1995 under a federal prescription after a twenty-year planning (intelligence and promotion) process that included consultation with the state of Wyoming and with the Wind River Reservation, among others (Jimenez, personal interview, July 2006). The federal government's prescription for wolf reintroduction set recovery goals of ten breeding pairs in Wyoming for three consecutive years, and it established conditions for delisting that included a requirement that the state create a management plan to provide for the continued existence of wolves in Wyoming, exclusive of the federally managed population in Yellowstone. To help mitigate potential conflict with livestock, reintroduced wolves were designated as an "experimental nonessential" population under section 10(j) of the Endangered Species Act (Federal Register 1994). The 10(j) classification meant that Wildlife Services, a branch of the US Department of Agriculture, could kill depredating wolves. This designation and a compensation program established by Defenders of Wildlife sought to address the concerns of ranchers and other opponents (Wilmot and Clark 2005).

This particular decision process, defined by wolf reintroduction and subsequent federal action, is ending. A new decision process, defined by state and tribal management of wolves, is beginning.

APPRAISING WOLF MANAGEMENT DECISION
PROCESSES IN THE TWO COMMUNITIES

Having described the overall decision process for wolf management in Wyoming, we now turn to a description and assessment of how the decision-making processes have functioned in the Wind River Reservation and the Upper Green River communities. Appendix 3.1 summarizes our evaluation of these decision processes. The appraisal contrasts the federal decision process for reintroduction with the tribal process through which the new management plan for the reservation was developed.

We organized appendix 3.1 to reflect the sequence of events that started with the decision (prescription) to reintroduce wolves to Yellowstone National Park, followed by repopulation of wolves in the park and surrounding areas, and concluding with the *next* decision process associated with managing wolves after delisting. By starting with prescription we are able to emphasize the repercussions of federal rules and guidelines for two small communities. The federal government set the prescription in early 1995, and stakeholders in Wyoming have spent the last eighteen years reacting. By ending our analysis with the intelligence and promotion functions, we are able to explore how the lessons learned from this former decision process might provide valuable information about how to approach the human side of the current decision process, in which wolves are being removed from the endangered species list and returned to tribal and state management.

Federal and State Decision Process for Wolf Reintroduction

Our evaluation revealed failures in both communities in each of the seven functions of the decision process for wolf reintroduction and management. Beginning with prescription, the stage at which rules and guidelines were set, both communities were disappointed, a sentiment that persisted throughout the process. To various degrees, each interest group has tried to increase its own grasp on desired values, but in so doing has deprived the other groups of values such as power, respect, wealth, rectitude, skill, and well-being. Rarely does anyone acknowledge the deprived values of other groups or work effectively to resolve disputes. In short, the debate about wolves has been reduced to a disrespectful discourse.

Both communities invested most of their efforts on the promotion function. As in most other politicized public arenas, each side feels a compulsion to speak out strongly on its own behalf. Predictably, this means that most of the discourse consists of each group making loud proclamations, while turning a deaf ear to what the other parties are saying. Thus other functions, such as application (which includes dispute resolution) and appraisal (review and evaluation), are largely ignored. In this particular case, the intelligence and promotion functions for the next decision process (to inform new prescriptions) have been inadequate and unsystematic because of the interference of litigation.

The federal and state decision process also fails the three tests developed by Brunner et al. (2002) for evaluating whether a decision process is operating in the common interest (appendix 3.2).

(1) The Procedural Test

A sense of exclusion from the process is a common sentiment in both the Native American and ranching communities, although some federal employees and environmentalists challenge this perception. One interviewee said that "some people may feel as if they have been excluded, but that's only because they didn't get their way" (anonymous conversation, January 2007). Even if this claim were true, it would not dismiss the validity of the test. When a controversial decision is handed down from the powers above, and people are left feeling as if they weren't heard, it is unlikely that the common interest has been served, and the "losing" voices may become louder over time. This is the case with wolves in Wyoming.

(2) The Substantive Test

Although a wealth of information about wolf reintroduction was disseminated, this decision process fails the substantive test because information alone has not been enough to compensate for the lack of reciprocal tolerance and mutual accommodation among the various parties engaged in the debate. The process itself has served to divide the parties further and has exacerbated the deprivation of respect among the groups. As questions about wolf management have dragged out through years of tedious litigation, access to information has also become more contentious. The USFWS, for example, maintains a code of secrecy concerning the locations of wolf packs in certain areas. This only adds to the sense of disrespect felt by many ranchers: "Why don't the Feds trust us to manage our own resources?" Withholding information, not to mention decision-making power, further undermines efforts to achieve common interest outcomes.

(3) The Practical Test

Wolf reintroduction has failed to live up to the hopes of any of the groups involved. Ranchers expected quick compensation for losses, timely delisting, and a small and highly controlled wolf population. Environmentalists expected a free-ranging wolf population, always protected and forever beyond the threat of guns. Native Americans expected to be treated as equal players in state and federal decision making. In implementation, none of these things has happened. From a biological perspective, wolf recovery was a success, but from a social perspective, all participants are frustrated.

Decision Process for the Tribal Wolf Management Plan

The Wind River Reservation Wolf Management Plan was developed after wolves returned to the reservation in 2005. Drafted in consultation with the USFWS, the plan was adopted in 2007. In contrast to the federal and state decision process for wolf reintroduction and recovery, the tribal decision process has performed relatively well in the decision activities. Although the intelligence and promotion activities were not fully inclusive—USFWS officials involved in preparing the plan interviewed tribal elders about how wolves should be managed, but did not consult the general population (Hnlicka, personal interview, July 2006)—the prescription activity broadened the range of participation. Since the plan had to pass both business councils and the general council and was therefore voted on by all enrolled tribal members, the plan may be seen as meeting the needs of much of the population.

The diversity of interests involved in prescription on the reservation necessitated a moderate approach in developing the plan. In our interviews some tribal members expressed hopes that the plan would serve as a vehicle for asserting tribal identity and cultural values in the face of Wyoming's white majority (anonymous interview, July 2006). The provisions of the plan dealing with wolves appear to be reasonably balanced. Tribal ranchers may be distressed by depredation and hunters may see wolves as competition for game resources, but other tribal members would prefer to see all carnivores protected. The involvement of the USFWS, the federal agency responsible for managing wolves under the Endangered Species Act, also helped to ensure that the Wind River Wolf Management Plan is less antagonistic to wolf supporters than the Wyoming State wolf management plan. It is too early to judge whether the decision process for the tribal plan meets all of the tests of the common interest, but it appears to allow flexibility to accommodate diverse values.

Moving Forward: Lessons and Recommendations

A land conservation organization in New Mexico describes its mission thus: "We believe saving relationships, not places, is the key to the future of con-

---

Box 3.1

Lessons for Managers from the Case of Wolves in Two Wyoming Communities

- Attention to social and community identities, as well as culturally appropriate relationships with wildlife, is necessary when managing carnivores, especially when planning and implementing reintroductions.
- Symbolic politics often mask or stand in for value demands, and these symbolic politics will escalate the longer the value demands remain unmet.
- Creating arenas for safely discussing value demands across identity groups could serve to de-escalate tensions around wildlife. To do this, however, all parties must be committed to the process and willing to put aside their own views, at least within the arena, in order to consider the views of others.
- Appeals to science-based management are likely to be met with limited success, even resistance, because the science on wolves is multifaceted, incomplete, acontextual, and open to varying interpretations and because the social discord over wolf management is so profound. The situation calls for a more robust cross-cultural dialogue, rather than a deeply entrenched legal battle over the science.
- Stop litigating and start talking!

---

servation" (Quivira Coalition 2006 Annual Report 2006, 4). The same could be said for wolf conservation in Wyoming. The solution does not lie with the number of wolves on the land or the dollar amount of a compensation check. Instead, the key to wolf conservation lies in honest and effective problem definition and movement toward a new decision process that is inclusive, factual, complete, rational, integrative, effective, balanced, prompt, constructive, ongoing, unbiased, practical, ameliorative, and that passes the three tests of the common interest. In order to reconstruct the existing decision process to meet these standards, efforts must focus on rebuilding the relationships that have broken down over the course of wolf reintroduction (Wilkinson, Clark, and Burch 2007).

Both the Native American and the ranching communities have expressed their desire for respect, economic self-sufficiency, and the power to make decisions about their own lives. As they and other groups involved in wolf management continue to make these demands through the polarizing filter of the media, rather than face-to-face in reasoned, respectful discussions, the chances of finding common ground diminish. Only when all sides commit to finding common ground and rebuilding relationships will the deeper governance problems be addressed (Clark, Miller, and Reading 2010; see chapter 8, this volume). The following are specific suggestions that will help these communities move toward these two important goals.

FROM KITCHEN TABLES TO WORKSHOPS:
NEW ARENAS FOR INTERACTION

Creating informal arenas around kitchen tables, over cups of coffee, and in hay meadows is an effective way to begin the process of rebuilding civility, human dignity, and skill among diverse communities. Common interest is realized by identifying, acknowledging, and pursuing shared interests. "Kitchen table" arenas are inexpensive to initiate and, in contrast to the polarized debate in the media, serve as informal settings in which conflicting groups can have respectful discussions about achieving shared goals. Conversations at a kitchen table arena may be facilitated by an impartial third party and should start with a focus on areas of common ground (e.g., preserving open space, improving land stewardship, managing natural gas development) as a medium through which to build social capital. Participants should be asked to define (and write down) the problem from their perspective. The facilitator's role would be to draw out the similarities in the various problem definitions and then lead the conversation in a way that acknowledges individual interests, while guiding the group toward integrated common goals.

At the same time, it is necessary to create more formal arenas and institutions (county/state/federal) to address identified weaknesses in decision processes. One possibility is stakeholder workshops, similar to the grizzly bear management workshops organized by Michael Gibeau of Parks Canada, which build participants' capacity and allow for the possibility of fostering respect and understanding through informal conversations as well as formal workshop activities (Rutherford et al. 2009; chapter 7, this volume). Workshops can also address the place-based nature and intricacies of a management issue and provide a forum for solutions that are tailored to a specific community. In this way, management strategies can benefit from local strengths and leadership.

Attendance at such workshops, conferences, and meetings by those groups that feel most alienated is more likely if the events are held on that group's territory. The California Rangeland Conservation Coalition, dedicated to protecting 13 million acres of oak woodland and grazing land, organized a panel discussion at the California Cattleman's Annual Conference entitled "Boots and Birkenstocks," which brought together ranchers and environmentalists to develop a strategy to prevent development of ranchland (Weiser 2007). A retired regional head of the US Fish and Wildlife Service, who is credited with inspiring the coalition, pushed hard to make ranchers and environmentalists identify common ground. "I kept saying, 'I understand what you're against. What are you for? . . . It turned out both the cattlemen and the environmental groups had a tremendous amount of overlap. It didn't surprise me, but I think it surprised them'" (Weiser 2007).

CHANGING AND BRIDGING: NEW NETWORKS

Change from within happens when an organization reaches beyond its established membership and invites in those who hold different views. Rancher Bill McDonald of the Malpai Borderlands Group in southern Arizona coined the term "Radical Center" to describe a neutral place where people can come together to explore their common interests rather than argue their differences. The exchange of information is important, but the relationships forged in the process of that exchange are equally important. As another example, *High Country News*, a pro-environmental publication, devoted a 2007 issue to the merits of "hook and bullet" conservation, and the editor commented, "As the West grapples with the forces of energy development and population growth, it will need the combined efforts of everyone who cares about wildlife and the wild places." The feature article was entitled "Predator Hunters for the Environment" (Larmer 2007). Such attempts to cross boundaries are challenging, but necessary in the process of establishing legitimacy among adversaries.

Established organizations can also create broader networks around common ground issues. The Greater Yellowstone Conservation Directory, an inventory and directory of conservation organizations in the Greater Yellowstone region, compiled by the Northern Rockies Conservation Cooperative in Jackson, Wyoming, is an example of an effort to build relationships among conservation groups and create arenas for policy discussions. The group surveyed the environmental community and convened formal and informal small-scale discussion groups about wolf conservation issues.

BEYOND LITIGATION: MORE SOPHISTICATED STRATEGIES FOR ENGAGEMENT

The environmental community frequently defaults to the "legality" or "illegality" of actions affecting endangered species by appealing, rhetorically and in court, to the law. Challenged on the social complexities of carnivore recovery, retorts about upholding the Endangered Species Act and forcing the USFWS to "do its job" are common from environmental advocates (anonymous, 2012). While upholding the integrity of the act is important, the wolf case illustrates the consequences of ignoring the social issues in favor of strict adherence to legal formalities. Following each challenge to a delisting attempt, the USFWS sought to address the concerns of the environmental groups and then delist again, and each time the environmental groups found another reason to sue to keep wolves on the list. Legally, the challenges were legitimate, but social and state-level political tolerance decayed to the point where senators from the western states attached a wolf delisting rider to a Congressional budget, creating a legally unassailable and entirely unscientific delisting and setting a disturbing precedent. Litigation is a powerful tool that should be used strategically and in concert with careful assessment of the social and political consequences as well as the legal outcomes.

MAKING A DIFFERENCE: NEW SOLUTIONS

Making on-the-ground improvements can attract new and diverse participants. Seth Wilson of the Blackfoot Challenge (chapter 6, this volume), Steve Primm the Madison Valley (Newsome et al. 2010), and Timmothy Kaminski of the Mountain Livestock Cooperative have helped reduce predation on livestock in several communities in northern Montana and in Alberta by building and maintaining strong relationships with ranchers and conservation groups while seeking innovative, practical, and scientifically-based measures to help reduce mortalities of predators and livestock. Using GIS (geographic information system) mapping of conflict hotspots in Montana, Wilson and his partners have implemented practical measures to reduce conflicts with bears, such as electric fencing of beehives and removal of livestock carcasses from the vicinity of herds (chapter 6, this volume). Kaminski's work in Alberta and Wyoming focuses on changing herding practices to account for the habits of wolves, by removing vulnerable animals from areas where wolves are likely to concentrate seasonal activity (Kaminski, personal interview, July 2007). Substantive improvements are a vital step beyond the forging of new relationships, but without a sound rapport with all the groups involved in a conservation effort, these improvements will be challenging.

Youth education is a powerful tool for effecting change. Environmental, ranching, and Native communities run numerous youth education programs, but few holistically combine conservation values with sustainable food production and spiritual respect for land. Given the diversity of players and issues, northwestern Wyoming might be a laboratory for a new type of youth program that could address landscape scale issues. Combining elements of animal husbandry from 4-H programs with environmental education curricula on carnivore biology and Native teachings about the relationship between humans and animals might capture the imaginations of young people from diverse backgrounds and inspire them to come up with more creative solutions than the policy experts in DC could imagine.

Finally, the process of rebuilding relationships should, if possible, take place at least partially on the literal common ground that all parties value—the landscape of Wyoming. Mending fences, putting up hay, or taking a challenging hike with a colleague or adversary offers a chance to connect and to understand differences and similarities in perspective across groups. The Green River Valley Land Trust uses activities like this to build relationships within the community. Spiteful letters to the editor and meetings attended only by like-minded people do nothing to advance the discussion about tough issues and, in many cases, serve as a polarizing force. The conversation must shift out of the newspapers and courtrooms and away from the conference table and move back outside where it is easier to gain perspective on the issues at hand.

All the abovementioned recommendations are based on empowering local leadership. Leaders "should have the ability to carry out more comprehensive, contextual, and rational wildlife management programs than currently exist. They must understand how to design and implement practice-based strategies. They must be skilled in integration and inclusion" (Clark and Rutherford 2005, 244–45). Change is not going to come from the outside. Instead, the role for outsiders is to identify the existing pillars of leadership in the community and ask, "how can I help?" Local leadership might be hidden in unlikely places, and it might not take the same form in all communities (McLaughlin, Primm, and Rutherford 2005).

We have not mentioned science as a means of improving the decision process. Science plays an important role in carnivore management, but will not by itself resolve looming value deprivation issues. Wolf management would be easy if the key lay in one number such as the size of the wolf population. Instead, there are other numbers involved: three sovereign nations (two Native and one federal); at least six government agencies (USFWS, National

Park Service, US Forest Service, WYGF, Montana Fish, Wildlife and Parks, Tribal Fish and Game Department); seven counties (Sublette, Teton, and Fremont in Wyoming, Gallatin and Madison in Montana, and Fremont and Teton in Idaho); twenty-one county commissioners; 6,728 Native Americans on the reservation; 7,359 residents of Sublette County; and some 8,000 cows roaming around on the Bridger-Teton National Forest. These other numbers are as important as the population of wolves.

Conclusion

Finding common interest happens, by definition, at the grassroots with individuals. Common interest is not something that can be legislated or dictated, but instead must come from the commitment of the community to adhere to an effective decision process. Our recommendations are drawn from conversations with real people, in real places, about real matters, and we believe we have laid out a workable plan for taking steps toward this type of process.

Canis lupus remains, in the end, an unfathomable creature, neither a noble representative of wilderness nor an agent of federal imperialism, but an animal intent, like all others, on survival in an uncertain world.

The problem, however, is not the wolf; it is the value demands and the lack of a means for integrating those value demands to identify and advance the common interest. This fact is rarely recognized. De-escalation of the conflicts will involve a commitment to looking honestly at values and the social and decision processes through which values are indulged or deprived. For the sake of all the communities involved, for the sake of the wildlife, and for the sake of the land in which all parties are deeply invested, a move toward common ground is essential.

Appendix 3.1
Decision Activities, Standards, and Effectiveness in Wolf Management in Two Wyoming Communities

Evaluation of the decision processes for wolf management in two Wyoming communities. Described below are the activities of the decision process, effectiveness standards for each, and the processes by which the two communities managed wolves. Our analysis reflects the sequence of events that started with the decision (prescription) to reintroduce wolves to Yellowstone National Park, followed by repopulation of wolves in the park and surrounding areas, and concludes with the *next* decision process associated with managing wolves after delisting.

PRESCRIPTION

> *Setting all of the rules and regulations governing how wolf reintroduction proceeds (i.e., 10(j) designation, compensation, removal of depredating animals, recovery goals, etc.).*

Stable expectations (effective)

Rational

Comprehensive

Wind River Indian Reservation Community

Federal Reintroduction Process

Initially, the prescription function was in line with respect and rectitude values of the reservation community and was generally viewed as lawful/ enforceable (*stable expectations*) and in the common interest (*rational*). Prescription did not, however, address how wolf reintroduction might exacerbate existing divides between the reservation and the wider Wyoming community and was therefore *not fully comprehensive*.

Tribal Management Plan

The prescription function associated with the creation of the tribal management plan in 2007 was effective in that it advocated for the common interest (*rational*) and left latitude to deal with all potential situations (*comprehensive*). Because it left latitude to deal with many different scenarios, however, some expectations are *unstable*.

Upper Green River Ranching Community

Federal Reintroduction Process

The ranching community was immediately suspicious of the prescription function of the decision process because, despite early promises that wolf reintroduction would not significantly affect their businesses (i.e., compensation programs, removal of depredating animals, etc.), they distrusted the decision makers and had *unstable expectations* for the outcomes of the process. Furthermore, the prescription function was viewed by ranchers as *irrational* (not in the common interest) because they perceived that it catered only to the interests of the scientific and environmental communities. Lastly, ranchers felt that the prescription function was *not fully comprehensive* because the decision did not have clearly stated goals, rules, contingencies, and assets.

From the ranchers' perspectives, the only clearly articulated items in the prescription were the sanctions, which were widely perceived as unfounded threats.

INVOCATION

Initiation of provisions within the prescription (i.e., claiming a wolf kill, asking for removal of wolves, requesting compensation, etc.).

Timely
Dependable
Rational
Nonprovocative

Wind River Indian Reservation Community

Federal Reintroduction Process

The low number of wolves and wolf conflict on the reservation for the first ten years after the reintroduction has meant that invocation of prescription has generally been seen as *timely*, *dependable*, *rational*, and *nonprovocative* on the reservation. Since members of the two tribes see themselves as disempowered within the wider context of Wyoming, and the traditional perspective of wolves as brothers and teachers has been overlooked in favor of addressing conflict in terms of economics and productivity for ranchers, some reservation residents feel that invocation was *not timely* in considering their concerns and *not consistently dependable*.

Tribal Management Plan

Tribal Fish and Game officials are responsible for implementation and enforcement of the tribal wolf management plan, which had just been adopted at the time of research. No data were available to assess invocation.

Upper Green River Ranching Community

Federal Reintroduction Process

The invocation function of the decision process for the UGRCA ranching community was initiated in 2000 when they had their first confirmed wolf kill. At that point, ranchers began the process of requesting the provisions for compensation and animal control laid out in the prescription. Ranchers perceived the invocation function as flawed and disappointingly reactive (rather than proactive) because depredation compensation and wolf control programs were *not timely*. UGRCA ranchers also perceived the invocation process as *undependable* because, in their view, the implementation was not consistent with prescriptions presented by the Feds before reintroduction (i.e., timely compensation, reliable control of depredating animals, and quick delisting after meeting biological recovery goals). Ranchers perceived the invocation process as *irrational* because the systems in place to deal with implementation were not equipped to help ranchers cope with depredation in an appropriate time frame. Lastly, ranchers felt that the invocation function caused maximum value loss (power, wealth, rectitude, and respect) because they felt that they had been lied to in the prescription process, and this further *provoked* ranchers to resist cooperation in implementation.

APPLICATION

Dispute resolution and enforcement of the prescription. How the action of invocation is resolved (i.e., payment of compensation owed, control of depredating wolves, litigation, etc.).
Rational and realistic
Uniformly applied

Wind River Indian Reservation Community
Federal Reintroduction Process
The reservation itself has seen few wolves and little wolf conflict. Looking at the wider Wyoming community, some members of the reservation felt that the application process has catered to the special interests of people who want fewer wolves and that therefore the application has been *nonuniform*. Some reservation inhabitants also felt that violations of prescription (e.g., illegal killing of wolves) were not addressed in surrounding non-Native communities and that the process beyond the reservation was *irrational* and *unrealistic*.

Tribal Management Plan
At the time of research, neither dispute resolution nor enforcement of prescription had become necessary.

Upper Green River Ranching Community
Federal Reintroduction Process
To date, disputes about wolves have been handled in the federal court system. Ranchers view the application function as *irrational* and *unrealistic* because the resolutions to date (achieved through litigation) have been only partially contextual, partially unbiased, and not sufficiently adaptive. Litigation takes a long time and costs a lot, and in the meantime no one is getting what they want. Lastly, unlike the Native community that has a trusted

third-party participant (Tribal Fish and Game) to mediate and enforce the process, ranchers feel that the USFWS is *nonuniformly* enforcing the original prescription. Not enough depredating wolves have been killed and the compensation checks are not arriving fast enough.

APPRAISAL

Internal and external review. How well have the various phases been able to evaluate themselves? What mechanisms for monitoring and evaluation were included in the prescription?

Dependable and rational

Comprehensive and selective

Independent

Continuous

Wind River Indian Reservation Community

Federal Reintroduction Process

Original appraisal plans for the biological success of the reintroduction were clear and straightforward, but failed to address social issues, leading most stakeholders, including many on the reservation, to feel that the process was *irrational*, *undependable*, and *not comprehensive*. Internal appraisals of the program have been supplemented by external appraisals, but ultimately the decision about wolf delisting was made by Congressional order, and with all of the stakeholders invested in the outcome, appraisal has *not been independent*. Litigation has led to a *discontinuous* appraisal process as the criteria for delisting perpetually change.

Tribal Management Plan

At the time of research, no data were available on perceptions of the appraisal process. The plan itself contains some measures for appraisal, which were generally agreed on (*rational and dependable*), but relatively vague to accommodate multiple scenarios (*comprehensive* but *not selective*). Some appraisers are embedded in the process in order to protect tribal sovereignty, but the USFWS serves as an outside appraiser, so appraisal is at least partially *independent*. Appraisal is likely to be *continuous* once evoked.

Upper Green River Ranching Community

Federal Reintroduction Process

The ranching community would argue that the appraisal function of this decision process has been *irrational* and *undependable* because there is no agreed on set of criteria for judging appraisal, and the data on which judgments are made is highly disputed. In addition, the UGRCA ranching community would argue that the appraisal process has been *neither comprehensive nor selective*. Mechanisms for meaningful monitoring and evaluation should have been imbedded in the prescription function and carried out within each successive phase of the decision process, but this was not the reality. Furthermore, all the interested parties are so imbedded in the heat of the debate that nobody is protected from threats or inducements, and nobody is acting as an external appraiser (i.e., *no independence*). Ranchers also feel strongly that appraisal has been *intermittent*. The process of litigation means that decisions are made from the top, passed down to the bottom, and then elicit reactions from the interested parties. This inherently creates a *discontinuity* in the appraisal process.

TERMINATION/SUCCESSION

Ending or moving on. Change of management regimes cued by success of biological recovery.

Timely

Dependable and comprehensive

Balanced

Ameliorative

Wind River Indian Reservation Community

Federal Reintroduction Process

The federal delisting process has *not been timely*, and in fact, since the delisting process started in 2007 we still have no clear resolution. The contention over the status of the wolf has made the process itself *undependable* and *nonameliorative* at the level of state-federal interaction, and this delayed a full return of management to the tribes.

Tribal Management Plan

The tribal management plan, representing termination of the federal prescription, was implemented in a *timely*, *dependable*, and *comprehensive* fashion. It is *balanced*, without unduly privileging special interest groups, and *ameliorative* in that it returns power over a resource to the local population.

Upper Green River Ranching Community

Federal Reintroduction Process

The termination function of this decision process should have begun when wolves achieved biological recovery and management regimes changed accordingly. Biological recovery of wolves occurred in 2000, but the delisting process is still underway (since 2007 = *not timely*), and thus the management regime has not yet changed in accordance with the prescription. Ranchers view the termination function as *undependable* and *noncomprehensive* because the policies set forth are the result of ever-shifting decisions from the courts. Ranchers feel that litigation has co-opted the decision process in a way that prevents progress toward state management of wolves (i.e., litigation is inhibiting change, and thus the system is *out of balance*). With the advent of termination, policies like third-party compensation for ranchers will come to an end, but the WYGF is likely to put *ameliorative* systems in place (similar to the control and compensation programs for grizzly predation) that replace the federal programs.

INTELLIGENCE

> *Gathering information and planning for what comes next. In this case, the process was entirely subverted because most intelligence gathered (to inform new prescriptions) has been collected in the process of litigation.*

Dependable
Comprehensive
Selective
Creative
Open

Wind River Indian Reservation Community

Federal Reintroduction Process
Not applicable.

Tribal Management Plan
As the reservation takes on independent management of reservation wolf populations, intelligence will focus on monitoring wolves on the reservation and keeping track of conflicts and public perception. The plan does not specify a predetermined number of wolves for the reservation, allowing for flexibility in management but also potentially compromising the dependability of the plan if the options are too broad. Whether the planning is dependable, comprehensive, and available to everyone remains to be seen.

Upper Green River Ranching Community

Federal Reintroduction Process
The intelligence function for the next decision process is already viewed as highly *undependable*, and that bodes badly for the rest of the process. The

next steps for wolf management will be dictated by the outcomes of current litigation. Congressional orders and court decisions are already replacing attempts to gather information in an unbiased and inclusive manner, and stakeholders are turning to the courts and the media as a vehicle through which to make their voices heard instead of respectful dialogue and collaboration. Even more so than the previous information gathering process, this new intelligence function is being completely subverted by a few well-funded interest groups/stakeholders using the promotion function and litigation to override the process of dutifully collecting information from all appropriate sources and affected people. The new intelligence function is therefore totally *noncomprehensive* in nature. It seems highly unlikely that Congressional orders or court decisions will be guided by adequate problem orientation, but will instead be unaware of what the "problem" really is and therefore *unable to be selective* in choosing appropriate solutions. Rather than correcting the (relatively) minor inadequacies imbedded in the last intelligence gathering process (prior to wolf reintroduction), this current decision process is barreling forward with a plan to base the next set of prescriptions on information gathered in current litigation and impassioned promotion. This is a mistake. Decision makers and those who are providing information to them should revisit the value of a properly conducted intelligence process and carry out the steps necessary to compare new objectives and strategies with the ones from the previous process. When the intelligence function is subverted by the promotion function and litigation, almost all parties (except those with lots of wealth and power) are excluded from access to intelligence/information. This *lack of openness* means that crucial input from affected stakeholders is overlooked.

PROMOTION

Open debate, in which various groups advocate for their interests or preferred policy.

Rational
Integrative
Comprehensive

Wind River Indian Reservation Community

Federal Reintroduction Process

Native communities feel that the promotion function following wolf reintroduction has been *irrational* because their values have not been adequately considered by decision makers, particularly at the state level. They also feel as if the process has been *nonintegrative* because their concerns and perspectives were put second to the concerns and perspectives of neighboring white communities. Surrounding non-Native communities have a strong interest in certain resource management issues (specifically, water) and successfully advocate for a position that will continue to allow them preferential access to disputed resources. Many Native Americans feel that the debate is *not comprehensive* because it fails to reflect the full range of community interests, especially the views of the disenfranchised.

Tribal Management Plan

Native Americans have been able to promote their view of wolf management by creating their own management plan, which bypassed the state of Wyoming. This promotion process was judged as largely *rational*, *integrative*, and *comprehensive* because it was in accordance with laws that grant sovereignty to tribes.

Upper Green River Ranching Community

Federal Reintroduction Process

Ranchers characterize the promotion function following wolf reintroduction as *irrational* because they feel that their proposed alternatives were disregarded (value deprivations of power, wealth, rectitude, and respect). Furthermore, ranchers feel that they were being asked to adopt a new set of husbandry practices to accommodate wolves on the landscape, but were not given any proactive resources or assistance to help implement the new practices (i.e., range riders, fencing, grazing plans, etc.). Ranchers also characterize the promotion function as *nonintegrative* because the debate after wolf reintroduction was bipolar (i.e., all or nothing, wolves with full protection or no deal). Lastly, ranchers characterize the promotion function as *noncomprehensive* because it has failed to encourage debate that includes people who feel the economic consequences of the proposed prescription.

Appendix 3.2
Tests of the Common Interest

Tests for determining whether a decision process is operating in the common interest (adapted from Brunner et al. 2002).

PROCEDURAL TEST

Is the process inclusive and open to broad participation, and are the participants responsible to the broader community?

Does the process involve all of the interested parties directly or through representatives?

Are participants responsible and accountable?

SUBSTANTIVE TEST

Does the process meet the reasonable expectations of the participants involved?

Is the process substantiated by good information that is available to all of the participants?

Are the expectations of participants about what will be accomplished warranted by the evidence available?

Are the demands of participants compatible with more comprehensive goals?

Is the process defined by reciprocal tolerance and mutual accommodation between and among the different special interest groups?

Have participants representative of the community as a whole signed off on the outcomes, indicating that they believe the outcomes are in the common interest?

PRAGMATIC (PRACTICAL) TEST

Do the outcomes of the process uphold the expectations of participants over time?

Is it adaptable to changing circumstances?

NOTE

We thank these individuals and groups for their help in preparing this chapter: Murray Rutherford, Denise Casey, Northern Rockies Conservation Cooperative, Jason and Kate Wilmot, Lydia Dixon, Timmothy Kaminski, David Mattson, Richard Baldes, Jason Baldes, Jolene Catron, Charles Price, V, John and Lucy Fandek, Gary Keen, Leif Videen, Lat and Jill Straley, Barb Franklin, Pinedale Forest Service Office, Bernie Holtz, Pinedale and Lander (Wyoming) Game and Fish Offices, Lander USFWS Office, Steve Primm, Mike Jimenez, Susannah Woodruff, Richard Noble, Bruce and Mary Wolford, Jenny and Lou Roberts, Linda Baker, Mark Bruscino, Kevin Fry, Greater Yellowstone Coalition, Quivira Coalition, Defenders of Wildlife, Norman Pape, Todd Stearns, Jonathan Schechter, and Todd Graham.

REFERENCES

Alliance for the Wild Rockies. 2011. "Conservation Groups Appeal Federal District Court Wolf Decision." Press Release, August 8. Retrieved June 10, 2013, from http://www.wildrockiesalliance.org/news/2011/0808wolfPR.shtml.

Associated Press. 2010. "Defenders of Wildlife End Compensation Program for Livestock." *The Missoulian*, August 31. Retrieved October 17, 2010, from http://missoulian.com/news/state-and-regional/article_d314d392-b57a-11df-a1be-001cc4c002e0.html.

Barringer, F., and J. M. Broder. 2011. "Congress, in a First, Removes an Animal from the Endangered Species List." *New York Times*, April 12. Retrieved June 13, 2011, from http://www.nytimes.com/2011/04/13/us/politics/13wolves.html.

Brown, M. 2007. "Wolf Plan Has Critics, Backers." *Casper Star Tribune Online*, July 18. Retrieved November 25, 2007, from http://www.trib.com/articles/2007/07/18/news/wyoming/be6002e0b053802c8725731c0010d973.txt.

Brunner, R. D., C. H. Colburn, C. M. Cromley, R. A. Klein, and E. A. Olson. 2002. *Finding Common Ground: Governance and Natural Resources in the American West.* New Haven, CT: Yale University Press.

Brunner, R. D., T. A. Steelman, L. Coe-Juell, C. M. Cromley, C. M. Edwards, and D. W. Tucker. 2005. *Adaptive Governance: Integrating Science, Policy, and Decision Making.* New York: Columbia University Press.

Casey, D., and T. W. Clark. 1996. *Tales of the Wolf: Fifty-One Stories of Wolf Encounters in the Wild.* Moose, WY: Homestead Press.

Chaney, R. 2010. "Molloy to Hear Wolf Lawsuit Argument June 15th; FWP Looks to Adjust Hunt." *The Missoulian*, April 28. Retrieved October 15, 2010, from http://missoulian.com/news/local/article_4951d04a-5280-11df-9b48-001cc4c03286.html.

Clark, S. G. 2002. *The Policy Process: A Practical Guide for Natural Resource Professionals.* New Haven, CT: Yale University Press.

———. "An Informational Approach to Sustainability: 'Intelligence' in Conservation and Natural Resource Management Policy." *Journal of Sustainable Forestry* 28:636–62.

Clark, S. G., B. J. Miller, and R. P. Reading. 2010. "Policy: Integrated Problem Solving as an Approach to Wolf Management." In *Awakening Spirits: Wolves in the Southern Rockies*, edited by R. P. Reading, B. J. Miller, A. L. Mascvhing, R. Edward, and M. K. Phillips, 147–60 . Boulder, CO: Fulcrum.

Clark, T. W., and R. D. Brunner. 1996. "Making Partnerships Work in Endangered Species Conservation: An Introduction to the Decision Process." *Endangered Species Update* 13(9):1–5.

Clark, T. W., and A. M. Gillesberg. 2001. "Lessons from Wolf Restoration in Greater Yellowstone." In *Wolves and Human Communities: Biology, Politics, and Ethics*, edited by V. A. Sharpe, B. Norton, and S. Donnelley, 135–49. Washington, DC: Island Press.

Clark. T. W., and M. B. Rutherford. 2005. "The Institutional System of Wildlife Management: Making it More Effective." In *Coexisting with Large Carnivores: Lessons from Greater Yellowstone*, edited by T. W. Clark, M. B. Rutherford, and D. Casey, 211–53. Washington, DC: Island Press.

Cronon, W. 1994. *Under an Open Sky: Rethinking America's Western Past*. New York: W. W. Norton.

Defenders of Wildlife webpage. n.d. "The Bailey Wildlife Foundation Wolf Compensation Trust." Retrieved November 29, 2007, from http://www.defenders .org/programs_and_policy/wildlife_conservation/solutions/wolf_compensation _trust/index.php?ht=.

———. 2007. Press Release, 5 July. Retrieved October 10, 2011, from http://www .defenders.org/newsroom/press_releases_folder/2007/07_05_2007_new_rule _expands_wolf_killing_in_northern_rockies.php.

———. 2008. "Defenders of Wildlife Helps Fund Montana Rancher Compensation Program." Press Release, February 13. Retrieved December 22, 2012, from http:// www.defenders.org/press-release/defenders-wildlife-helps-fund-montana-rancher -compensation-program.

——— 2010a. "Defenders Shift Focus to Wolf Coexistence Partnerships." Press Release, August 20. Retrieved October 17, 2010, from http://www.defenders.org /newsroom/press_releases_folder/2010/08_20_2010_defenders_shifts_focus_to _wolf_coexistence_partnerships.php.

———. 2010b. "Federal Protections Restored for Northern Rockies' Wolves." Press Release, August 5. Retrieved October 15, 2010, from http://www.defenders.org /newsroom/press_releases_folder/2010/08_05_2010_federal_protections_restored _for_northern_rockies_wolves.php.

Earthjustice. 2010. "Wolf Recovery under Attack in the Northern Rockies." Retrieved October 17, 2010, from http://www.earthjustice.org/features/campaigns/wolf -recovery-under-attack-in-the-northern-rockies.

Eastern Shoshone Tribe website. 2007. Retrieved September 16, 2007, from http:// www.easternshoshone.net/EasternShoshoneHistory2.html.

Farquhar, B. 2007. "A Predator Hot Spot." *Casper Star Tribune Online*, May 30. Retrieved November 10, 2007, from http://www.trib.com/articles/2007/05/30 /news/wyoming/fc87b5975390c236872572ea008059bb.txt.

Federal Register. 1994. "Endangered and Threatened Wildlife and Plants: Proposed Establishment of a Nonessential Experimental Population of Gray Wolf in Yellowstone National Park, Wyoming, Idaho, and Montana." *Federal Register*, vol. 59, no. 157, August 16. Retrieved December 19, 2007, from http://ecos.fws.gov /docs/federal_register/fr2657.pdf.

Gearino, J. 2009. "Wolf Lawsuits Grow." *Casper Star Tribune Online*, April 9. Retrieved May 23, 2009, from http://casperstartribune.net/articles/2009/04/12/news /wyoming/b9b3bfd0d1fb368087257593007ddf00.txt.

Jacquet, J. 2007. Sublette County Socioeconomic Analysis Advisory Committee website. Retrieved November 15, 2007, from www.sublette-se.org.

Justia US Law. 2011. "808 F.2d 741: The Northern Arapahoe Tribe, in Its Own Right and on Behalf of All Members of the Northern Arapahoe Tribe, Plaintiff-Appellant, v. Donald P. Hodel, Secretary of the Interior, Kenneth L. Smith, Assistant Secretary, Indian Affairs, Richard C. Whitesell, Bureau of Indian Affairs, Billings Area Office

Director, L.W. Collier, Wind River Agency Superintendent, Defendants-Appellees, and the Shoshone Tribe of the Wind River Indian Reservation, Wyoming, Intervening Defendant-Appellee." Retrieved October 8, 2011, from http://law.justia .com/cases/federal/appellate-courts/F2/808/741/173971/.

Koltko-Riveral, M. E. 2004. "The Psychology of Worldviews." *Review of General Psychology* 8:3–58.

Larmer, P. 2007. "Editor's Note." *High Country News.* 39(12):2.

Lasswell, H. D. 1971. *A Pre-View of Policy Sciences.* New York: American Elsevier.

Lasswell, H. D., and M. S. McDougal. 1992. *Jurisprudence for a Free Society: Studies in Law, Science and Politics.* 2 vols. New Haven, CT: New Haven Press.

Loomis, B. 1995. "One Bullet Prompted Regret." *High Country News.* Retrieved November 1, 2007, from http://www.hcn.org/servlets/hcn.Article?article_id=789.

Lozier, R., and D. Platts. 1982. *Cowboys on the Green River Circa 1918.* Jackson, WY: Pioneer Press.

Mattson, D. J., K. L. Byrd, M. B. Rutherford, S. R. Brown, and T. W. Clark. 2006. "Finding Common Ground in Large Carnivore Conservation: Mapping Contending Perspectives." *Environmental Science and Policy* 9:392–405.

Mattson, D. J., and S. G. Clark. 2011. "Human Dignity in Concept and Practice." *Policy Sciences* 44:303–20.

Mattson, D. J., H. Karl, and S. G. Clark. 2012. "Values in Natural Resource Management and Policy." In *Restoring Land—Coordinating Science, Politics, and Action: Complexities of Climate and Governance*, edited by H. A. Karl et al., 239–60. Dordrecht: Springer Science.

Maughan, R. 2007. "New Rule Would Lower the Bar on Killing Endangered Wolves in Northern Rockies." *Ralph Maughan Online Press Release*, July 5. Retrieved November 28, 2007, from http://wolves.wordpress.com/category/wolves-wisconsin/.

McLaughlin, G. P., S. Primm, and M. B. Rutherford. 2005. "Participatory Projects for Coexistence: Rebuilding Civil Society." In *Coexisting with Large Carnivores: Lessons from Greater Yellowstone*, edited by T. W. Clark, M. B. Rutherford, and D. Casey, 177–210. Washington, DC: Island Press.

Meaney, C., C. Sullivan, and S. G. Clark. 2010. "Water Management on the Wind River Indian Reservation, Wyoming: A Rapid Assessment and Recommendations." Yale School of Forestry and Environmental Studies, Working Paper 24:79–97.

Miller, J. 2007. "When Will Wyo Gain Wolf Control?" *Casper Star Tribune Online*, June 4. Retrieved November 10, 2007, from http://www.trib.com/articles/2007 /06/04/news/wyoming/fdeb1000133cedd2872572ef00211805.txt.

———. 2010. "Idaho Won't Manage Wolves under Endangered Species." *Seattle Pi*, October 18. Retrieved October 18, 2010, from http://www.seattlepi.com/local /6420ap_id_wolf_management_idaho.html.

Newsome, D., J. Hoyle, E. Alcott, J. Siegal, T. Rosen, S. Kamal, and R. Wynn-Grant. 2010. "Large Scale Conservation in the Greater Yellowstone Ecosystem: A Field Assessment and Recommendations." Yale School of Forestry and Environmental Studies, Working Paper 24:99–123.

Northern Arapaho Tribe website. 2007. Retrieved November 30, 2007, from http:// www.northernarapaho.com/node/3.

Northern Rockies Conservation Cooperative. 2009. *Greater Challenges, Fewer Resources: Preliminary Findings of Greater Yellowstone Conservation Organization Survey, 2009.* Retrieved October 15, 2010, from www.gycoi.org.

O'Gara, G. 2000. *What You See in Clear Water: Indians, Whites, and a Battle over Water in the American West.* New York: Knopf.

Pielke, R. A., Jr. 2007. *The Honest Broker: Making Sense of Science in Politics.* Cambridge: Cambridge University Press.

Platts, D. 1989. *Wolf Times in the Jackson Hole County—A Chronicle.* Jackson, WY: Bearprint Press.

Quivira Coalition. 2006. *Transitioning to a New Model of Conservation.* 2006 Annual Report. Santa Fe: Quivira Coalition.

Rentfrow, P. J., S. D. Gosling, and J. Potter. 2008. "A Theory of Emergence, Persistence, and Expression of Geographic Variation in Psychological Characteristics." *Association for Psychological Sciences* 3:329–69.

Robbins, P. 2004. *Political Ecology.* Malden, MA: Blackwell.

Rutherford, M. B., M. L. Gibeau, S. G. Clark, F. Edwards, and E. Chamberlain. 2009. "Interdisciplinary Problem Solving Workshops for Grizzly Bear Conservation in Banff National Park, Canada." *Policy Sciences* 42:163–87.

Sanders, K. 2001. *Interior Secretary Bruce Babbitt Bids Farewell to Yellowstone National Park.* Retrieved May 20, 2009, from http://www.yellowstone-bearman.com/babbitt .html.

Shoshone and Arapaho Tribal Fish and Game Department. 2007. *Wolf Management Plan for the Wind River Reservation.* Shoshone and Arapaho Tribal Fish and Game Department. Ethete, WY.

Sommers, A. P., C. C. Price, C. D. Urbigkit, and E. M. Peterson. 2010. "Quantifying Economic Impacts of Large-Carnivore Depredation on Bovine Calves." *Journal of Wildlife Management* 74:1425–34.

Steelman, T., and M. E. DuMond. 2009. "Serving the Common Interest in US Forest Policy: A Case Study of the Healthy Forest Restoration Act." *Environmental Management* 43:396–410.

Steelman, T. A., and J. Rivera. 2006. "Voluntary Environmental Programs in the United States." *Organization and Environment* 19(3):1–21.

Taylor, D., and T. W. Clark. 2005. "Management Context: People, Animals, and Institutions." In *Coexisting with Large Carnivores: Lessons from Greater Yellowstone*, edited by T. W. Clark, M. B. Rutherford, and D. Casey, 28–68. Washington, DC: Island Press.

Urbigkit, C. 2010. "Defenders Ends Wolf Compensation Program." *Pinedale Online*, August 25. Retrieved December 22, 2012, from http://www.pinedaleonline.com /news/2010/08/Defendersendswolfcom.htm.

Volz, M. 2010. "Judge Orders Wolf Relisting." *Casper Star Tribune Online*, August 6.

Retrieved October 15, 2010, from http://trib.com/news/state-and-regional/article
_4453aa54-4f87-5d59-bf27-59d07376d15f.html.

Weiser, M. 2007. "Guardians of the Range." *Sacramento Bee Online*, May 8. Retrieved
November 10, 2007, from http://www.sacbee.com/378/story/171250.html.

Weiss, J. 1989. "The Powers of Problem Definition: The Case of Government
Paperwork." *Policy Sciences* 22:92–121.

Western, S. 2002. *Pushed off the Mountain, Sold down the River: Wyoming's Search for Its
Soul*. Moose, WY: Homestead Publishing.

Westernwolves.org. 2009. *Managing Wolves*. Retrieved May 22, 2009, from http://www
.westernwolves.org/index.php/managing-wolves.

———. 2009. *Casper Star Tribune Online*, February 18. Retrieved February 18, 2009,
from http://www.trib.com/articles/2009.

Wetzler, A. 2010. "New Wolf Bills: Bad Science, Bad Policy, and Bad Legislation."
Natural Resource Defense Council's "Switchboard" blog post, September 28.
Retrieved October 16, 2010, from http://switchboard.nrdc.org/blogs/awetzler
/new_wolf_bills_bad_science_bad.html.

Wild Earth Guardians. 2011. "Conservation Groups Challenge Wolf Delisting
Rider: Lawsuit Seeks to Restore Federal Protection to Gray Wolves in Northern
Rockies." Press Release, May 5. Retrieved January 8, 2012, from http://www
.wildearthguardians.org/site/News2?news_iv_ctrl=-1&page=NewsArticle&id=6797.

Wilkinson, K. M., S. G. Clark, and W. R. Burch. 2007. "Other Voices, Other Ways,
Better Practices: Bridging Local and Professional Knowledge." Yale School of
Forestry and Environmental Studies, Report 14:1–58.

Wilmot, J., and T. W. Clark. 2005. "Wolf Restoration: A Battle in the War over the
West." In *Coexisting with Large Carnivores: Lessons from Greater Yellowstone*, edited
by T. W. Clark, M. B. Rutherford, and D. Casey, 138–74. Washington, DC: Island
Press.

Wood, K. n.d. "Gray Wolf/Timberwolf Threatens Human Life." *All American Patriot*.
Retrieved October 17, 2010, from http://www.allamericanpatriot.com/content
/gray-wolftimberwolf-threatens-human-life.

Wyoming Game and Fish. 2006. "Wyoming Responds to the US Fish and Wildlife
Service's Refusal to Delist Wolves." *Wyoming Game and Fish Press Release*, October 6.
Retrieved November 25, 2007, from http://gf.state.wy.us/services/news
/pressreleases/06/10/06/061006_5.asp.

4 Science-Based Grizzly Bear Conservation in a Co-Management Environment The Kluane Region Case, Yukon

DOUGLAS CLARK, LINAYA WORKMAN, AND D. SCOTT SLOCOMBE

Introduction

The biological argument for collaborative, multi-stakeholder, and regional-scale grizzly bear conservation has been well established and is widely accepted by management agencies (Craighead, Sumner, and Mitchell 1995; Herrero 1994; Mattson et al. 1996; Raufflet, Vredenburg, and Miller 2003). This approach was attempted in Kluane National Park and Reserve (KNP) in the southwestern corner of the Yukon Territory, Canada, but collapsed in a remarkably short time. During the 1990s KNP conducted an intensive ecological study on grizzlies in the park that suggested the population was likely in decline (McCann 1998). In response a multi-stakeholder conservation planning exercise for grizzlies was begun in the region, where wildlife and protected areas have been co-managed by federal, territorial, and Aboriginal governments since the settlement of comprehensive Aboriginal land claims in 1993. Led by the region's two main co-management organizations, a multi-agency planning team began work in 2000. Local residents, however, distrusted the park's research methods, felt that bears were abundant, and resented what they saw as an inaccurate and imposed problem definition. By June 2001 the planning team had decided to put the planning process on hold. Two subsequent attempts to restart the plan were unsuccessful, and the resistance that greeted a 2007 proposal for a grizzly bear population survey adjacent to the park, which also foundered, indicated that substantive governance problems had remained unaddressed in the southwest Yukon's bear management regime.

This chapter concentrates on a detailed analysis of events between 2000 and 2005, focusing specifically on the regional grizzly bear conservation planning process. We also examine the continuation of trends since then and some specific recent events in the context of those trends. We describe this case, analyze how it unfolded and why, make some recommendations for the Kluane grizzly bear situation specifically, and extract some relevant lessons for large carnivore conservation efforts more broadly.

The Kluane situation was one of four case studies of grizzly bear conservation examined by Clark (2007). Data to construct this case study came from several sources. Semi-structured interviews (n = 24) were conducted from 2003 through 2005 with participants who were involved in or affected by grizzly bear management programs in the southwest Yukon. Also in 2004 we held a series of five focus groups in Haines Junction. Our objective was to gain greater understanding of local residents' values and perspectives on historic and current bear management; this endeavor was also intended to assist with integrated wildlife management planning for the Champagne and Aishihik First Nations' (CAFN) Traditional Territory. Interviews and focus groups were recorded, transcribed, and then coded using HyperRESEARCH v.2.6, running on Apple OS X. A preliminary set of a priori codes was based on the elements of the policy sciences' problem analysis framework, and other emergent codes were developed from the data (Lasswell 1971; Miles and Huberman 1994).

Our standpoints are multifaceted. Clark worked as a park warden in Kluane National Park from 2000 to 2002 before returning to graduate school and undertaking the research described here. Workman has had long involvement with wildlife and fishery co-management in the southwest Yukon and has worked on these issues for the CAFN since before the settlement of their land claim in 1993. Both these authors were involved in many of the events described in this chapter. Slocombe is a professor at Wilfrid Laurier University and has over two decades of research experience in the region, including supervising Clark's doctoral research.

Context and Problem Definition

The Kluane region has been inhabited for at least 8,500 years and is part of the traditional territory of the Southern Tutchone people. From south to north the Champagne and Aishihik First Nations, Kluane First Nation, and White River First Nation have overlapping traditional territories. Non-Aboriginal people only arrived in the area in the 1890s, and since then there have been massive changes in Southern Tutchone society (Cruikshank 1991, 1998, 2005; McClellan et al. 1987).

SOCIETIES AND INSTITUTIONS

When the Alaska Highway was built during World War II, concern about over-hunting in the southwest Yukon by the influx of American military personnel led to the creation of the Kluane Game Sanctuary in 1942 (Lotenberg 1998; McCandless 1985). Legislation creating the game sanctuary also prevented Aboriginal people from carrying out traditional hunting and trapping activities within it. In 1976 much of the sanctuary became Kluane National Park and Reserve (Theberge 1980). In 1980 KNP, along with the adjacent Wrangell-St. Elias National Park and Preserve in Alaska, was designated by the United Nations Educational, Scientific and Cultural Organization (UNESCO) as a World Heritage Site. Later, Glacier Bay National Park and Preserve in Alaska and Tatshenshini-Alsek Park in British Columbia were added to the site's designation, recognizing this as the largest internationally protected area in the world. When the park was established, control of the resources was transferred to the federal government, but the First Nations' perceptions that their people remained excluded from practicing traditional activities in the park did not change (Lotenberg 1998; Nadasdy 2003). Early park policies didn't discourage this perception; park establishment proceeded without local consultation and with considerable subsistence and economic impacts on local Aboriginal and non-Aboriginal people. The memory and legacy of this process remains strong in the region, despite a steady flow of economic benefits from park visitors and the region's growing tourism sector.

In 1973 Aboriginal people in the Yukon began negotiating land claims with the federal and territorial governments (Yukon Native Brotherhood 1973). This resulted in the signing of the Yukon-wide Umbrella Final Agreement in 1993 and, more locally, the CAFN Final Agreement, also in 1993. Regionally, the Alsek Renewable Resource Council (ARRC) was created by the Final Agreement as the primary instrument for renewable resource management in the CAFN traditional territory. The council makes recommendations about management of fish, wildlife, and their habitats to the territorial government, CAFN government, and the Yukon Fish and Wildlife Management Board. Similarly, the Kluane Park Management Board (KPMB) was created by the agreement to make recommendations to the park superintendent and ul-

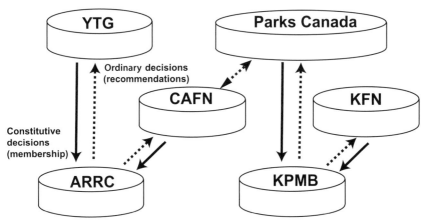

Figure 4.1. Institutional relationships in the southwest Yukon. Dashed arrows indicate the flow of recommendations and consultations, and solid arrows indicate constitutive decisions that determine how the influenced institutions will make their decisions. Acronyms used are as follows: Alsek Renewable Resource Council (ARRC), Champagne and Aishihik First Nations (CAFN), Kluane First Nation (KFN), Kluane National Park Management Board (KPMB), Government of the Yukon (YTG is the common acronym in use there).

timately to the responsible federal minister. The Kluane First Nation ratified its final agreement in 2003, and the White River First Nation claim is still being negotiated. The signing of Kluane First Nation's claim has resulted in a new resource council for their traditional territories and the addition of new members to the KPMB. Aboriginal co-management of Kluane National Park is contentious, not least because of the fifty-year history of exclusion of First Nations from the park area after having occupied it for millennia (Lotenberg 1998; Weitzner 2000). Figure 4.1 illustrates this complex, interconnected institutional arrangement.

GRIZZLY BEAR MANAGEMENT IN THE KLUANE REGION

Grizzly bears have always received considerable research and management attention in the southwest Yukon. One of the first in-depth studies of grizzlies in Canada took place there in the 1960s prior to park establishment (Pearson 1975). National park managers initially considered grizzlies primarily in the context of their conflicts with park visitors, and managing bear-human conflicts remains a high priority both inside and outside the park (e.g., MacHutcheon and Wellwood 2002a, 2002b; Leonard, Breneman, and Frey 1990). Although KNP measures 22,015 km^2, just 18 percent of the park is vegetated, and only 7 percent is forested: the majority of the park is largely high-altitude glaciers and rock (Parks Canada Agency 2004). So despite its large size the park is relatively poor quality grizzly habitat, and regionally the biotic linkages between protected areas and adjacent jurisdictions are often stronger than those between the protected areas themselves (Danby and Slocombe 2005).

Historically, bear-human conflicts have tended to occur in communities such as Haines Junction and Burwash Landing, in the Slims River Valley inside KNP, and at Shäwshe (Dalton Post) and Klukshu—traditional CAFN salmon fishing sites in the Tatshenshini River system (Leonard, Breneman, and Frey 1990; Maraj 2010). The territorial government manages grizzly bears as a game animal, with spring and fall seasons for Yukon residents. Notably, this regime applies to CAFN members because grizzlies aren't considered a subsistence species under their final agreement. Harvest by nonresidents was initially regulated through a system of points allocated to outfitters, which was replaced by a flexible sex-ratio quota system in 1997 (Smith 1990; Yukon Dept. of Renewable Resources 1997).

Protecting regionally significant wildlife populations is one of the major goals of KNP, and with a growing institutional emphasis on ecological science through the 1990s (Parks Canada Agency 2004; Canadian Parks Service 1994, 1992), grizzlies came to be seen as an indicator of the park's wilderness character and ecological integrity. A major study on them ran from 1992 to 1997, with some monitoring activities continuing to 2004 (McCann 1998, 2001; Wielgus, McCann, and Bunnell 1992). McCann (2001) estimated that approximately 225 bears used the park at any one time during that study. Significantly, 43 percent of the bears monitored spent time outside the park, and the population appeared to be declining at 3 percent per year, though there was a wide confidence interval associated with that point estimate (McCann 1998, 2001). Grizzly bears are difficult animals to census, so not only is there substantial uncertainty about the estimates of regional grizzly bear population size and growth rate, but also those estimates are not of a closed population with clear boundaries (Boyce et al. 2001; Schwartz et al. 2006).

PERSPECTIVES: CONTRASTING NARRATIVES OF GRIZZLY CONSERVATION

Aboriginal people and non-Aboriginal people generally have strikingly different concepts of bear-human relationships, which conflict over the appropriate methods and goals of bear research and management (Hallowell 1926; Rockwell 1991; Georgette 2001; Loon and Georgette 1989; Van Daele et al. 2001). The word "respect" almost inevitably arises in discussions in northern Canada about how to interact with grizzly bears, but as Nadasdy (2003) noted, a First Nations definition of respect is quite different from a non-Aboriginal one and is rooted in a worldview based on kinship and reciprocity with animals. Even today a traditional concept of respect for bears is widespread among CAFN members and, for many, profoundly influences their behaviors on the land (Clark and Slocombe 2009; similar to Watters and Anderson's description of the Shoshone and Arapaho peoples' relationships with wolves in western Wyoming, chapter 3, this volume). It appears that the Aboriginal concept of respect is also difficult for some non-Aboriginals to understand accurately. For example, some interviewees, all wildlife and park managers, saw shooting of bears by people who profess respect for them as hypocrisy. Not all did so, however. In terms of people coexisting with bears on a daily basis, local residents in the pro-resource-use focus groups expressed strikingly similar ideas to those in both First Nations focus groups. Clearly though, the potential consequences of such misunderstandings could be significant, justifying marginalization of Aboriginal co-management partners, eroding trust and (ironically) interpersonal respect (Clark and Slocombe 2009).

The fundamental differences between these Aboriginal perspectives and institutionalized resource management approaches led to strong negative reactions to the KNP grizzly study. The study was conceived and implemented in the early 1990s, prior to land claim settlement and following the then-current organizational emphasis on ecosystem management and better science (Canadian Parks Service 1992; Henry and Lieff 1992). Within Parks Canada this study was "leading edge" when it was designed. However, once land claims were settled and co-management bodies formed, other participants' expectations of the park were not just for passive consultation, but rather for direct involvement in all stages of decision making.

A mid-1990s attempt at making the public more aware of the park's bear research had unintended effects when a television documentary on the project offended a number of CAFN elders, who felt that the research methods showed a lack of respect for the bears. Park staff had used a First Nation-owned film company for the documentary and apparently did not want some of the handling procedures filmed, but felt pressured into allowing it. In 2003 park interpreters developed an outreach program for local school children based on playing the role of a radio-collared grizzly bear in the park, which was also poorly received in the community. The interpreters involved were baffled by this outcome, but another former park interpreter remarked that she found this event unsurprising since in the park bureaucracy they worked in separate "silos": wardens always dealt with management issues as their prerogative, but they never substantively informed the interpreters about them. Much of the concern focused on "disrespectful" research methods, which were common wildlife study techniques such as chemical immobilization and radio-collaring, all consistent with federal animal care standards (Wielgus, McCann, and Bunnell 1992). It is intriguing to note that, though Pearson's (1975) earlier study used much the same methods, his work is much less criticized by local residents, many of whom mentioned it approvingly in interviews. However, Pearson's work predated the park, and community involvement was not demanded of scientists during the time of his study (Yukon Native Brotherhood 1973). He also remained in the Yukon after his research was finished.

Understanding others' myths and perspectives has proven to be a real challenge with serious consequences in this case. Besides the difficulty with Aboriginal expressions of respect for bears, there is the question of understanding the viewpoint of researchers. Negative reactions toward the park's bear study were strong and not limited to First Nations people. This outcome reduced the bear study's legitimacy in the local communities, making it an unlikely foundation for policy decisions by the relatively new community-based institutions. Diminished legitimacy of the study became important because, while its results suggested that the grizzly population was in decline, the local perception was that the results were not only obtained with questionable methods, but also were very much at odds with their own perceptions of grizzlies being abundant and simply not in any imminent danger. A common explanation by local people was that if they believed that grizzlies were threatened in the region they would certainly support conservation measures. Since they were not convinced that was the case they saw no need for such action.

Faced with a growing crisis of legitimacy, park managers and biologists made genuine efforts to communicate their perspectives to other participants, but most still seemed reluctant to question publicly their own individual and collective standpoints. Their perspectives hardened in defense of their research, and their collective institutional reaction was reflected in McCann (2001, 2): "The frequent appearance of subadult grizzly bears along highways and near communities strongly influences perceptions as to the grizzly bear densities throughout the entire ecosystem." This characterization attempts to explain local viewpoints contrary to the study results, but it is speculative and insufficient since the elements and processes involved in the formation of local perceptions about grizzly bear numbers are still very poorly understood.

Unfortunately, no common ground has yet been found between these divergent standpoints, and two competing narratives currently exist in the region: a local perception of abundant bears and governments with protectionist agendas versus a pessimistic narrative of inevitably-declining grizzly populations and an ill-informed or unobservant populace (Schwartz 2001; appendix 4.1). Each narrative has a distinct camp of followers, with the conservationist narrative expressed most strongly by KNP staff, the KPMB (in actions, if not by individual members), newer locals who had moved into the region from down south, and some Government of the Yukon (YTG, the acronym commonly used in the study area) staff. In contrast, the localist counternarrative is espoused by the ARRC (again in actions, if not by individual members), some CAFN members, some YTG staff, many locals, and particularly by an identifiable group of more conservative, established, and influential men. In 2005 a visiting European scientist used the term *stammtisch* to describe this group. The word literally means "table reserved for regulars" (source: www.babelfish.com, accessed March 21, 2006), but idiomatically refers to a German institution in small towns in which local men gather at a particular tavern table, where local-level political and business decisions are made.

In many ways these two narratives are very similar to those identified by Clark and Rutherford (2005) in Wyoming, but these narratives feed off each other and are not independent. It may therefore be more productive to view them not as two separate narratives but as a dialectic, the contrasting elements of which are summarized in appendix 4.1. Though this dialectic is a useful heuristic for understanding perspectives and values, the situation can't be entirely reduced to a simple set of two underlying myths because of variability among individuals, change over time, and the complexity introduced by Aboriginal and non-Aboriginal perspectives and dynamics that are not completely reflected within the two narratives.

At the most straightforward level the conservationist narrative espouses principles of conservation biology such as population viability and connectivity. However, more deeply, this reflects a faith in science and governmental conservation institutions. Moving downward in the appendix, more fundamental value positions (e.g., rectitude) and core myths emerge. The national park mandate is privileged. There is clear normative priority given to ecological values, rationalization of preferred alternatives, and dismissal of "parochial" countering positions using the rhetoric of conservation biology. The myth of the Yukon as a frontier is debunked, and it becomes conceptualized as a symbolic "last best place." Grizzlies are emblematic of this rugged fragility, and their fates/conservation strategies employed down south are well known and often mentioned as justification for similar approaches in the Yukon. Interestingly, this "nonfrontier" is not primarily conceived as an Aboriginal homeland, as it is in the more commonly invoked dualism about the Canadian north (Berger 1977).

In contrast, the localist counternarrative doesn't focus on bears in depth. Instead, it is more of a response to the conservationist narrative and at all levels reflects a desire for independence and freedom from governmental control. Base values concerned are well-being and respect. To some, bear hunting isn't so much an end in itself as a means to get outside and experience the land. Such experiences are deeply cherished, especially in springtime after a long, cold winter. Where bears are discussed, it is in the context of deprivation of these values: fear of loss of hunting opportunities, regulations that make defense kills problematic, or their own observations on the land being discounted by government biologists. The park has a central role but not a positive one, and it is a deeply distrusted institution. Proponents of this narrative place more faith in the resilience of local ecosystems and informal systems of resource conservation through local stewardship than the other camp does. The independent, entrepreneurial Yukoner is an operative symbol here, as opposed to the depersonalized place-based symbology of the conservationist narrative. This relative priority is often mentioned in the context of environmentalists valuing bears more than people, a perceived position that this camp finds deeply offensive. The fear of change also appears to influence this narrative: one interviewee mused philosophically that we all outlive our tolerance for change. Finally, there is much local-level skepticism about "managing" grizzlies.

As much as the two narratives differ from each other, neither are they wholly homogenous positions within the two camps. Where this appears to be strongest is in how the varying perspectives on science and local knowledge differ within these groups (Berkes 1999). For example, both appeal to science at times and both consider such knowledge valid but doubt the other side's use or interpretation of that knowledge. Also, one participant in focus group 4, in which the conservationist narrative was strongest, perceptively questioned the uncritical application of "southern park" management approaches such as the core/buffer idea in Kluane. Subsequently, however, that study participant did change his mind in response to what he saw as the growing influence of the localist narrative in land-use decisions.

The base values that underlie these two narratives are largely power, re-spect, and rectitude. Enlightenment is present, but not as widely sought as one might expect with a science-based issue. It appears that knowledge is used strategically and is not objectively determinative in forming view-points—not an uncommon situation (Healy and Ascher 1995). Additionally, the sentiment that Yukoners aren't naïve about development and don't need to be saved from themselves isn't limited to the localist counternar-rative. It was eloquently expressed by a participant in the conservationist-oriented focus group. The examples from that group, described by partici-pants who had come to the Kluane region from "outside" (though at very different times), suggest that some people may modify their ideas as they come to learn about a place and how it differs from their own previous expe-riences. Indeed, some of the most forceful arguments for a top-down, scien-tific management approach came from a participant who had moved there only recently from Alberta.

All study participants seemed to agree on general "motherhood" goals such as human safety and the desire to not see grizzly bears disappear lo-cally. The details, perceptions of relative importance, and means of achieving those goals are where disagreements occur. The devil is very much in those details though, since they are fundamental to the conservationist and lo-calist narratives. Human injuries in the region have fortunately been rare, but according to several interviewees those maulings apparently resulted in reduced tolerance for bears and consequently more bears shot immediately afterward. Bear attacks in the Kluane region have evidently shaped inhabi-tants' perceptions of risks over the long term.

The Kluane Grizzly Decision-Making Process

Overall, the current picture is of a recently developed network of institutions with CAFN occupying a central "hub" position (fig. 4.1). Within this network there are strongly expressed value positions (particularly power, well-being, and wealth), and in the focal time period there were a number of assertive individual actors, mostly male, for whom respect appeared to be an impor-tant scope value.

PARTICIPANTS, STRATEGIES, AND OUTCOMES

The relationships between institutions are strongly asymmetrical, despite the intent and rhetoric of community-based management. Governments exert considerable influence on the two main regional co-management or-ganizations, the ARRC and KPMB, which only have nonbinding advisory authority on substantive issues. In the case of the KPMB, it appears that the board has a long history of being a marginal institution. CAFN and Parks Canada have preserved a government-to-government relationship at senior levels and, unlike the ARRC, the board had considerable difficulty defining its role. Consequently, the KPMB may not have functioned effectively as an institution of public government during the policy process discussed here. Although CAFN is a hub in the institutional network, its members felt left out of the park's grizzly project. The park's numerous and well-intentioned attempts to involve First Nations and community members (even inviting First Nations onto the peer review committee) were generally considered to be too little, too late and so were not viewed as terribly meaningful. CAFN and its members also still feel unable to talk about "managing" bears, meaning controlling bear populations to benefit ungulates, as is done with wolves. This is a contentious topic in the Yukon, and CAFN is well aware of this (Todd 2002).

Adaptation to the post-land claim environment has been difficult for Kluane National Park. Turnover of field staff has been extremely low, especially in contrast to relatively rapid rotation in senior management positions. On the ground, this has reinforced status quo policies and procedures, making change difficult (Clark and Slocombe 2005). Additional factors hindering adaptation include ongoing agency-wide reorganizations and budget cuts and the need to follow standardized national policies and legislation (Dearden and Dempsey 2004). Nevertheless, since 2004 there have been several changes in senior management that produced rapid, substantial change in how the park approaches the task of co-management. First was the arrival of a new superintendent, paired with a First Nations trainee superintendent, and a new young visitor service manager. By 2005 park staff (including the previously marginalized interpreters) were actively involved in events such as a week-long First Nations youth science camp held within the park. This constructive trend continues.

An important but controversial role was that of a particular YTG conservation officer who was widely praised throughout the region for closing numerous small dumps and helping many lodges, camps, and businesses along the Alaska Highway to bear-proof their facilities. However, in many circles he was also criticized, sometimes harshly, for frequently shooting bears in response to bear-human conflicts. Nonetheless, such criticism was often tempered by the sympathy that many people had for the high demands on him to respond to bear-human conflicts over a large area and with relatively few resources compared to the national park. As such, these criticisms were perhaps more institutional than personal. His strategies had mixed effects. Garbage control efforts were clearly very positive, but after he charged a CAFN member for shooting a bear at Klukshu in 2002, he lost community trust about reporting bears shot in defense of life and property there, which has probably been damaging in terms of both conservation and interagency cooperation.

Despite this prominence in the field, YTG played a less significant role in the regional grizzly planning process than might have been expected. This history is important because YTG went through its own substantial crisis with wolf control for moose and caribou population recovery throughout the 1980s and 1990s (Hayes et al. 2003; Todd 2002). Throughout that time there was much internal conflict on those issues, in which bears were implicated as major predators of moose calves, although manipulating bear populations was apparently less preferable than controlling wolves (Larsen, Gauthier, and Markel 1989). As a result there was no support for YTG to conduct bear research of its own, the bear biologist position had remained intermittently occupied or unstaffed for many years, and the points-based system for allocating grizzly harvests to outfitters deteriorated from a rich dialogue to a form letter advising them of the government's quota decisions before being replaced (Smith 1990). Altogether, YTG was in a position with interests to consider in the planning process but limited ability to substantively contribute at the time.

INTELLIGENCE

The Kluane grizzly bear study was a prolonged and costly intelligence collection effort. It was highly regarded within the agency and academia because it was designed and led by academic scientists and was guided by a peer review committee of eminent North American bear biologists (Kluane National Park and Reserve 1998; Wielgus, McCann, and Bunnell 1992). Their belief was that good science would provide park managers with the necessary information to make good decisions about bear conservation in Kluane. However, the park staff's perspective on the role of science in decision making differed from that of the university-based biologist who was principal investigator for the study. For example, at times that biologist felt pressure to provide scientific justification where none existed for management decisions that park wardens wanted to make, such as regulating the timing and number of river rafting groups. This institutional arrangement also made ownership of the project unclear in many peoples' minds. Community members perceived Parks Canada as controlling the study, yet the wardens generally deferred to the biologist as project spokesman. The study was never clearly terminated either. Although major data collection efforts finished in 1997, four years later there was still much internal debate among park staff about whether or not to continue monitoring bears by radio-collaring (ultimately some bears were), and a final project report has yet to be published.

Although considerable resources were expended on biological research the intelligence function was far from comprehensive. There was no organized attempt to collect traditional ecological knowledge for the grizzly conservation plan, yet by that time such efforts were considered a normal, appropriate, and expected part of the intelligence function in northern Canadian policy processes (Armitage and Clark 2005). Managers and biologists interviewed expressed a common frustration that they were expected to accomplish this task yet felt poorly prepared to do so. Nevertheless, their responses often still suggested an instrumental view of traditional ecological knowledge, that it was simply one more source of data to be collected and analyzed through conventional positivistic means, instead of another whole knowledge system that needed to be brought into the policy process, with an awareness of all that that might entail (Nadasdy 2003; Natcher et al. 2005).

Different perspectives, especially the long-term knowledge and observations of First Nations people, are arguably necessary for coping with the dynamic character of the southwest Yukon's biophysical environment, an attribute invoked by a number of different focus group participants (Cruikshank 2005). For example, focus group 5 reported that in 2000 there was still a high snowpack in spring and consequently an abundance of bears seen along the Haines Road (which parallels the park boundary) grazing the new growth on the roadsides. Though this area is well known to locals, bear hunters, and wildlife-viewing guides for that reason, the abundance of bears that spring was remarkable even four years later. The ARRC's public meeting where local people objected to the bear plan and to changes in bear hunting regulations took place later that same year, and such recent local observations may very well have influenced the meeting's outcomes. If so, the value of having mechanisms to integrate, assess, and discuss knowledge from a variety of sources becomes apparent.

INVOCATION AND APPLICATION

Implementation of bear management procedures began in the late 1980s and carried on independently of the concurrent intelligence function and planning processes. Despite the failure of the formal planning process there have been a number of notably successful strategies applied in the region.

Cumulatively, these localized efforts have probably had region-wide benefits for bear conservation and human safety. Most of these initiatives were aimed at controlling bears' access to human garbage, a major cause of bear-human conflicts and human injuries by bears throughout North America (Herrero 1970, 1985; Gilbert 1989; Hastings, Gilbert, and Turner 1989). In the southwest Yukon, these loosely coordinated (but mutually endorsed) efforts involved YTG conservation officers closing many small landfills along the Alaska Highway, municipal installation of an electric fence around the Haines Junction landfill, and bear-resistant garbage containers installed in all jurisdictions throughout the region, including CAFN's custom-built steel garbage trailer enclosed by an electric fence at Klukshu. Techniques to assess and mitigate the risk of grizzly bear interactions with river rafters were pioneered in Kluane and subsequently implemented throughout other neighboring protected areas (Wellwood and MacHutchon 1999; MacHutchon and Wellwood 2002a, 2002b). This type of outcome is similar to other regional conservation efforts that were informally implemented at the field level, even though they were never officially approved by senior levels of government (Danby and Slocombe 2002).

PROMOTION AND PRESCRIPTION: THE REGIONAL GRIZZLY BEAR PLAN

This policy process was initiated during concurrent established planning processes, specifically, YTG's annual harvest allocations (with pressure from outfitters to increase limits adjacent to the park), the 1990 Kluane park management plan revisions, and the Greater Kluane Land Use Plan. The first attempt to reconcile these differing park and territorial agendas took shape as the Kluane Ecosystem Bear Working Group (Breneman and Smith 1989).

Following the park's study, the KPMB, ARRC, KNP, YTG, CAFN, the British Columbia ministries responsible for wildlife and parks, and the Yukon Fish and Wildlife Management Board initiated a planning process that was intended to provide consistent management of grizzly bear populations among jurisdictions at a regional ecosystem scale. This policy prescription was first proposed by a YTG biologist at the park grizzly research project's peer review meeting in October 1998, and subsequent to these discussions the KPMB and ARRC agreed to lead the preparation of the plan jointly (Kluane National Park and Reserve 1998). The process began with an all-day meeting in Haines Junction on September 13, 2000, with a professional facilitator and representatives of all participating organizations present.

From the outset, leading the process proved challenging for the two co-management institutions. The KPMB was not in a strong position to lead the grizzly bear planning process because of its history of disempowerment and its ongoing search for a substantive role in decision making. At the time the ARRC was also very busy with two other major planning processes (forestry, and fishing in Dezadeash Lake) and had little time or energy to spare. Further, there was no single champion for the process, but simply individuals administering different aspects of it. During meetings and teleconferences over the next ten months, participants laid out the elements of the plan: its geographic scope, the importance (and sensitivity) of collecting local and traditional ecological knowledge and input and the methods for gathering it, obtaining KNP's final study results, comparisons with Pearson's earlier work, writing-up, and logistics and financial support. A written "terms of reference" that defined the terms of cooperation between the participants was seen as a key part of the co-management process. This document became increasingly problematic as participants were encouraged to circulate lists of "desired outcomes" and "political/policy considerations," and different standpoints surfaced. Despite these differences, however, participants remained optimistic that the usual procedures for community-based wildlife planning in the Yukon would work in this situation.

FOCUSING EVENTS AND TERMINATION

Two interrelated focusing events likely had much to do with the planning process's collapse. The first was the July 2000 "Bear 66" incident, in which a female grizzly bear obtained improperly stored food from a hiking group in the park and intensive efforts were directed toward keeping that bear alive and out of trouble for the rest of the summer (D. Clark 2001). This required closing areas of the park, radio-collaring the bear, and capturing and moving her several times. Additionally, once soapberries (*Shepherdia canadensis*) ripened in the valley bottoms and Bear 66 left the park seeking them, park wardens followed her daily to prevent conflicts along the Alaska Highway and at Congdon Creek campground. Though park wardens are ex officio territorial conservation officers and thus empowered to work outside the park, park management exchanged letters with YTG to legitimize this arrangement and ensure that both agencies were comfortable with the procedures to be followed. Such intensive management of Bear 66 by park wardens occurred every summer thereafter, until her death following a collar replacement in autumn 2003.

The second focusing event was the December 12, 2000, ARRC public meeting that led to termination of the grizzly bear planning process. The *stamm-tisch* was there in force. The presentation on the regional bear planning process by the chair of the park management board provoked a harsh round of criticism of the park's bear management actions and the demand that the park not participate in the bear plan (Alsek Renewable Resource Council 2000). The Bear 66 incident, which was seen by park staff as a success story, was described by some present as direct evidence of the park expanding its mandate outside its jurisdiction and overriding YTG policy (Alsek Renewable Resource Council 2000).

Besides the clash of the two prevailing narratives, several contingent elements likely affected that meeting's outcome. The presentation of the bear plan immediately followed discussion of unpopular changes to territorial bear hunting regulations, and the territorial bear biologist was absent from the meeting so he was unable to explain those recommendations directly. Over the next six months little progress was made on the plan. At a meeting on June 21, 2001, the decision was made by those involved with the planning process to put it on hold until autumn 2002. Although there were attempts in 2003 and 2005 to revive the plan, nothing came of them, so this chain of events effectively ended with a rapid and premature termination of the planning process, with no substantive appraisal.

Moving Forward: Lessons and Recommendations

At heart, the conflicts described really aren't about bears, and therefore seeking a solution to "the Kluane grizzly problem" would have little chance of success. Fundamentally, this situation reflects a broader power struggle for self-determination and local independence from traditionally hierarchical, colonialist government institutions in the Yukon. In this way it resembles the paradigm shift of Brunner et al. (2005) from scientific management to adaptive governance. The Kluane case involves competing myths and problem definitions, and on this specific issue the localist narrative successfully displaced the conservationist. Because of the symbolic value of grizzlies in the conservationist narrative and the accumulated history of conflict over them, Clark (2007) predicted that targeted bear conservation efforts justified by that narrative would be very likely to exacerbate conflict and diminish the stock of social capital in the region's communities and institutions (see Putnam 2000; Pretty 2003; Pretty and Smith 2003). Indications are that recent promotional efforts rooted in the scientific management approach did indeed have such an effect.

Box 4.1

Lessons for Managers from the Kluane Grizzly Bear Case

- Work hard at understanding the social process of a given issue as deeply and comprehensively as possible, particularly other participants' perspectives, myths, and values (plus how your own standpoint fits with theirs, or doesn't). Patience is an important virtue since such understanding is never absolute and often accumulates slowly, especially when working across cultures.
- In any organization the trade-off between long-term, experienced personnel and new staff with new ideas can be difficult to manage. Ideally, an organizational culture would respect and use the hard-won knowledge gained through experience while also honestly supporting innovation, experimenting with new approaches and locally-developed rules, and creating opportunities for collaborative learning within and between agencies and other institutions.
- Enable early and substantive involvement of all participants in all seven decision functions.
- Individual champions can play important roles in shaping conservation project outcomes. However, they need to be committed for considerable time and must actively engage with all other participants to avoid being seen as "outsiders" with agendas.
- Empower and build democratic decision-making capacity in local co-management institutions.
- Create space for self-organization (e.g., ways for dealing with bear–human conflicts) and support such efforts without attempting to control or direct them.
- Work with senior managers and elected officials to create a receptive political context for emergent community-based initiatives.
- Accept that application of many of these lessons will challenge your professional and organizational norms and procedures and, in some cases, even your professional identity and personal beliefs. Realize that you're not alone in facing such challenges. Seek out resources and other people to assist you with your thinking process (e.g., peers coping with comparable situations, elders, or even experienced researchers).

IT'S ABOUT GOVERNANCE, NOT GRIZZLIES

Rather than focusing on bears, a more beneficial strategy would likely be to build social capital and governance capacity within the Kluane region's co-management institutions (appendix 4.2). Trust and respect—arguably two components of social capital that are particularly important for co-management regimes and for adaptive governance—are exhaustible and slow to replenish, particularly in a relatively remote region with a small population (Manseau, Parlee, and Alyes 2005; Weitzner and Manseau 2001; Brunner et al. 2005). "Volunteer burnout" is a widely recognized phenomenon in Haines Junction, where a limited number of people are depended on for many community functions, from sitting on boards and councils to coaching hockey and running the ambulance and fire department.

In northern communities where resource management institutions are made up of local people, many of them lifelong residents, social capital is very likely a prerequisite for developing healthy, functional, and adaptive governance institutions. Putnam (2000) identified two forms of social capital: bonding (creation of exclusive groups, which can have drawbacks) and bridging (inclusive connections among individuals). These forms provide a useful way to consider how to move past the polarized dialectic on bears and potentially other matters. There are numerous institutions that create bonding social capital in the Kluane region (e.g., First Nation potlatches, cultural events and Village Bakery music nights, the end-of-day convergence on the post office to collect mail, and perhaps even the *stammtisch*), as well as bridging (e.g., school- and youth-oriented activities, hockey leagues, potlatches, and the Society for Education and Culture, now Junction Arts and Music). Further bridging opportunities might be useful though. At the very least they would make it easier for more people to find common ground or just to confine their polarized confrontations to the constituted governance arenas. Such compartmentalization would be a particularly important outcome in a small community, where civility is vital in day-to-day life.

With reasonably abundant social capital in Haines Junction it's unsurprising that governance through regional co-management institutions, particularly the ARRC, has evolved considerably in the years since land claim settlement in 1993. The outcomes of this case study suggest that resource management regimes in the southwest Yukon are shifting slowly away from scientific management as the sole approach, but are they actually shifting toward adaptive governance? Likely they are, but such regimes are dynamic, evolving systems and the outcomes of this transition are not yet clear. An affirmative argument could be made since there is now a more inclusive and bottom-up policy process overall with co-management institutions than was the case in previous regimes (McCandless 1985; Lotenberg 1998; Weitzner 2000). Problem definitions clearly can be redefined based on peoples' input, and they are all attempting to integrate knowledge from different sources. However, much emphasis remains on quantitative management targets (e.g., wildlife quotas, salmon escapement, ecological integrity indicators within KNP). Further, those institutions still display some discomfort with uncertainty in intelligence and a general desire to minimize it.

Ultimately the co-management institutions are limited by land claim legislation to making recommendations to government ministers, who retain the power to decide whether or not to implement them. Such an arrangement isn't necessarily a bad thing in principle, since outcomes in practice are what count (Bryan 2004). Unfortunately, such practices may still have a long way to go toward achieving the ideals of polycentric and adaptive governance. In 2005 an apparent example of ministerial overriding of an ARRC recommendation on allocating part of the Aishihik caribou herd's quota to hunting outfitters caused considerable local resentment and mistrust. Clearly, the movement toward adaptive governance is vulnerable to interruption by special interests and is far from an inevitable outcome (Brunner et al. 2005). Such actions could also have corrosive effects on the social capital that co-management institutions depend on. As a former council member complained, if that's what's going to happen, why bother getting involved and wasting your time?

Besides the grizzly bear case, there have been other events in the Kluane region that reinforce the argument that a paradigm shift toward adaptive governance may be underway. Clark and Slocombe (2005) considered the Aishihik wolf control project in the 1990s as another possible example of this transition, but the decision process details, effects, and outcomes of that episode are not clear. Subsequent development of the Yukon wolf management plan by a citizen committee is perhaps a clearer example of progress toward adaptive governance (Todd 2002). The ARRC's development of numerous management plans for wood bison (*Bison bison athabascae*), fisheries in Dezadeash Lake, integrated wildlife management plans, and the Strategic Forest Management Plan are also examples of bottom-up decision processes, although many, such as the forest plan, have complex cross-scale linkages with other institutions (Alsek Renewable Resource Council 2004). Nadasdy's (2003) critical assessment of the Ruby Range Dall sheep (*Ovis dalli*) recovery does run counter to this trend in terms of its outcomes, yet overall these geographically-clustered examples generally do suggest at least a slow regional shift. More comprehensive review and assessment of trends in co-management in the territory or among northern national parks would no doubt be revealing, but the number of countervailing cases from which one could selectively argue conclusions is large (Smith 2004). Despite such indeterminacy, however, the fundamental governance paradigm shift that we may be seeing here shouldn't be expected to be rapid, complete, or uneventful, let alone perfect.

Regional co-management institutions are evolving and still learning how to "do" governance. The broad challenge of integrating scientific, local, and traditional knowledge, as well as different value sets, is widely recognized and being approached through a variety of specific endeavors such as regional forestry planning and the CAFN/Parks Canada "Broken Connections" project, aimed at reestablishing CAFN members' ties with the land in the park. There are more specific procedural challenges as well. One noteworthy example was raised by members of focus group 4, who commented that they much preferred the small focus group format to the more common "town hall shout-down," such as the December 2000 meeting where the bear plan was rejected. Members appreciated having a "safe" opportunity to talk, and in 2005 one of those people mentioned that focus groups now seem to be more common in the region than before the first author's fieldwork. If so, that may represent a successful diffusion and adaptation of a particular technique by local participants (Brunner et al. 2005).

More substantively, however, this outcome suggests that those proponents of the conservationist narrative who aren't government personnel may be marginalized, partially because they are often newcomers and few in number so far. Also, they are often mobile entrepreneurs or professionals, and so may move on more frequently than other groups if they see better livelihood or lifestyle prospects elsewhere. It is possible that they may be ignored by (or simply less likely to be nominated to) co-management bodies that favor the other, more established narrative. People who hold this narrative are a growing part of the community in the long term, drawn by the area's wilderness and relaxed rural lifestyle. Ensuring that their voices are heard and respected will be necessary in order to keep them engaged in local decision making and not drive them toward divisive or coercive political strategies—or just leaving. To that end, it will be important to maintain inclusiveness and improve the general level of facilitation skills locally, as well as the ability to deal respectfully but effectively with coercive or abusive individuals in public settings.

POSTSCRIPT: A SHIFT TO COMMUNITY-BASED BEAR MANAGEMENT

In 2007 YTG proposed a grizzly bear population survey for the area south of Haines Junction and immediately adjacent to Kluane National Park. Discussed as early as 2001 (D. Clark, unpublished observation), such an inventory project has apparently long been a priority for the Fish and Wildlife Branch. Like the many other wildlife inventory projects also undertaken in 2007, it was likely made possible by the Yukon Government's increases in revenue from energy exploration (Mason 2008; Yukon Dept. of Environment 2009). This particular proposal was met with some concern by the local co-management bodies, but the intention was that the project would be developed through community consultations and be steered by a local working group. Concurrently, the community and the ARRC in particular expressed concerns about an apparent abrupt increase in bear encounters in Haines Junction and expressed their priority for attention to dealing with bear-human conflicts over a population survey (CBC 2007). It is widely supposed in Haines Junction that forest thinning in the 1990s to reduce wildfire risk in the wake of a bark beetle epidemic has increased soapberry production around town, attracting bears into close proximity with people. Given the perception of changing berry crop distribution as the cause for increased bear encounters, it is unsurprising that there was little community support for what was perceived to be a top-down scientific project that neither acknowledged local observations nor addressed residents' concerns.

Plans for the population survey were canceled in 2009, and participants agreed instead to a very deliberate strategy of identifying and implementing feasible, focused ways to prevent bear-human conflicts in Haines Junction (Yukon Dept. of Environment 2009). These measures have included soapberry bush removal, remote-camera monitoring of those treated berry patches, distribution of educational materials, and encouraging residents to report bear observations (Yukon Dept. of Environment 2009, 2010). The current situation is thus another example in which community-based approaches to bear management, such as the prototypes discussed by Wilson et al. elsewhere in this volume, can be developed and implemented where larger-scale, top-down endeavors wouldn't succeed. From an ecological point of view, the clustered nature of bear-human conflicts in the region makes the southwest Yukon an eminently suitable site for such local-scale interventions (Maraj 2010). Though conflict hot spots have shifted and will continue to shift over time, managing them can reduce bear mortality. This approach can benefit bear populations at regional scales since individual bears have large home ranges, and it offers a resolution to the characterization of the main valley of the region as a mortality sink (Primm and Wilson 2004; Wilson et al. 2005; Maraj 2010).

Our data suggest that one specific aspect of this community-based program could be usefully expanded, effectively engaging an even broader proportion of Haines Junction residents. Education and knowledge about bears were widely seen as important by focus group participants and should be part of any community-based initiative, as they rightfully are in the current situation. However, the sort and source of information to be trusted can differ. As discussed earlier, scientific information was often described as less relevant or useful than experiential knowledge. This was not universally so, however. Many focus group participants and interviewees expressed regret that the results of local bear studies weren't more accessible to them. The knowledge of First Nations elders was widely expressed as valuable and necessary to transmit to younger generations, preferably on the land through shared experience. The consistent message from all study participants was to emphasize contextually appropriate and grounded knowledge about how to interact with bears safely and respectfully and to make it available to everyone who might encounter bears within the region, residents and tourists alike. In the long term, there will be a need to maintain real opportunities for people and bears to observe, interact with, and learn about each other (Herrero et al. 2005; Smith, Herrero, and Debruyn 2005; Clark and Slocombe 2009). Clearly, there would be widespread support within the community for ways that local residents' own knowledge and experience could be acknowledged and incorporated into the content and structure of the program in the future.

Conclusions

The conflicts described here extend far beyond bears, yet bears remain potent symbols of those deeply rooted controversies. Direct attempts to solve "the Kluane grizzly problem" through scientific management approaches have systematically failed and simply built an edifice of controversy over grizzlies and protected areas that well-intentioned but unwary new participants (like the first author in 2000) have repeatedly stumbled over. Changing such deeply entrenched strategic narratives as now exist in Kluane is a daunting task. Röling and Maarleveld (1999) suggest that in such cases it may be more fruitful to construct a compelling new narrative through collaborative learning. Fortunately, the elements are largely in place to create just such a new, shared narrative through the recent community-based bear management initiatives, although time will tell what ultimately does evolve from that effort. The emergence of community-based bear management in Haines Junction is an obvious extension of current governance trends and is consistent with the larger region's progress toward adaptive governance. This situation is also consistent with one of the more intriguing theoretical concepts about adaptive forms of co-management, that top-down imposition of "community-based" management approaches (or any local-scale form of governance) can be counterproductive and increase the odds of failure (Ruitenbeek and Cartier 2001). Such initiatives have the greatest odds for success if they self organize, which is essentially what took place in Haines Junction in response to the 2007 population survey proposal. The co-management bodies and First Nation clarified and communicated their priorities to the territorial and federal governments, which ultimately responded in a constructive fashion. It is too early to appraise these initial outcomes thoroughly, but it is probably safe to speculate that participants will need to maintain this sort of respectful, adaptive relationship if they are to continue to clarify and secure the common interest in grizzly bear co-management in the southwest Yukon.

Appendix 4.1
The Grizzly Bear Management Dialectic in the Southwest Yukon

GRIZZLY POPULATION
Conservationist Narrative
The regional grizzly population is precarious, if not in trouble already. McCann's findings are an authoritative source that says the population is probably declining.

Localist Narrative
There's no problem: if promoters thought there was, they would support conservation actions. There may even be too many bears, and if so they should be thinned out, as was done in the past. The park study's methods were disrespectful to bears and its conclusions are wrong.

VISIBILITY OF BEARS
Conservationist Narrative
Locals see lots of bears because they see the same bear all the time or see just subadults (which indicates an unbalanced age ratio, a sign of overharvesting).

Localist Narrative
Promoters are confident that there are lots of bears ("we know what we see on the land") and insulted that scientists discount their observations.

HUNTING
Conservationist Narrative
The precautionary principle should be followed. Promoters are not against hunting (and take pains to point that out), but rationalize bear hunting restrictions because so few people hunt bears anyway.

Localist Narrative
Promoters want to be left alone and don't want to see their hunting opportunities taken away for no good reason, especially when hunters aren't the problem. For example, more bears are shot by YTG than by hunters, and lots has been done already (dump closures and fencing).

DEFENSE KILLS
Conservationist Narrative
Defense kills need to be minimized; nonlethal alternatives (e.g., aversive conditioning) are preferred.

Localist Narrative
People need to be able to defend themselves and their property, and government regulations are making this more difficult. Nonlethal alternatives are ineffective and, to First Nations people, disrespectful.

KLUANE NATIONAL PARK
Conservationist Narrative
Kluane National Park protects critical habitat for grizzly bears (a "protected core" surrounded by a territorial game preserve buffer) and Parks Canada's legal mandate to preserve ecological integrity requires conserving grizzlies outside the park as well.

Localist Narrative
Bears aren't exclusively park bears and the bear plan was an attempt by the park to expand its jurisdiction. The park is run by people with their own environmental/exclusive agenda.

Conservationist Narrative

Science (conservation biology) will provide the answers, bear numbers need to be monitored, and a coordinated, interjurisdictional plan is required to prevent grizzlies going down a "slippery slope" to extinction. Promoters invoke traditional ecological knowledge, but often say they don't know how to use it in conservation planning.

Localist Narrative

Bears can't be managed because they're wild and essentially uncontrollable. Promoters are skeptical that bears can be counted accurately (though not anti-science more generally) and often say that studies don't include local and traditional knowledge but should.

PLANNING AHEAD
Conservationist Narrative

The Yukon is no longer a frontier and requires careful (and mostly governmental) management because northern ecosystems are fragile and unproductive. There's still a chance to avoid the mistakes made down south. Yukoners need to see the big picture.

Localist Narrative

There's plenty of space, resources, and wilderness in the Yukon. People should be able to make a living here, but government over-regulates. Such planning only ever seems to affect the private sector, and it kills the entrepreneurial spirit. Yukoners aren't naïve and don't need to be protected from themselves.

Appendix 4.2
Problems with the Decision Process from the Kluane Grizzly Bear Case and Recommended Improvements

INTELLIGENCE
Malfunctions

Despite detailed ecological research, intelligence gathering efforts weren't sufficiently comprehensive.

Recommendations

Efforts to collect local and traditional ecological knowledge must receive comparable treatment to scientific investigations, and efforts must be made to understand participants' values, interests, and positions in potentially controversial situations.

PROMOTION

Malfunctions

The problem definition that was promoted was based on a single contested narrative and worldview.

Recommendations

Build an arena where different problem definitions can be discussed and negotiated in the absence of strong power differentials between the participants. Here this means further empowering the co-management institutions and enhancing their decision-making capacity.

PRESCRIPTION

Malfunctions

As a prescription, the grizzly bear conservation plan was inadvertently provocative.

Recommendations

1. Practitioners should clearly and comprehensively understand a situation's social process before making even initial prescriptions.
2. Failed prescriptions should be evaluated and not simply repeated.

INVOCATION AND APPLICATION

Malfunctions

Disconnected from the formal intelligence function.

Recommendations

Despite the disconnect, concrete local successes emerged (e.g., garbage control), so this wasn't necessarily a problem here.

APPRAISAL

Malfunctions

There was none until this study.

Recommendations
Create authoritative and acceptable appraisal mechanisms, ideally through the co-management institutions in order to ensure broad ownership of the results and trust in them.

TERMINATION
Malfunctions
Premature termination of the regional grizzly conservation planning process.

Recommendations
1. More and better tools are needed for co-management institutions to obtain representative public input democratically, so that coercive "town hall shout-downs" can be avoided.
2. Research projects must be clearly terminated (and new projects must be differentiated from them) and not prolonged as "monitoring" programs.
3. The park's final grizzly bear research report still must be completed and publicly distributed to bring closure to the project.

NOTE
Financial support for this research was provided by the Canon National Parks Science Scholars Program, the Social Sciences and Humanities Research Council of Canada, Wilfrid Laurier University, Mountain Equipment Co-op's Environment Fund, the Northern Scientific Training Program of the Canadian Department of Indian Affairs and Northern Development, a TransCanada Pipelines Graduate Award, and Yukon College's Northern Research Institute. Lyn Hartley professionally facilitated our focus groups in 2004, which were also actively supported by the Alsek Renewable Resource Council. We are grateful to all the study participants, our student transcribers, and the many other people and organizations who supported this work.

REFERENCES
Alsek Renewable Resource Council. 2000. *Minutes of Public Meeting December 12, 2000*. Haines Junction: Alsek Renewable Resource Council.
———. 2004. *Strategic Forest Management Plan for the Champagne and Aishihik Traditional Territory*. Haines Junction: Alsek Renewable Resource Council.
Armitage, D., and D. Clark. 2005. "Patterns, Currents, Boundaries and Scales: Framing an Applied Research Agenda for Integrated Oceans Resource Management in Canada's North." In *Breaking Ice: Renewable Resource and Ocean Management in the Canadian North*, edited by F. Berkes, H. Fast, M. Manseau, and A. Diduck, 337–62. Calgary: University of Calgary/Arctic Institute of North America.

Berger, T. R. 1977. *Northern Frontier, Northern Homeland: The Report of the Mackenzie Valley Pipeline Inquiry, vol. 1.* Toronto: James Lorrimer and Co.

Berkes, F. 1999. *Sacred Ecology: Traditional Ecological Knowledge and Resource Management.* Cambridge: Cambridge University Press.

Boyce, M., B. M. Blanchard, R. R. Knight, and C. Servheen. 2001. "Population Viability for Grizzly Bears: A Critical Review." In *International Association of Bear Research and Management Monograph Series* No. 4. Knoxville: University of Tennessee.

Breneman, R., and B. Smith. 1989. *Proceedings, Kluane Ecosystem Bear Working Group, October 25–26, 1989.* Haines Junction: Parks Canada and Government of the Yukon, Department of Renewable Resources.

Brunner, R., T. A. Steelman, L. Coe-Juell, C. M. Cromley, C. M. Edwards, and D. W. Tucker. 2005. *Adaptive Governance: Integrating Science, Policy, and Decision Making.* New York: Columbia University Press.

Bryan, T. 2004. "Tragedy Averted: The Promise of Collaboration." *Society and Natural Resources* 17:881–96.

Canadian Broadcasting Corporation (CBC). 2007. "Haines Junction Gets Double Dose of Grizzly Trouble." Accessed September 26 from http://www.cbc.ca/canada/north /story/2007/09/26/yk-grizzly.html.

Canadian Parks Service. 1992. *Toward Sustainable Ecosystems: A Canadian Parks Service Strategy to Enhance Ecological Integrity.* Final Report of the Ecosystem Management Task Force, Canadian Parks Service, Western Region. Calgary: Canadian Parks Service.

———. 1994. *Guiding Principles and Operational Policies.* Ottawa: Minister of Public Works and Government Services.

Clark, D. 2001. "The Bear #66 Incident: A Creative Approach to Law Enforcement in Kluane National Park." Parks Canada Agency, *Regulatory News* 2000:10–12.

———. 2007. *Local and Regional-Scale Societal Dynamics in Grizzly Bear Conservation.* PhD Dissertation. Waterloo: Wilfrid Laurier University.

Clark, D., and D. S. Slocombe. 2005. "Re-Negotiating Science in Protected Areas: Grizzly Bear Conservation in the Southwest Yukon." In *Presenting and Representing Environments*, edited by G. Humphrys and M. Williams, 33–53. Dordrecht: Springer.

———. 2009. "Respect for Grizzly Bears: An Aboriginal Approach for Co-Existence and Resilience." *Ecology and Society* 14 (1):42. Available from http://www .ecologyandsociety.org/vol14/iss1/art42/.

Clark, T. W., and M. B. Rutherford. 2005. "The Institutional System of Wildlife Management: Making It More Effective." In *Coexisting with Large Carnivores: Lessons from Greater Yellowstone*, edited by T. W. Clark, M. R. Rutherford, and D. Casey, 211–53. Washington, DC: Island Press.

Craighead, J. J., J. S. Sumner, and J. A. Mitchell. 1995. *The Grizzly Bears of Yellowstone: Their Ecology in the Yellowstone Ecosystem, 1959–1992.* Washington, DC: Island Press.

Cruikshank, J. 1991. *Reading Voices. Dän Dhá Ts'edenintth'é. Oral and Written Interpretations of the Yukon's Past.* Toronto: Douglas and McIntyre.

————. 1998. *The Social Life of Stories: Narrative and Knowledge in the Yukon Territory.* Vancouver: University of British Columbia Press.

————. 2005. *Do Glaciers Listen? Local Knowledge, Colonial Encounters, and Social Imagination.* Vancouver: University of British Columbia Press.

Danby, R. K., and D. S. Slocombe. 2002. "Protected Areas and Intergovernmental Cooperation in the St. Elias Region." *Natural Resources Journal* 42(2):247–82.

————. 2005. "Regional Ecology, Ecosystem Geography, and Transboundary Protected Areas in the St. Elias Mountains." *Ecological Applications* 15(2):405–22.

Dearden, P., and J. Dempsey. 2004. "Protected Areas in Canada: Decade of Change." *The Canadian Geographer* 48(2):225–39.

Gilbert, B. K. 1989. "Behavioral Plasticity and Bear-Human Conflicts." In *Bear-People Conflicts: Proceedings of a Symposium on Management Strategies*, edited by M. Bromley, 1–8. Yellowknife: Government of NWT.

Georgette, S. 2001. *Brown Bears on the Northern Seward Peninsula, Alaska: Traditional Knowledge and Subsistence Uses in Deering and Shishmaref*, No. 248. Juneau: Alaska Department of Fish and Game, Division of Subsistence.

Hallowell, I. A. 1926. "Bear Ceremonialism in the Northern Hemisphere." *American Anthropologist* 28(1):1–175.

Hastings, B. C., B. K. Gilbert, and D. L. Turner. 1989. "Effect of Bears Eating Campers' Food on Human-Bear Interactions." In *Bear-People Conflicts: Proceedings of a Symposium on Management Strategies*, edited by M. Bromley, 15–18. Yellowknife: Government of NWT.

Hayes, R. D., R. Farnell, R. M. P. Ward, J. Carey, M. Dehn, G. W. Kuzyk, A. Baer, C. L. Gardner, and M. O'Donoghue. 2003. "Experimental Reduction of Wolves in the Yukon: Ungulate Responses and Management Implications." *Wildlife Monographs* 152:1–35.

Healy, R. G., and W. Ascher. 1995. "Knowledge in the Policy Process: Incorporating New Environmental Information in Natural Resources Policymaking." *Policy Sciences* 28:1–19.

Henry, J. D., and B. Lieff. 1992. *Ecosystem Management of National Parks, Western Region, Canadian Parks Service: Seminar Proceedings.* Calgary: Canadian Parks Service.

Herrero, S. 1970. "Human Injury Inflicted by Grizzly Bears." *Science* 170:593–98.

————. 1985. *Bear Attacks: Their Causes and Avoidance.* Piscataway, NJ: Winchester Press.

————. 1994. "The Canadian National Parks and Grizzly Bear Ecosystems: The Need for Interagency Management." *International Conference on Bear Research and Management* 9:7–21.

Herrero, S., T. Smith, T. D. DeBruyn, K. Gunther, and C. A. Matt. 2005. "Brown Bear Habituation to People—Safety, Risks, and Benefits." *Wildlife Society Bulletin* 33:362–73.

Kluane National Park and Reserve. 1998. *Grizzly Bear Research Project: Review and Workshop, October 21–22, 1998.* Haines Junction: Parks Canada.

Larsen, D. G., D. A. Gauthier, and R. L. Markel. 1989. "Causes and Rate of Moose Mortality in the Southwest Yukon." *Journal of Wildlife Management* 53(3):548–57.

Lasswell, H. D. 1971. *A Pre-View of Policy Sciences*. New York: American Elsevier.

Leonard, R. D., R. Breneman, and R. Frey. 1990. "A Case History of Grizzly Bear Management in the Slims River Area, Kluane National Park Reserve, Yukon." *International Conference on Bear Research and Management* 8:113–23.

Loon, H., and S. Georgette. 1989. *Contemporary Brown Bear Use in Northwest Alaska*, No. 163. Kotzebue: Alaska Department of Fish and Game, Division of Subsistence.

Lotenberg, G. 1998. *Recognizing Diversity: An Historical Context for Co-Managing Wildlife in the Kluane Region, 1890–Present*. Whitehorse: Parks Canada.

MacHutchon, A. G., and D. W. Wellwood. 2002a. "Assessing the Risk of Bear-Human Interaction at River Campsites." *Ursus* 13:293–98.

———. 2002b. "Reducing Bear-Human Conflict through River Recreation Management." *Ursus* 13:357–60.

Manseau, M., B. Parlee, and G. B. Alyes. 2005. "A Place for Traditional Ecological Knowledge in Resource Management." In *Breaking Ice: Renewable Resource and Ocean Management in the Canadian North*, edited by F. Berkes, H. Fast, M. Manseau, and A. Diduck, 141–64. Calgary: University of Calgary/Arctic Institute of North America.

Maraj, R. 2010. "Bears and Humans: How Canadian Park Managers Are Dealing with Grizzly Bear Populations in a Northern Landscape." *Park Science* 27 (2). Available from http://www.nature.nps.gov/ParkScience/index.cfm?ArticleID=415,andPage=1.

Mason, G. 2008. "Progress Is Bittersweet in Gold Rush Country." *The Globe and Mail*, July 26. Available from http://www.theglobeandmail.com/servlet/story/RTGAM .20080726.wbcmason26/BNStory/National/?query=.

Mattson, D. J., S. Herrero, R. G. Wright, and C. M. Pease. 1996. "Designing and Managing Protected Areas for Grizzly Bears: How Much Is Enough?" In *National Parks and Protected Areas: Their Role in Environmental Protection*, edited by R. G. Wright, 133–64. Cambridge, MA: Blackwell Science.

McCandless, R. G. 1985. *Yukon Wildlife: A Social History*. Edmonton: University of Alberta Press.

McCann, R. 1998. *Kluane National Park Grizzly Bear Research Project. Interim Report to Accompany the Project Review, October 21 and 22, 1998*. Center for Applied Conservation Biology, UBC. Vancouver: University of British Columbia.

———. 2001. *Grizzly Bear Management Recommendations for the Greater Kluane Ecosystem and Kluane National Park and Reserve*. Haines Junction: Parks Canada.

McClellan, C., L. Birckel, R. Bringhurst, J. A. Fall, C. McCarthy, and J. R. Sheppard. 1987. *Part of the Land, Part of the Water: A History of the Yukon Indians*. Toronto: Douglas and McIntyre.

Miles, M. B., and A. M. Huberman. 1994. *Qualitative Data Analysis: An Expanded Sourcebook*, 2nd ed. Thousand Oaks, CA: Sage Publications.

Nadasdy, P. 2003. *Hunters and Bureaucrats: Power, Knowledge, and Aboriginal-State Relations in the Southwest Yukon*. Vancouver: University of British Columbia Press.

Natcher, D. C., S. Davis, and C. G. Hickey. 2005. "Co-Management: Managing Relationships, Not Resources." *Human Organization* 64:240–50.

Parks Canada Agency. 2004. *Kluane National Park and Reserve of Canada Management Plan*. Ottawa: Parks Canada.

Pearson, A. M. 1975. "The Northern Interior Grizzly Bear *Ursus arctos L.*" *Report Series* No. 34. Ottawa: Canadian Wildlife Service.

Pretty, J. 2003. "Social Capital and the Collective Management of Resources." *Science* 302:1912–14.

Pretty, J., and D. Smith. 2003. "Social Capital in Biodiversity Conservation and Management." *Conservation Biology* 18:631–38.

Primm, S., and S. M. Wilson. 2004. "Re-Connecting Grizzly Bear Populations: Prospects for Participatory Projects." *Ursus* 15:106–16.

Putnam, R. D. 2000. *Bowling Alone: The Collapse and Revival of American Community*. New York: Simon and Schuster.

Raufflet, E., H. Vredenburg, and P. S. Miller. 2003. "Uneasy Guests: The Grizzly Bear PHVA in the Central Canadian Rockies." In *Experiments in Consilience: Integrating Social and Scientific Responses to Save Endangered Species*, edited by F. R. Westley and P. S. Miller, 185–202. Washington, DC: Island Press.

Rockwell, D. 1991. *Giving Voice to Bear: North American Indian Rituals, Myths and Images of the Bear*. Niwot, CO: Roberts Reinhardt.

Röling, N., and M. Maarleveld. 1999. "Facing Strategic Narratives: An Argument for Interactive Effectiveness." *Agriculture and Human Values* 16:295–308.

Ruitenbeek, J., and C. Cartier. 2001. "The Invisible Wand: Adaptive Co-Management as an Emergent Strategy in Complex Bio-Economic Systems." *CIFOR Occasional Paper* No. 34. Jakarta: Center for International Forestry Research.

Schwartz, C. C. 2001. "The Paradigm of Grizzly Bear Restoration in North America." In *Large Mammal Restoration*, edited by D. S. Maehr, R. Noss and J. L. Larkin, 225–29. Washington, DC: Island Press.

Schwartz, C. C., M. A. Haroldson, G. C. White, R. B. Harris, S. Cherry, K. A. Keating, D. Moody, and C. Servheen. 2006. "Temporal, Spatial, and Environmental Influences on the Demographics of Grizzly Bears in the Greater Yellowstone Ecosystem." *Wildlife Monographs* 161:1–68.

Smith, B. 2004. *Applying the Knowledge, Experience and Values of Yukon Indian People, Inuvialuit, and Others in Conservation Decisions: Summaries of 55 Yukon Projects, 1985–2003*. Report MR-04-01. Whitehorse: Government of the Yukon, Department of Environment.

Smith, B. L. 1990. "Sex Weighted Point System Regulates Grizzly Bear Harvest." *International Conference on Bear Research and Management* 8:375–83.

Smith, T. S., S. Herrero, and T. D. Debruyn. 2005. "Alaskan Brown Bears, Humans, and Habituation." *Ursus* 16:1–10.

Theberge, J., ed. 1980. *Kluane: Pinnacle of the Yukon*. Ottawa: Doubleday Canada.

Todd, S. 2002. "Building Consensus on Divisive Issues: A Case Study of the Yukon Wolf Management Team." *Environmental Impact Assessment Review* 22:655–84.

Van Daele, L. J., J. R. Morgart, M. T. Hinkes, S. D. Kovach, J. W. Denton, and R. H. Kaycon. 2001. "Grizzlies, Eskimos, and Biologists: Cross-Cultural Bear Management in Southwest Alaska." *Ursus* 12:141–52.

Weitzner, V. 2000. *Taking the Pulse of Collaborative Management in Canada's National Parks and National Park Reserves: Voices from the Field*. Winnipeg: Natural Resources Institute, University of Manitoba.

Weitzner, V., and M. Manseau. 2001. "Taking the Pulse of Collaborative Management in Canada's National Parks and National Park Reserves: Voices from the Field." In *Crossing Boundaries in Park Management: Proceedings of the 11th Conference on Research and Resource Management in Parks and on Public Lands*, edited by D. Harmon, 253–59. Hancock, MI: George Wright Society.

Wellwood, D., and G. MacHutchon. 1999. *Risk Assessment of Bear-Human Interaction at Campsites on the Alsek River, Kluane National Park, Yukon*. Haines Junction: Parks Canada.

Wielgus, R., R. McCann, and F. L. Bunnell. 1992. *Study Design for Kluane National Park Reserve Grizzly Bear Research Program*. Winnipeg: Canadian Parks Service.

Wilson, S. M., M. J. Madel, D. J. Mattson, J. M. Graham, J. A. Burchfield, and J. M. Belsky. 2005. "Natural Landscape Features, Human-Related Attractants, and Conflict Hotspots: A Spatial Analysis of Human-Grizzly Bear Conflicts." *Ursus* 16:117–29.

Yukon Dept. of Environment. 2009. *Fish and Wildlife Inventory Project Summaries*. Whitehorse: Government of the Yukon, Department of Environment.

———. 2010. *Fish and Wildlife Branch Highlights 2009–2010*. Whitehorse: Government of the Yukon, Department of Environment.

Yukon Dept. of Renewable Resources. 1997. *Grizzly Bear Management Principles—July 1997*. Whitehorse: Government of the Yukon, Department of Renewable Resources.

Yukon Native Brotherhood. 1973. *Together Today for Our Children Tomorrow. A Statement of Grievances and an Approach to Settlement by the Yukon Indian People*. Brampton: Charters Publishing.

5

Wolf Management on Ranchlands in Southwestern Alberta Collaborating to Address Conflict

WILLIAM M. PYM, MURRAY B. RUTHERFORD, AND MICHAEL L. GIBEAU

Introduction

The ranching communities of southwestern Alberta have been managing livestock for generations on the grasslands, foothills, and lower slopes along the eastern edge of the Rocky Mountains. These areas also provide habitat for a variety of wildlife species, including the grey wolf (*Canis lupus*). An unfortunate consequence of the overlap between livestock operations and wolf habitat is that wolves occasionally prey on livestock, and the losses to ranchers can be substantial. To remove the threat, ranchers sometimes kill wolves or encourage others to kill them. Typically, the wolf population recovers within a few years and the cycle of depredation and wolf removal is repeated. This ongoing conflict has continued for many years despite the 1991 adoption of a comprehensive wolf management plan for the province (Alberta Forestry Lands and Wildlife 1991).

In this chapter, we discuss two innovative local collaborations that attempted to break the cycle of depredation and eradication in southwestern Alberta. The first started in 1998, when members of two nongovernmental organizations, the Central Rockies Wolf Project (CRWP) and the Southern Alberta Conservation Cooperative (SACC), began working together to learn more about wolf ecology, ranching practices, and ranchers' concerns, and to test wolf deterrence techniques and alternative ranching practices. The collaboration lasted about five years before disbanding because of internal dissent and attrition of key participants. The second initiative was the multi-stakeholder Oldman Basin Carnivore Advisory Group

(OBCAG 2004), established by the Alberta government in 2003 to operate in the same general area as the CRWP/SACC. The OBCAG was formed to advise the province about all large carnivores in the region, but in practice it focused on wolves and livestock management (OBCAG 2004). The OBCAG lasted until 2007, when the facilitator took a job elsewhere, after which the group stopped meeting. There were no formal links between the CRWP/SACC and the OBCAG, but after the CRWP/SACC dissolved one of its members continued with wolf research and eventually became involved with the OBCAG. Both projects sought to engage ranchers, scientists, wildlife managers, and conservation organizations in order to improve understanding about wolf-livestock interactions and develop alternative management practices to reduce conflicts. Both groups made good early progress in engaging with participants and testing strategies for deterring depredation, but neither was able to maintain an active presence in the region for more than a few years or to effect substantial long-term improvements. Our objective here is to evaluate the strengths and weaknesses of these two initiatives and identify lessons for similar efforts in the future.

In our evaluation we used the decision process framework and criteria developed by Harold Lasswell and his colleagues (Lasswell 1971; Clark 2002). We began by reviewing the published and unpublished literature, reports, websites, and other information available on the two projects. We also interviewed many people involved with or knowledgeable about the projects, including scientists, ranchers, environmentalists, and others. Five interviewees were members of the CRWP/SACC or had worked directly with the group. Nine were members of the OBCAG. The interview protocol consisted of sixty-four primary questions supported by additional probing questions in an open-ended format (see Pym 2010 for the interview protocol and description of methods).

Our observational standpoint in this evaluation reflects our personal histories and perspectives. Pym grew up in a small Ontario town in a county that boasts of its beef industry, but he has no personal ranching experience. His academic training is in biology, education, and natural resource management, and he has worked as a planner for government and a project manager for an environmental consulting firm. Rutherford has an academic background in science, law, and policy. As a student he spent several summers working in Alberta, first as a geologist and then as a policy analyst for a private firm. He has researched and written about conservation policy, including large carnivore conservation, and has participated in several workshops targeted at improving local decision making for conservation. Gibeau is a wildlife biologist who recently retired from a position as the large carnivore specialist for the mountain national parks in Alberta and is now working for a conservation organization. He grew up on a ranch in southeastern Alberta and has participated in several conservation-oriented projects with ranchers, but was not directly involved with the CRWP/SACC or OBCAG. All three authors strongly believe that decision-making processes should be transparent, should offer people a genuine opportunity to participate in decisions that affect their lives, and should strive to identify and advance the common interest.

Context and Problem Definition

An estimated fifty to sixty thousand grey wolves inhabit Canada in about 80 percent of their original range (Environment Canada 2008). Alberta's 1991 wolf management plan estimated that the wolf population in the province ranged from "a late winter low of 3,500 to an early-summer high of about 5,500 following the birth of pups," with a midwinter average of about 4,200 (Alberta Forestry Lands and Wildlife 1991, xi). A more recent assessment by Robichaud and Boyce (2010) estimates Alberta's total winter population at about 5,100 wolves. The Committee on the Status of Endangered Wildlife in Canada considers grey wolves in Alberta "not at risk" and the Alberta government classifies the species as "secure" (Canada 2009; Alberta Environment and Sustainable Resource Development 2013).

The modern era of wolf management in Alberta began in the 1860s when European settlers killed wolves for pelts and in retaliation for robbing food caches (Alberta Forestry Lands and Wildlife 1991). Depredation on livestock was recorded in the late 1870s and 1880s, and by 1899 a wolf bounty was being administered by the Western Stock Growers' Association (Alberta Forestry Lands and Wildlife 1991; Alberta Sustainable Resource Development 2002a). Since then management has focused on controlling wolf numbers (Gunson 1992; Alberta Sustainable Resource Development 2002a; Musiani and Paquet 2004). Methods have included poisoning, trapping, snaring, and shooting from the air or ground (Alberta Sustainable Resource Development 2002a; Musiani and Paquet 2004; Alberta Wilderness Association 2008). Wolves were nearly extirpated from the province twice, but recovered because of high reproductive rates and relaxation of large-scale control efforts (Alberta Sustainable Resource Development 2002a; Alberta Wilderness Association 2008). Since the last province-wide cull in the 1950s, the province has attempted to address wolf-livestock conflicts through population studies, monitoring disease, instituting a compensation program for livestock losses, and continuing to control the wolf population through hunting, trapping, and area-specific killing (Gunson 1992; Alberta Sustainable Resource Development 2002a; Alberta Sustainable Resource Development 2002c; Alberta Wilderness Association 2008).

The 1991 wolf management plan, still in effect twenty years later, promotes trapping and hunting of wolves and also authorizes the killing of wolves to protect property, reduce depredation on livestock, control disease (primarily rabies), and increase ungulate populations (Alberta Forestry Lands and Wildlife 1991). The plan aims to maintain a winter population of 4,000 wolves in Alberta, with fifty located in the southern region of the province, including the area in which the CRWP/SACC and OBCAG operated. Under current Alberta regulations, wolf hunting is allowed for nine months of the year in most areas, and landowners may kill wolves at any time without a license if the animals are on or within 8 kilometers of their land or their authorized public grazing land.

FACTORS CONTRIBUTING TO THE CONFLICT
Several factors have contributed to the high level of conflict between wolves and livestock producers in southwestern Alberta and the controversy over potential solutions. First, the eastern slopes of the Rockies are slowly losing their pristine nature with the expansion of industry (forestry, oil and gas, coal mining, and power production), recreational development, and human communities (Alberta Wilderness Association 2002; Southern Alberta Land Trust Society 2007; Janusz 2008; Wearmouth 2009). A recent report on land use trends in the Chief Mountain area of southwestern Alberta predicts that the human population there will more than double within fifty years, with more people living on residential than agricultural acreages and increasing linear features and forest fragmentation (Silvatech Consulting Ltd. 2008). The report warns that "growth in settlements and transportation networks represent[s] significant threats to grassland integrity" (1). Livestock numbers in the region are also expected to grow substantially, and with wolves and livestock squeezed together in smaller and more fragmented areas, the potential for conflict will likely rise. The predictions in the Chief Mountain report are especially troubling given the current high intensity of conflict between wolves and livestock. Morehouse and Boyce (2011) studied three wolf packs in southwestern Alberta and found that on average each pack killed about seventeen cattle per year, with the wolves' diets shifting from wild prey to cattle during the grazing season. The wolves also fed on livestock carcasses in boneyards on ranches. The authors describe the region as

"an area of intense overlap between wolf territories and cattle grazing areas" (Morehouse and Boyce 2011, 444).

A second factor contributing to the problem is that difficult market conditions are reducing ranchers' tolerance for losses to predators. Markets for Canadian beef were badly affected by the discovery in 2003 of a cow in Alberta with bovine spongiform encephalopathy (BSE). The United States and many other countries banned imports of Canadian beef products (Le Roy, Klein, and Klvacek 2006). Although the bans were partially lifted in the following years, Canadian beef exports to the United States in 2008 remained below the level of exports prior to the BSE discovery (US Department of Agriculture Foreign Agricultural Services 2008). Losses to the Canadian cattle industry related to BSE from 2003 to 2005 alone are estimated at over four billion dollars (Le Roy et al. 2006). Weak market conditions continue to challenge the profitability of cattle ranching, and new costs associated with tighter Canadian policies and regulations are adding to the financial pressures (Le Roy et al. 2006; USDAFAS 2008).

A third factor is the recovery of wolf populations in American states south of Alberta. In the 1970s wolves in the Lower 48 were listed as endangered under the US Endangered Species Act (US Fish and Wildlife Service 2012). Recovery efforts included the relocation of sixty-six wolves from Canada to Yellowstone National Park and other locations in the northwestern United States. Wolf numbers have since climbed to more than 1,700 in the northwestern United States, with the majority in Idaho, Wyoming, and Montana (US Fish and Wildlife Service et al. 2012). The foothills of southwestern Alberta function as a corridor connecting these US wolves with protected areas in the Canadian Rockies. It is likely that populations just south of the border not only provide support for populations in southern Alberta, but also reduce the available territory for Alberta wolves that might disperse south (Alberta Sustainable Resource Development 2002b).

A fourth factor in the Alberta wolf-livestock problem is that public attitudes are changing. An excerpt from a 1907 report from Alberta's chief game guardian illustrates historical perceptions: "As the depredations of timber wolves and coyotes cause a loss to the settlers of this province amounting to many thousands of dollars annually a few remarks as to the best and most successful manner of capturing or destroying them will no doubt prove of interest. . . . Poisoning is a very common, as well as successful, way of destroying these pests" (Alberta Forestry Lands and Wildlife 1991, 87–88). Today, however, many people in Canada and the United States consider it unacceptable to poison wolves or kill them on a large scale (Boitani 1995; Gilbert

1995; Minta, Karieva, and Curlee 1999; Musiani and Paquet 2004; Alberta Wilderness Association 2008). Even in the rural ranchlands of southwestern Alberta there is no longer broad support for completely eradicating wolves. In a 2011 survey of rural landowners near Waterton Lakes National Park, almost 90 percent of respondents agreed that "whether or not I encounter large carnivores, it is important to know that they exist in the region," and more than 60 percent agreed that "it is important for me to know that there are healthy populations of large carnivores in southwestern Alberta" (Quinn and Alexander 2011). Even so, the impacts of these animals are considered problematic: more than 65 percent of respondents agreed that "the current rate of livestock loss to large carnivore depredation is unacceptable," about one-third agreed that "I would be happier if there were no large carnivores on or near my property," and almost 80 percent agreed that "it is acceptable for people to kill a large carnivore if they think it poses a threat to their property (including pets and livestock)." Of particular interest here with respect to the CRWP/SACC and the OBCAG is that more than 70 percent of respondents in the Quinn and Alexander (2011) survey agreed that "people and large carnivores can successfully share a landscape if properly managed."

THE CRWP/SACC COLLABORATION

The CRWP/SACC collaboration of six wildlife scientists and one business-person began in 1998 and ended in 2003. The group adopted three main strategies. First, in order to build better relationships with ranchers and government staff, while gathering local knowledge of wolves and livestock management practices, members of the group "went down and talked to people about what the issues were and asked people what techniques they were using and . . . what the wolf numbers were, . . . the population and distribution" (all quotes in this chapter, unless otherwise attributed, are from interviewees, whose identities are being kept anonymous). A second strategy involved trapping, radio collaring, and tracking wolves to "learn more about what motivates wolves to kill cattle and how we may prevent that, or at least reduce losses." The third strategy was to conduct research on techniques to deter wolves from ranches and livestock. The group focused on fladry: hanging flagging on fence lines so that it flaps and deters wolves from approaching or crossing the flagged fence (Musiani et al. 2003). One interviewee observed that CRWP/SACC "landed on fladry as being a really good one to test, very innovative, but there was also discussion of shock collars, . . . but fladry was the one, and also management, right, from the human end, if the ranchers could manage their cattle in a certain way then there would be less risk."

The CRWP/SACC was initially quite successful in building good relationships within the group and with several ranchers with whom they interacted regularly. Researchers with CRWP/SACC also completed two fladry experiments on separate ranches in successive years, and the results made an important contribution to knowledge about this deterrence method (Musiani et al. 2003). However, the group's efforts were hampered because it had no direct representation from government or the ranching community, and only a few ranchers were sufficiently interested and supportive to collaborate. Other ranchers did not trust the group and some opposed its efforts.

The first fladry experiment exacerbated the problem of trust. During the experiment neighboring ranchers became concerned that wolves were denning on the experimental property and killing cattle on their nearby ranches. After the experiment the neighbors requested support from CRWP/SACC to locate and kill the problem wolves. Some members of CRWP/SACC were willing to help the ranchers in this way, but two members were strongly opposed because they felt that killing wolves went against the mandate of CRWP. These two individuals imposed a decision to withhold resources. As one participant described it, "a couple of members from our group said, 'let's go down there and kill some wolves,' and [name omitted] said, 'I don't think so, I don't think we're doing that.'" Most of the blame for this decision, however, seemed to fall on the research scientists tied to the CRWP rather than the field staff of the CRWP/SACC (the fact that one of the field staff subsequently was involved with the OBCAG also suggests that he continued to maintain a positive relationship with ranchers).

The CRWP/SACC collaboration eventually dissolved for a variety of reasons, including two researchers leaving the region, the value-based disagreement about helping ranchers locate and kill wolves, deteriorating trust and frustration about perceived inconsistencies in decision making, and accusations of poor management of resources.

THE OBCAG

The OBCAG was established in 2003 by the Alberta government "to provide input and advice on the management of large carnivores in the Oldman Basin area" (OBCAG 2004). The group met intermittently until 2007. Participants included representatives of conservation groups, the Alberta Beef Producers organization, the Alberta Trapper's Association, local municipalities, and Alberta's Sustainable Resource Development agency (SRD), along with individual ranchers and wildlife biologists (OBCAG 2004). Funding came from the province, Defenders of Wildlife, and the Alberta Beef Producers.

As the name "advisory group" indicates, the OBCAG offered guidance and recommendations to the provincial government, but did not have formal authority to make and implement decisions on behalf of the province (OBCAG 2004, 2005).

Our interviewees felt that the OBCAG was successful in building internal trust and respect among its own members. The group appeared to have a good understanding of important trends and conditions and elected to focus its efforts on wolf-livestock conflict cycles, ranching practices, and the economic viability of ranching. Field staff radio-collared and tracked wolves to study wolf movement and habitat use patterns. The OBCAG also discussed protocols for wolf removal and agreed on a lethal removal protocol for individual wolves or packs responsible for repeated depredation, which the province adopted. The OBCAG examined alternative ranching practices and tested a night rider program. Finally, the group discussed and recommended changes to the provincial compensation program for livestock depredations.

Along with its successes the OBCAG encountered challenges that hampered its ability to continue effectively: it was unable to secure sufficient long-term resources, it could not enforce consistent and timely implementation of its recommendations, and it did not communicate well beyond its own constituents (the group worked on a broad communication plan but it was not finalized). In 2007 the local provincial biologist, who had acted as facilitator for the OBCAG, moved to another job outside Alberta, and, although someone filled the biologist position, the role of facilitator was not taken up and the group did not reconvene.

PROBLEM ORIENTATION FOR
THE CRWP/SACC AND THE OBCAG

Our interviews reveal how the CRWP/SACC and the OBCAG defined the problem of wolf-livestock conflict in southwestern Alberta.

Goals

According to participants, the primary goal of CRWP/SACC was to work with local ranchers and improve understanding of wolf-livestock interactions in order to reduce the killing of wolves and obtain a more stable and viable wolf population:

> "We really tried to focus on the coexistence component, we really felt that there was a way for ranchers to do business in southern Alberta and wolves to exist there, and not just exist there but fulfill their role."

"The belief that wolves have a place in southwestern Alberta and as did ranching, . . . knowing that wolves can impact the ranching community and that certain individuals are hit fairly hard at times, that we would try to figure out some way of achieving a balance where wolves aren't wiped out willy-nilly."

"To gather knowledge about wolf predation on livestock and also to help the ranchers practically."

These observations correspond with the formally prescribed goals of the collaborating organizations, CRWP—to conduct research and education to support a healthy population of wolves in the Central Rocky Mountains (Wadlow, Callaghan, and Musiani 2003; Parks Canada 2009)—and SACC— to gather information on conflicts, improve understanding of carnivore behavior in relation to livestock, work with ranchers on depredation avoidance, improve livestock compensation policy, and provide information about practical approaches to conservation on private lands (Kaminski et al. 2003). The SACC was initially formed as a practice-based extension of CRWP to distinguish the efforts of field staff working with the ranching community (SACC) from the "wolf-friendly" reputation of the research organization (CRWP).

For the OBCAG, the primary goals were to bring together interested and affected groups in order to make recommendations to the provincial government to reduce wolf-livestock conflicts while maintaining a stable wolf population. One interviewee said that the group was trying "to minimize conflict between livestock and wolves . . . and . . . do some actual basic studies to try to understand the behavior of the interaction more and to see if there were ways to co-exist." Another commented that "normally, I think it was to get people together and talk about the issue and try to come up with some solutions other than just killing wolves." According to its formal terms of reference, the OBCAG sought to promote cooperation in large carnivore management and conservation, provide a means for effective local representation and participation, identify potential carnivore management principles, goals, and actions, develop and manage a model work plan for achieving carnivore management goals, and report to and provide recommendations to the SRD toward creating a local carnivore management framework (OBCAG 2004).

Trends

Interviewees identified five important trends for the CRWP/SACC and the OBCAG. First was the growing number of wolves: "We have more wolves . . .

between the Bow River and the Montana border than in living memory." "Wolves weren't a real problem until the late 1990s." A second trend was the historical cycle of increased depredation followed by indiscriminate killing of wolves: "Historically, the wolf numbers would build up and the population would start preying on livestock and the wolves would be killed off, the whole pack, and then it would be quiet for two to three years or maybe four, and then the wolf numbers start again and then the depredations would start and then they would increase and you kill them all again."

The third trend was increasing commercial and residential development on lands once used mainly by livestock and wolves. Fourth, interviewees stressed that the urban public has grown to accept and respect large carnivores and is interested in preserving these animals and their habitat. "We set aside large areas to be protected as wilderness and national parks, . . . but they are not effective places for housing this complement of ungulates and carnivores that now preoccupy people's vision of what they want from public lands." One interviewee commented on opposition to culling: "Public opinion seems to think that that's not an appropriate way to deal with wolves." Another observed, "There is an element at least in the urban population which is the majority that would rather have the wolves and kick the ranchers out." The final trend identified by our interviewees was the challenging economic climate for the ranching industry. Some interviewees saw this as a factor underlying other trends.

Conditioning Factors

Eight factors were recognized as causing or contributing to the trends just described. First, provincial policy enabled killing of wolves, particularly in response to depredation. Some interviewees referred to the province's "liberal" wolf management policies: "There is the regulations to factor, the hunting regulations are even more liberal, you know, no limit, virtually no season. Well, they can shoot them right to . . . June 15 and . . . any landowner can shoot a wolf." "Wolves can be shot on sight, by any rancher for any reason, at any time on or within 8 km of their ownings." Second and third, interviewees felt that wolf populations of southern Alberta were being influenced by the increase in wolf numbers south of the border and also by declining trapping efforts: "Wolves recovering in the Yellowstone region and in Montana and Idaho were a source population for southern Alberta wolves. So there were more wolves coming in." "As long as you are running cattle you'll have issues because wolves are moving up and down all the time from Montana to north of Jasper." "A real increase in the wolf population. I'm not sure why it is exactly. . . . I think it might be that nobody is trapping wolves right now. . . . In the US, I think they are protected . . . and wolves move back and forth." "Trapping in the old days kept the wolf numbers much lower. Since the decline of the fur industry . . . the wolf is not subjected to as intense a trapping effort as it was around about fifty years ago, or even twenty years ago."

We also heard about the overlap of grazing lands, wild prey habitat, and wolf habitat, and how increasing numbers of wolves along with more recreational activity and oil and gas development in the forest/grassland interface and backcountry have led to more wolf-livestock encounters and depredation. "There's not a place in this whole region where wolf territories are not overlain by cattle leases." Fifth, some interviewees identified the cyclical history of depredation and eradication as a factor. A sixth factor identified by interviewees is that ranching is a traditional way of life that has been passed down through generations, and livestock management practices have changed little. Seventh, interviewees mentioned the challenging economic circumstances for ranching. "Part of the problem was that we were going through a time of BSE and a lot of the ranching was very poor." Compensation from the provincial government for depredation was not considered adequate. Finally, we heard about the emotional effects of seeing cattle that have been maimed or killed: "You just have to be at a few depredation events. . . . [Wolves] create a huge amount of emotion. That is what you are dealing with and people who are confronted with that. . . . It's the emotional part of how they do it and the upfront. It is not clean, there is suffering, and it is all these things that create, that trigger an emotional response, and for those ranchers it is real."

Projections

Participants felt that without intervention the depredation and cull cycle would continue indefinitely. Many believed that their efforts could reduce wolf-livestock conflicts and encourage tolerance of wolves. One commented, "Well, if [the OBCAG] hadn't happened, there would have been ad hoc killing of wolves by government and ranchers, including poisoning and indiscriminate shooting, the same old cycle." Others were more pessimistic: "As long as wolf numbers stayed down and we had helicopters in the air and the collars on wolves, that was good enough for them, and that would be an ongoing continuous process." And "I don't know that wolf management even ten years later in the province of Alberta is all that more enlightened."

Alternatives

As described earlier, the strategies adopted by the CRWP/SACC group focused on researching wolf ecology and wolf-livestock interactions, talking with ranchers to learn about and address their concerns, and testing mitigative management practices such as fladry. For the OBCAG, the strategies selected were radio collaring and monitoring to study wolf ecology and wolf-livestock interactions, lethal control of wolves when needed based on a "three-in-three" threshold, a communication strategy, and alternative management practices, including night riding, deterrents such as rag boxes, and different feeding arrangements for livestock. The "three-in-three" threshold refers to three depredation events in three months, the level permitted for a particular pack before lethal control is implemented. Night riding is monitoring a pasture overnight (camping out) with telemetry gear, waiting for collared wolves to approach, and then intercepting them and scaring them away before they reach the livestock. One interviewee spoke of the alternative management options considered by the OBCAG, saying, "We talked about many, many things, non-lethal control, husbandry practices, what pasture do you put your cattle on, what mix of cattle, do you put yearlings in the back pasture in the spring of the year when the wolves are there."

Wolf Management Decision Process
CRWP/SACC AND THE COMMON INTEREST

To judge whether the CRWP/SACC decision process served the common interest, we applied the procedural, substantive, and practical tests outlined by Brunner (2002), Brunner and Steelman (2005), and Steelman and DuMond (2009) (appendix 5.1). The CRWP/SACC process failed the procedural test, only partially satisfied the substantive test, and failed the practical test.

The CRWP/SACC lacked representation from the ranching community, governments (provincial, local, First Nations), and other important interests such as hunters and recreationists. Interviewees indicated that the group actively communicated with some ranchers, but was not generally accountable to the community for its decisions and actions. Thus it did not meet the procedural test of inclusive participation.

The expectation that CRWP/SACC could reduce conflicts between wolves and livestock was reasonable, and reducing conflicts could potentially benefit both ranchers and conservationists. But the group's lack of representation from government and community meant that support for its decisions was largely limited to its own members and the few ranchers who chose to be involved in its efforts: "Some ranchers liked and were interested in what the research group had to offer and were willing to try the group's approach, but most weren't willing." Many ranchers considered the CRWP/SACC to be "wolf-loving" environmentalists. One interviewee said that the ranching community's view of the CRWP/SACC's efforts was "about 50/50." Another said, "I think not having government involvement and real commitment made a huge difference." Since the outcomes of the SRWP/SACC process did not meet reasonable community expectations and only partly satisfied community needs and goals we judged the substantive test as partly met.

The efforts of the CRWP/SACC to understand and reduce conflicts aligned well with broad community goals, but the group's approach to depredating wolves was controversial. The inability to address ranchers' concerns about depredation during the first fladry experiment, especially when a primary objective of the CRWP/SACC was to build rapport with ranchers, could be interpreted as a failure to adapt to changing circumstances. At the same time, the fact that the second fladry experiment was conducted the following winter on a ranch further north where there were fewer surrounding landowners was an example of adaptation. Unfortunately, the credibility of CRWP/SACC in the ranching community had been damaged by the initial experiment. Overall, the CRWP/SACC did not pass the practical test, which requires successful practical outcomes and adaptation to changing circumstances.

CRWP/SACC DECISION PROCESS

The CRWP/SACC performed very well in the intelligence function and moderately well in the invocation function, but poorly in the remaining functions and on the criteria for the overall decision process (appendix 5.2). For the pre-decision functions, the main strength of the group was in gathering and generating information. The participants were knowledgeable people, mainly wildlife biologists, who with good intentions and strong field experience conducted research on wolf-livestock interactions and ranching practices. Interviewees recalled occasions when participants questioned the validity of information to ensure that it was reliable. The group seemed to understand important trends and conditions and the social values inherent in the struggle over wolf-livestock conflicts.

Although members of the CRWP/SACC communicated with ranchers and government on an ongoing basis, actual decisions were made with little community or government involvement. As a consequence, many of our interviewees felt that the CRWP/SACC's decisions did not adequately reflect the interests of the wider community. Moreover, proposed strategies were not fully evaluated based on costs and benefits or impacts to social values. When asked to help kill depredating wolves, group members disagreed about values and what actions were appropriate to attain their goals and objectives. The group seemed inadequately prepared for this contingency, which suggests that they had not sufficiently clarified their collective values and goals, agreed in advance about what actions would be acceptable, or mapped out a plan to deal with foreseeable social consequences of the experiment.

The CRWP/SACC group also did not perform well in the prescription function or in implementation of its strategies. The group was not recognized by community members as having authority and had no real control to enforce its decisions. Existing provincial policies made it relatively easy to kill wolves, contradicting the group's efforts to limit wolf removal and encourage alternative livestock management practices. Although the CRWP/SACC cannot be blamed for provincial policies, the group's prescriptions did not include sufficient sanctions (negative or positive) to alter the behavior of many ranchers, and enforcement was very difficult because people who worked with the group did so solely out of interest. To encourage participation and compliance, the group could have offered additional incentives, such as helping to remove problem wolves, locating cattle that had been killed by wolves (to assist compensation claims), or instituting a program to supplement provincial compensation payments for losses. In spite of these difficulties, the group was reasonably successful at developing and implementing several positive strategies in a timely manner. CRWP/SACC members made quick decisions in the field when needed and made deliberate and thoughtful decisions for longer-term research such as collaring and fladry experiments.

Unfortunately, the fladry project encountered major problems because of poor communication. Although CRWP/SACC members were working or talking with some ranchers, the group had not encouraged feedback on their plans from others, nor had they contacted all of the potentially affected ranches prior to the start of the experiment. Had CRWP/SACC members engaged better with neighboring ranchers early on they might have identified concerns that could have been addressed or perhaps even selected a different location for the experiment. As a result of the perceived increase in depredation and the decision not to help remove wolves, some of our interviewees felt that the project imposed unnecessary costs on nearby ranchers. Others disagreed, suggesting that the depredation would have occurred in any event and that it was the province's responsibility to deal with wolf control and livestock compensation. One member of the CRWP/SACC told us that the depredation on neighboring ranches was expected and required to validate the research, because it showed that wolves were still in the area and killing livestock although they were not doing so on the experimental property: "The fact that wolves would kill livestock on a neighboring property to me was always a given." This individual also suggested that the level of depredation on the neighboring properties had not increased as a result of the experiment. Neighboring ranchers clearly did not agree.

Another challenge for the CRWP/SACC group was that its process for choosing among alternatives was not always consistent, leading eventually to a decline of internal trust and a sense of frustration over allocation of the group's limited resources. The group normally operated by consensus, but participants told us that some decisions about media releases and allocation of resources were made and implemented without debate and consensus approval and without adequate consideration of the interests of all members. Examples included purchasing equipment and collaring wolves in a national park outside the group's main working area and deciding not to assist in removing wolves after the fladry experiment. These concerns about decision making were not discussed by the group to clarify norms and rectify the perceived problems.

The CRWP/SACC group also performed poorly in the evaluation function and in planning for termination. There were no specific timelines or termination policies for decisions, with the exception of a two-year period established for the fladry experiment. One interviewee noted that timelines for the group's work were simply dictated by resource availability. Prior to the present study, the CRWP/SACC's process and strategies had not been formally evaluated, with the exception of peer review of a research paper from the fladry experiment (Musiani et al. 2003).

Finally, the CRWP/SACC did not satisfy many of the standards for the overall decision process. The group was generally efficient in its use of resources and appeared to be committed to the idea of supporting human dignity, but some participants questioned the honesty of others in light of decisions that were not conducted through the normal decision-making process. The group did not adjust well to concerns about the fladry project, and powerful members imposed a decision on those who disagreed. The process did not incorporate differentiated structures for each function of decision making.

OBCAG AND THE COMMON INTEREST

The OBCAG performed fairly well on tests of the common interest. With regard to the procedural test, the group's membership was reasonably representative, with the exception of a few interests such as the Alberta Fish and Game Association (which represents anglers and hunters), and some local leaders. First Nations were notably absent. Despite these omissions, the inclusiveness of the OBCAG benefited its decision making, as a range of community values was considered. A few interviewees, however, felt that some OBCAG members did not communicate well with the constituents they represented, and the majority of interviewees indicated that the OBCAG did not sufficiently discuss its work beyond its constituents. One interviewee observed that "it was never established how it was to be communicated to the people they represent, and to a large extent I don't think it was." Another said "there was no evidence that they took answers, solutions, proceedings, or progress from the committee to the rest of the community." A rancher, who was not a member of the OBCAG, commented: "You just can't fly your helicopter any time you want to shoot a wolf and not tell anybody when . . . or why you shoot a wolf or anything else, . . . and that's the way it's been." Even a member of the OBCAG said, "I just honestly never heard that much of what they were trying, . . . you know, what direction they were going." In contrast, another OBCAG member commented that "there were good reports coming back from representatives that were on the board, particularly from the producers. The producers were diligent in keeping the community informed of what was going on, and members of their groups were really diligent in keeping their member on the committee well informed."

The OBCAG satisfied the substantive test of the common interest. The group's expectations, interests, and longer-term goals were reasonable for the context. A few interviewees, including members of the OBCAG and ranchers not involved with the group, did question whether its efforts really advanced the interests of the ranching community. One suggested, "If this process went away, ranchers would probably be happier," and another, "not that the ranchers would change the way they ranch to accommodate wolves," and a third, "the ranchers see this as somewhat of a burden." The majority of our interviewees, however, including ranchers inside and outside the OBCAG, felt the overall goal of reducing wolf-livestock conflict was in everyone's best interest. One stated, "If [ranchers] didn't feel it was important they wouldn't participate . . . . We found that the group was pretty much in agreement that wolves had a place on the landscape, . . . and everyone agreed that wolves should not be present on the landscape at the expense of the cattle producer." Another commented, "The conservation community at the table made it a point, repeatedly, to say we want to reduce predation, we're not here to feed cows to wolves, and the ranchers would say we don't want to kill every wolf, we just don't want them eating all of our cows."

Interviewees indicated that the government representatives on the OBCAG were dedicated to the group's efforts, but that the group lacked support from senior levels in the provincial government, and the resources provided by the province were not sufficient. One interviewee said, "The OBCAG was not something that necessarily had the minister's ear." Another observed, "Carrie [a provincial representative on the OBCAG] was committed, . . . but you get one step above that and I think the commitment just declines, declines, declines."

The OBCAG's decision process was mainly consensus based or by majority vote on the rare occasions when consensus could not be achieved. Thus all members had a direct voice in decisions: "You try to go on consensus. . . . There has certainly been some compromises made." "I really can't say that I remember a vote. Perhaps you could say, though, that we operated by consensus."

Concerning the practical test, interviewees indicated that the OBCAG's decision process was representative and responsible, but that over time the outcomes lost support from the ranching community and were not always compatible with community interests. Although the OBCAG's membership experienced several changes, the membership was not adjusted to include representatives of any additional interests and the group did not devote much time to discussing individual roles and responsibilities despite concerns about communicating with constituencies and the wider community.

Interviewees indicated that after the OBCAG stopped meeting, support for the wolf control program declined as a result of increased depredation. One rancher observed, "It is getting worse and worse every year. Our number of wolf kills are up drastically. . . . There is tons of room for improvement. . . . The 'three-in-three,' that's gone by the wayside, we don't even get that anymore." An agency employee commented, "That is what we are going through, a major period of [livestock depredation] right now." A member of the OBCAG told us that the "three-in-three" rule was still in place but that the SRD had not been able to deliver as a consequence of funding cuts. According to this individual, "One of the things that the ranching community has been frustrated by on this committee is that instead of removing a full pack when there is trouble we've tended to reduce it and given it benefit . . . in the hope that things would stop, and it hasn't." However, most of the ranchers on the OBCAG that we interviewed indicated they supported the recommended wolf control protocol.

OBCAG DECISION PROCESS

The OBCAG's decision process performed moderately well in comparison with recommended standards (appendix 5.2). The group was particularly adept in gathering information and promoting and debating ideas (the intelligence and promotion functions), but had more difficulty with the rest of the decision functions.

The OBCAG included knowledgeable representatives of most interests, who generated and gathered valuable information used to evaluate and debate alternative strategies. The participants developed trust and respect for one another and felt that they were able to express their opinions and ideas freely. A comfortable setting for group meetings facilitated debate and development of strategies that respected the group's goals and the varied interests of its members.

Choices among alternatives were made through consensus, compromise, and occasionally reluctant agreement. The group reached consensus on radio collaring programs for both wolf research and wolf control. The "three-in-three" threshold for lethal wolf control was a compromise among conservation groups, biologists, and ranchers. Representatives of the ranching community reluctantly agreed to the development of a communication plan to share the work of the OBCAG with people outside the advisory group despite concerns about potential backlash. The ranching representatives were hesitant because they had received negative public attention in the past over wolf killing: "When the word got out, it resulted in wolf lovers actually targeting some local ranchers with letters or a presence at their gate, . . . a spotlight that the ranchers didn't want."

The OBCAG was successful in choosing and implementing several strategies that furthered the common interest in a timely and equitable manner. They adopted or recommended protocols such as the "three-in-three" threshold for wolf control and a limit of two radio collars per pack, which directed allocation of scarce resources and encouraged unbiased implementation. The group acted quickly when immediate decisions were needed about wolf control or radio collaring to minimize losses to ranchers while maintaining two collars per wolf pack for ongoing research and monitoring. In contrast, the group labored over preparing a communication plan and the plan was never implemented. Interviewees suggested that progress on the communication plan was slow because of the sensitive nature of wolf-livestock conflicts and lethal control. The slow progress and accommodation of these concerns may have improved trust and perhaps would have eventually resulted in a thoroughly considered plan that addressed concerns. However, at least one member of the OBCAG believed that the slow progress was a consequence of two other members who did not want publicity for the group or its efforts and purposely delayed the completion and implementation of the plan.

Four main weaknesses were evident in the prescription, invocation, and application functions. First, the OBCAG was only an advisory body to the provincial government and had no direct control over the implementation of its own decisions. If Alberta SRD did not act on the OBCAG's recommendations, it is unlikely that the advisory group could have forced action by approaching other government agencies, and the OBCAG's lobbying potential was limited as a result of its policy of not publicizing its efforts to people and groups beyond its own members.

The second challenge was that the group's prescriptions, including norms and contingencies, were not clearly and formally specified. Some group members were unaware of some decisions, and other members expressed concern that the details about when and how strategies would be implemented were not clear. Poorly defined prescriptions possibly contributed to confused expectations about what the SRD was supposed to accomplish with the group's recommendations and made it difficult to hold the government accountable for inaction or poor performance. Interviewees commented that the SRD was not properly following the OBCAG's recommendations, but at least one felt it was probably because the recommendations were not clear.

The third challenge was that resources for the OBCAG were inadequate and declined over the life of the group. Several members of the OBCAG suggested that senior levels in the provincial government did not support the process, as evidenced by the limited budget. One government representative said that relations between the group and senior levels of government could have been improved with more effort on promotion and working together with upper government staff. Several interviewees suggested that one of the chairpersons of the OBCAG acted in ways that did not support public relations and that this reduced support from members of the ranching community as well as government.

The OBCAG's lack of authority and control, poorly specified prescriptions, and limited resources all contributed to the group's fourth challenge, inconsistent implementation. The "three-in-three" protocol and the "two radio collars per pack" rule were designed to improve consistency, but other prescriptions were less clearly specified, and problems with resources and authority became more pronounced with time.

The OBCAG performed poorly in the evaluation function and in planning for termination. The group did not develop specific timelines or termination policies for its decision process or recommendations, nor did it evaluate performance other than through informal discussions at meetings (our own evaluation was not instigated by the OBCAG).

On the overall decision process criteria, the OBCAG performed well. The group's members and facilitator were considered honest and devoted to supporting human dignity. Decisions were deliberate and well debated. All OBCAG interviewees commented that the OBCAG was very efficient in its use of limited resources. However, three interviewees (two OBCAG members and one rancher) said that, although resources were used efficiently for particular tasks, perhaps the resources could have been allocated elsewhere for improved overall benefits or particular tasks could have been completed differently in order to preserve resources. One rancher on the OBCAG observed that if the objective were to remove wolves, poisoning would be a cheaper alternative to helicopters and euthanasia. Another rancher (not on the OBCAG) said that reducing livestock losses from poisonous vegetation would have a greater positive impact on livestock survival than managing wolves. Another OBCAG member commented that funding research on alternative livestock management practices might provide longer-lasting improvements for livestock survival than funding shorter-term wolf control strategies. Finally, like the CRWP/SACC, the OBCAG did not incorporate differentiated functions for decision making.

Moving Forward: Lessons and Recommendations

Both the CRWP/SACC collaboration and the OBCAG sought to reduce conflicts between wolves and livestock in southwestern Alberta. There were several notable similarities in the two initiatives: both operated mainly by consensus, encouraged feedback from participants, and were open to considering and discussing a variety of ideas and strategies. Both groups were very good at monitoring wolves and gathering information about wolf behavior, and both completed successful research on deterrence strategies. Neither group had clear process rules, structured evaluations, or termination policies. Finally, resources for both groups were very limited and were administered by a single authoritative entity so that most participants had little control over the allocation of funds. In the CRWP/SACC process, members of the CRWP dictated how resources would be allocated despite the otherwise collaborative approach adopted by the group. For the OBCAG, the provincial government controlled the resources and made decisions about whether or not to fund recommended strategies.

There were also some substantial differences between the CRWP/SACC

and the OBCAG. The OBCAG included representation from most of the interests in the community, whereas membership in the CRWP/SACC was limited
to biologists and a businessperson. Only the actual members of the CRWP/
SACC participated in its decision making, and this may have contributed
to the high emphasis that CRWP/SACC placed on preventing wolf mortality. The dominance of scientists on the CRWP/SACC may also explain another key difference between the two groups: the CRWP/SACC conducted its
main tests of deterrence strategies as formal scientific experiments, using
one ranch as the experimental treatment and adjoining ranches as controls.
Although this approach may have strengthened the validity of the experiments, the predictable value-based negative reactions of adjoining ranchers
damaged the reputation of the researchers in the ranching community. A related difference between the groups was that the OBCAG was successful at
maintaining trust among its member participants, but for the CRWP/SACC
internal trust deteriorated over time.

Finally, the CRWP/SACC members were responsible not only for making
decisions but for implementing those decisions, whereas the OBCAG was an
advisory group that only recommended strategies, leaving the responsibility
for successful implementation in the hands of the provincial government.
For the CRWP/SACC any failure in implementation was attributed to the
group itself, while the OBCAG was to some extent isolated from blame as the
provincial government was perceived to be primarily responsible.

DESIGNING EFFECTIVE DECISION PROCESSES

This evaluation of the CRWP/SACC and the OBCAG highlights several key
lessons for designing effective decision processes in such local initiatives.

Inclusive participation and effective communication are critical for identifying concerns and developing strategies that reflect the perspectives, interests, and goals of the community. The tests of the common interest stress
broad representation because decisions developed by inclusive processes
are more likely to advance the common interest and to be widely supported
(Brunner 2002; Brunner and Steelman 2005). These tests embody Lasswell's
"procedural" approach to ensure that common interests prevail over special
interests: "First, involve all interests actively all the time or on particularly
important occasions; second, involve third-party representatives of the common interest in other situations" (Lasswell 1971, 105). Inclusive decisions
may also aid implementation because community members are more likely
to respect the process through which the decision was made and accept that
the decision serves common interests. The experiences of the CRWP/SACC

and the OBCAG not only demonstrate the importance of inclusive representation, but also show that in addition to clarifying goals participants need to explore and understand each other's perspectives and think about how their perspectives will affect the strategies that are considered appropriate.

For an advisory body such as the OBCAG, with limited funding and little direct control over implementation, engaging with the community can help to develop a supportive constituency of interested, knowledgeable, and involved people to lobby for funding and implementation of recommendations. The OBCAG is also a good example of the tension between the desire for confidentiality in initial discussions of sensitive or controversial issues (such as killing wolves) and the openness that is needed to ensure good decisions and public support. Lasswell's (1971) criteria for the intelligence function emphasize openness, but recognize constraints: "The output is closed to unauthorized persons and organizations for appropriate periods.

Restrictions on dissemination are reasonable when lawful goals would be compromised, or when avoidable deprivations are imposed on third parties" (p. 88).

Decision makers need *sufficient authority and control* to make and implement publicly acceptable decisions. "To be authoritative is to be identified as the official or agency competent to act; to be controlling is to be able to shape results" (Lasswell 1971, 99). When a group finds itself without adequate authority and control, it needs to consider strategies to address this deficiency, including the possibility of adding participants with authority and control to the decision-making process. This may also increase the probability of effecting change in formal laws or policies.

It is also important to develop *clear norms and expectations about decision processes*, including the roles and responsibilities of each participant. A clearly described process helps to create unambiguous expectations so that people can be held accountable for not contributing or for working against the group's efforts. Process rules and participants' responsibilities should be formally specified in "terms of reference" or similar documentation and available to all participants.

Prescriptions should be clearly and comprehensively specified. Effective prescriptions set out the goals that the decision is designed to advance, the rules or other mechanisms being instituted to achieve those goals, the contingencies in which the rules will apply, sanctions (positive and negative) to encourage performance, and sufficient resources for implementation (Clark 2002). Decisions should be formally documented so that the decision group and those affected have clear expectations and the formal prescription can be referenced during implementation and evaluation.

Evaluation of decision processes and outcomes is essential to ensure that they meet goals. Evaluations can identify strengths and weaknesses in decision making and warn when prescriptions need to be adjusted or terminated to deal with error or changing circumstances. Decision makers should adopt evaluative criteria and evaluation policies, including regular internal and external reviews. The evaluation process should be supported by plans for adaptation and termination.

Plans for adaptation and termination of prescriptions should be developed to ensure that effective strategies end when they have achieved their goals, that those that fail are not continued, and that all strategies are adapted with changing circumstances. For the OBCAG, support for the group's wolf control strategy appears to be eroding in recent years because of perceived increases in depredation. One possible reason is that resources have declined so that the wolf-control protocol is no longer being applied consistently. If these trends and conditions continue, the protocol should be revisited.

Collaborative groups should consider hiring or at least consulting an *experienced facilitator*, who can help them to develop clear process and outcome standards and guide them through the intellectual tasks of problem orientation (Rutherford et al. 2009).

Stable resources for the long term are required to ensure that groups continue to operate and their prescriptions are implemented consistently and effectively. This, of course, is often a major challenge, but building engaged and supportive constituencies can be an asset.

Develop *plans for succession*. The loss of key participants was an important factor in the demise of both the CRWP/SACC collaboration and the OBCAG. Succession planning is difficult, but it is important for groups to make plans to develop or recruit candidates to take on leadership roles when others withdraw.

BUILDING TRUST AND SOCIAL CAPITAL

Local initiatives that seek to reduce conflicts between wolves and livestock on private lands cannot succeed without the willing engagement and participation of landowners. Landowners will not engage and participate unless they believe in and trust the process and the people involved. Thus, local initiatives must generate the trust and social capacity for these and other participants to work together to overcome conflicts and identify and pursue their common interests. It is very difficult for an external intervention to construct social capacity (Ostrom 1999). Local knowledge is required, and participants must be "strongly motivated to facilitate the growth and empowerment of others" (Ostrom 1999, 182). Working with a few community members may not be enough, as broader cultural and institutional dynamics are critical. The experiences of the OBCAG show that it is possible to build trust among ranchers, conservationists, wildlife scientists, and provincial government staff in dealing with wolves. The CRWP/SACC process provides another example of how trust can be built, but also shows how easily that trust can erode.

We identified six factors that contributed to a high level of trust and capacity in the OBCAG. The first was the level of commitment of participants to the process. They not only gave their own time and financial support, but also were willing to make real compromises to advance the common interest. The wildlife specialists and conservation members understood that by keeping wolves on the landscape the ranchers would be the people who would experience the most deprivation as cattle would continue to be killed. The ranchers understood that some amount of wolf depredation would continue indefinitely and that there was growing public support for having large carnivores on lands outside protected areas. The ranchers settled for a reduced level of wolf removal, while the biologists and conservationists agreed to a moderate level of lethal wolf control as long as it targeted wolves that repeatedly preyed on livestock.

The second factor was that the group isolated its sensitive deliberations from the outside community and media, which created a sense of freedom to speak openly, especially when discussing controversial topics such as killing wolves. Although general information about the group and its efforts was available to anyone who inquired, the specific information discussed by the group, its recommendations, and the implementation of its strategies were not openly shared with the media. Knowing that sensitive and potentially controversial discussions would not be shared with the public at large likely permitted OBCAG members to speak more freely. One participant commented, "Well it's had a pretty low profile, and that has been on purpose because we've done some things that aren't particularly popular like collaring wolves and not destroying an entire pack when there's some predation." A study by Lachapelle, McCool, and Patterson (2003) of planning groups in the western United States found that in one case concerns over trust developed following the release of a newsletter that "generated negative media attention, angered the public, and intensified a sense of mistrust." In contrast, the OBCAG refrained from releasing information to the media because of the sensitive nature of wolf management and concerns over negative media attention. Unfortunately, the communication plan for the OBCAG was never finalized. As discussed earlier, the trade-off for the group's confidentiality was that better engagement with external interests might have helped build support, identify outside concerns, find additional funding sources, and encourage the province to commit additional long-term resources.

The third factor that promoted trust was the consensus decision rule adopted by the OBCAG. This ensured that each participant's concerns and interests were considered. The fourth factor was the informality of the group's meetings and decision processes, which made participants more comfortable in taking part in its deliberations and voicing their interests. A fifth factor was the limited responsibility of the OBCAG, as the province was responsible for implementing decisions. Failures in implementation were mainly attributed to the government, rather than the advisory board or its members. This may also have created an OBCAG-versus-the-government mentality that fostered cohesion within the OBCAG. Unfortunately, blaming another group, the province, may have resulted in some apathy among participants about whether their recommendations were implemented correctly. Another disadvantage is that group members may have tried to redirect blame for failure on the implementation of recommendations rather than addressing potential problems with the recommendations themselves.

The sixth factor regarding trust was the willingness of the Alberta SRD agency to act quickly to implement the OBCAG's early recommendations, which were supported by consensus among the advisory group. The "three-in-three" livestock depredation threshold was adopted immediately by the SRD after recommendation by the OBCAG. Other recommendations, including radio collaring for research, were also adopted quickly. The willingness of the SRD to act on the initial recommendations validated the meaningfulness of the process for the group, provided a sense of accomplishment, and fostered trust in the government representatives on the OBCAG.

In contrast, the CRWP/SACC collaboration was not successful in maintaining lasting trust among its participants. An important lesson from this case is that nonconforming actions need to be addressed when they occur. Interviewees commented that at times decisions were made without consensus, and funding was applied to equipment and work not agreed on by all CRWP/SACC members. Allowing these actions to continue unchecked created frustration and distrust. Noncompliant actions may arise from a misunderstanding of rules, which can be rectified, but sometimes they are not misunderstandings and need to be addressed before the group's process deteriorates further.

Practical Methods for Reducing Conflicts between Large Carnivores and People

Can local initiatives develop effective mechanisms to reduce conflicts with large carnivores that are also acceptable to the broader community? The answer from the southwestern Alberta cases seems to be . . . possibly. Through the CRWP/SACC process the use of fladry was shown to be effective at reducing depredation (Musiani et al. 2003). The OBCAG also demonstrated that night riding could reduce livestock losses. One difficulty with both strategies is that ranchers may not have the resources or interest to implement them. This difficulty is not insurmountable: the Blackfoot Challenge and partners in Montana have operated a successful range riding program since 2008 (People and Carnivores 2013), and the Mountain Livestock Cooperative has had positive results with a range riding program in Alberta in which ranchers cooperate to ride each others' ranches (Minbashian 2012).

In addition to testing deterrence methods the OBCAG attempted to address conflicts between wolves, livestock, and people by recommending changes to the provincial government's wolf control strategy and livestock compensation program. The OBCAG convinced the province to change the wolf control program in southern Alberta from relatively indiscriminate wolf removal, sometimes using strychnine, to an approach that targeted "problem" wolves using net gunning, trapping, and euthanasia. The revised strategy was successful, but required diligent attention since the wolf population was no longer being eliminated from time to time. According to our interviewees, wolf removal activities have become less effective since the OBCAG stopped meeting.

The OBCAG also recommended an increase in provincial compensation for livestock losses to better reflect the true costs to ranchers. With higher compensation it is possible that ranchers might be more willing to accept occasional depredation events. In 2012 a community-based group in southwestern Alberta developed a prototype for a revised provincial compensation scheme that is now being considered by the provincial government (Morrison 2013).

Conclusion

This evaluation of the CRWP/SACC collaboration and the OBCAG shows that both initiatives were very successful in generating information for decisions, but both struggled with other aspects of the decision process. Particular weaknesses included poor goal definition, lack of clarity in process, unclear responsibilities of participants, lack of evaluation and termination policies, and poor external communication. Despite these weaknesses, the experiences of these two innovative projects offer hope that such collaborative endeavors can develop trust among diverse stakeholders and that conflicts between wolves and livestock can be managed more effectively.

The members of the CRWP/SACC, in particular, faced numerous difficulties in their efforts to reduce conflicts. However, this should not be taken as a wholesale indictment. Maintaining wolves and livestock in a shared landscape is a complex challenge. The CRWP/SACC field staff initiated a good working relationship with some ranchers, and the group demonstrated the potential of fladry as a deterrence strategy. Their efforts provided a foundation for further work in southwestern Alberta and may have fostered the interest of conservationists, government, and ranchers in working together to reduce wolf-livestock conflicts. As one interviewee commented, "I don't think we'd be near as far along at this point in time had we not started together back in 2000." Some of the field staff from the CRWP/SACC collaboration have continued to conduct research and work on reducing conflicts with livestock in Alberta and elsewhere.

The OBCAG collaborative process also faced difficulties in its decision making, but it built a solid foundation of trust among diverse participants, and it recommended several management strategies that were implemented effectively. Notably, it successfully altered the provincial wolf removal practice in the short term. Recent increases in depredation and other challenges in the region indicate a pressing need for further collaboration to develop effective policies for wolves, livestock, ranchers, and other interests in southwestern Alberta.

Appendix 5.1
Tests of the Common Interest

Tests to determine whether a decision process serves the common interest (adapted from Brunner et al. 2002; Brunner and Steelman 2005; and Steelman and DuMond 2009).

PROCEDURAL TEST

Is the decision-making process *inclusive* (i.e., are the participants representative of the community as a whole)?

If some interests are not directly represented in the process, are those interests reflected in the outcomes?

Are the participants *responsible* (i.e., are they willing and able to serve the community as a whole, and can they be held accountable for the consequences of their decisions)?

SUBSTANTIVE TEST

Are the expectations of the participants about what will be accomplished reasonable?

Have all valid and appropriate concerns been taken into account?

Have the outcomes been approved by participants representative of the community as a whole, indicating that they believe the outcomes are in the common interest?

Are the outcomes compatible with broad societal goals (e.g., democracy, equity, timeliness)?

Do the outcomes ostensibly solve the problem?

Do the outcomes work in practice, and do they uphold the reasonable
expectations of those who participated?
Are decisions adapted over time to deal with changing circumstances?

Appendix 5.2
Decision Activities, Standards, and Effectiveness in the Central Rockies Wolf Project/Southern Alberta Conservation Cooperative and the Oldman Basin Carnivore Advisory Group

Appraisal of decision-making processes, specifically decision activities and
effectiveness standards, for the CRWP/SACC and the OBCAG.

INTELLIGENCE
Recognizing the problem and gathering information.
Dependable
Comprehensive
Selective
Creative
Open

CRWP/SACC
Participants had expertise in wildlife biology, but limited expertise in
other areas.
Good understanding of important trends and conditions.
Well-designed experimental research, selectively focused on deterrence.
Critically examined the dependability of information.
Many interests unrepresented.
Insufficient exploration of perspectives of participants.

OBCAG
Participants had diverse expertise.
Good understanding of important trends and conditions.
Worked together to generate valuable information.
Critically examined the dependability of information.
Inclusive representation of the community, with the exception of a few
interests.
Poor communication of information beyond membership (openness).

PROMOTION

Open debate, in which various groups advocate for their interests or preferred policy.

Rational

Integrative

Comprehensive

CRWP/SACC

Insufficient debate and assessment of costs and benefits to community interests.

Many interests unrepresented.

Some alternatives did not advance the common interest.

OBCAG

Thorough debate and assessment of costs and benefits to community interests.

Developed high levels of respect and trust, freedom to express opinions without fear of persecution.

Comfortable setting for the exchange of ideas.

Alternatives reflected diverse interests of members.

Slow in developing communication plan.

PRESCRIPTION

Setting the policy, rules, or guidelines.

Effective (stable expectations)

Rational

Comprehensive

CRWP/SACC

Inconsistent decision rule; normally consensus-based, but occasionally arbitrary.

Little authority or control.

Insufficient sanctions (positive or negative) and resources.

Many interests unrepresented; decisions didn't always reflect the common interest.

Decisions not always brought to the attention of potentially affected parties.

OBCAG

Consistently applied consensus-based decision rule; participants showed willingness to compromise.

Some authority, but little control.

Some prescriptions not sufficiently clear about norms and contingencies.

Insufficient sanctions (positive or negative) and resources.

Wide representation; decisions generally reflected the common interest.

Decisions not always brought to the attention of potentially affected parties.

INVOCATION

Implementation
Timely
Dependable
Rational
Nonprovocative

CRWP/SACC

Timely implementation of selected strategies, particularly field decisions.

More deliberative when time allowed.

Fladry studies highly provocative for neighboring ranchers, perceived to impose unfair deprivations.

OBCAG

Advisory body, little control over implementation.

Limited resources.

Timely implementation of some prescriptions, but others not implemented.

More deliberative when time allowed.

Standardized protocols encouraged unbiased invocation.

APPLICATION

Dispute resolution and enforcement of the prescription.
Rational and realistic
Uniform

CRWP/SACC

Goals of CRWP favored in application; complaints by neighboring ranchers considered, but no remedy provided.

No review of complaints about deviations from normal decision rules.

OBCAG

Little control over application.

Nondiscriminatory.

Nonconformance not always rectified.

APPRAISAL

Review and evaluation of the activities so far.

Dependable

Rational

Comprehensive

Selective

Continuing

Independent

CRWP/SACC

No explicit evaluation processes or standards, other than peer review of academic publication.

OBCAG

No explicit evaluation processes or standards.

TERMINATION/SUCCESSION

Ending or moving on.

Timely

Comprehensive

Dependable

Balanced

Ameliorative

CRWP/SACC

No explicit timelines or termination policies, other than a two-year period for the fladry experiments.

OBCAG

No explicit timelines or termination policies.

OVERALL STANDARDS

Honest
Economical
Technically efficient
Loyal and skilled personnel
Complementary and effective impacts
Differentiated structures
Flexible and realistic in adjusting to change
Deliberate
Responsible

CRWP/SACC

Loyal and skilled personnel, but with narrow expertise.
Expressed commitment to the common interest.
Efficient use of resources.
Honesty and responsibility of some participants questioned because of
    decisions about allocation of resources.
Mixed record of adjusting to change.
No differentiated structures for decision-making functions.
Deterioration of internal trust over time.

OBCAG

Loyal and skilled personnel.
Expressed commitment to the common interest.
Efficient use of resources.
Participants considered honest.
Limited opportunity to adjust to change.
No differentiated structures for decision-making functions.
Maintained high levels of internal trust over time.
Deliberate and responsible decision making.

NOTE

We thank the Northern Rockies Conservation Cooperative and the Social Sciences and Humanities Research Council of Canada for providing funding for this research. We also thank our interviewees and Susan Clark, Denise Casey, and several anonymous reviewers for their comments.

REFERENCES

Alberta Environment and Sustainable Resource Development. 2013. *Gray Wolf* (Canis lupus). Accessed January 8, 2013, from http://www.srd.alberta.ca/FishWildlife /WildSpecies/Mammals/WildDogs/GrayWolf.aspx.

Alberta Forestry Lands and Wildlife. 1991. *Management Plan for Wolves in Alberta*. Edmonton: Alberta Forestry Lands and Wildlife.

Alberta Sustainable Resource Development. 2002a. *Wolves in Alberta: The Evolution of Wolf Management*. Accessed May 26, 2008, from http://www.srd.gov.ab.ca /fishwildlife/wildlifeinalberta/wolvesalberta/evolutionmanagement.aspx.

———. 2002b. *Wolves in Alberta: Present Status of the Wolf*. Accessed May 26, 2008, from http://www.srd.gov.ab.ca/fishwildlife/wildlifeinalberta/wolvesalberta /presentstatus.aspx.

———. 2002c. *Wolves in Alberta: Problems with Wolves*. Accessed May 26, 2008, from http://www.srd.gov.ab.ca/fishwildlife/wildlifeinalberta/wolvesalberta/problems .aspx.

Alberta Wilderness Association. 2002. "Alberta's Development Agenda Puts Wildlife in Danger and Ignores Public Opinion." News Release, November 18, 2002. Accessed January 16, 2009, from http://news.albertawilderness.ca/2002NR/NR021218 /NR021218.htm.

———. 2008. *Wolves History*. Accessed May 26, 2008, from http://issues.alberta wilderness.ca/WL/wolveshistory.htm.

Boitani, L. 1995. "Ecological and Cultural Diversities in the Evolution of Wolf-Human Relationships." In *Ecology and Conservation of Wolves in a Changing World*, edited by L. N. Carbyn, S. H. Fritts, and D. R. Seip, 3–12. Edmonton: Canadian Circumpolar Institute, Occasional Publication No. 35.

Brunner, R. D. 2002. "Problems of Governance." In *Finding Common Ground: Governance and Natural Resources in the American West*, edited by R. D. Brunner, C. H. Colburn, C. M. Cromley, R. A. Klein, and E. A. Olson, 1–47. New Haven, CT: Yale University Press.

Brunner, R. D., and T. A. Steelman. 2005. "Toward Adaptive Governance." In *Adaptive Governance: Integrating Science, Policy, and Decision Making*, edited by R. D. Brunner, T. A. Steelman, L. Coe-Juell, C. M. Cromley, C. M. Edwards, and D. W. Tucker, 268– 304. New York: Columbia University Press.

Canada. 2009. *Species at Risk Public Registry, Species Profile, Northern Grey Wolf*. Accessed June 5, 2010, from http://www.sararegistry.gc.ca/species/speciesDetails _e.cfm?sid=613.

Clark, S. G. 2002. *The Policy Process: A Practical Guide for Natural Resource Professionals*. New Haven, CT: Yale University Press.

Environment Canada. 2008. *Non-Detriment Finding for Canada—Grey Wolf*. Accessed June 5, 2010, from http://www.ec.gc.ca/cites/default.asp?lang=En&n=BB314F25-1.

Gilbert, F. F. 1995. "Historical Perspectives on Wolf Management in North America with Special Reference to Humane Treatments in Capture Methods." In *Ecology and Conservation of Wolves in a Changing World*, edited by L. N. Carbyn, S. H. Fritts, and D. R. Seip, 13–17. Edmonton: Canadian Circumpolar Institute, Occasional Publication No. 35.

Gunson, J. R. 1992. "Historical and Present Management of Wolves in Alberta." *Wildlife Society Bulletin* 20:330–39.

Janusz, B. D. 2008. "Carnivore Corridors Threatened in the Crowsnest." *Wild Lands Advocate* 16 (2):10–12.

Kaminski, T., C. Mamo, C. Callaghan, and M. Musiani. 2003. *Southern Alberta Conservation Cooperative: A Proposal for Conserving Large Carnivore and Rural Communities in Southern Alberta.* Accessed June 26, 2008, from http://www.y2y.net /science/grants/carn7.asp.

Lachapelle, P. R., S. F. McCool, and M. E. Patterson. 2003. "Barriers to Effective Natural Resource Planning in a 'Messy' World." *Society and Natural Resources* 16:473–90.

Lasswell, H. D. 1971. *A Pre-View of Policy Sciences.* New York: American Elsevier.

Le Roy, D., K. K. Klein, and T. Klvacek. 2006. *The Losses in the Beef Sector in Canada from BSE.* Commissioned Paper CP 2006-5, Canadian Agricultural Trade Policy Research Network. Accessed January 16, 2009, from http://www.uoguelph.ca /catprn/publications_commissioned_papers.shtml.

Minbashian, J. 2012. "Wolves and Livestock: Can They Share a Landscape?" *Conservation Northwest* 90:4–5.

Minta, S. C., P. M. Karieva, and A. P. Curlee. 1999. "Carnivore Research and Conservation: Learning from History and Theory." In *Carnivores in Ecosystems: The Yellowstone Experience,* edited by T. W. Clark, S. C. Minta, A. P. Curlee, and P. M. Karieva, 323–404. New Haven, CT: Yale University Press.

Morehouse, A. T., and M. S. Boyce. 2011. "From Venison to Beef: Seasonal Changes in Wolf Diet Composition in a Livestock Grazing Landscape." *Frontiers in Ecology and the Environment* 9:440–45.

Morrison, C. 2013. *Carnivores and Conflict: A Community Approach to Carnivore Compensation; Report 2, Proposed Amendments to the Alberta Wildlife Predator Compensation Program.* Accessed February 10, 2013, from http://www .watertonbiosphere.com/uploads/biosphere-resources_36_3809318697.pdf.

Musiani, M., C. Mamo, L. Boitani, C. Callaghan, C. Gates, L. Mattei, E. Visalberghi, S. Breck, and G. Volpi. 2003. "Wolf Depredation Trends and the Use of Fladry Barriers to Protect Livestock in Western North America." *Conservation Biology* 17:1538–47.

Musiani, M., and P. C. Paquet. 2004. "The Practices of Wolf Persecution, Protection, and Restoration in Canada and the United States." *BioScience* 54:50–60.

Oldman Basin Carnivore Advisory Group (OBCAG). 2004. *Community Oriented Wolf Strategy: Year 1 Progress Report.* Pincher Creek: Alberta Sustainable Resource Development.

———. 2005. *Community Oriented Wolf Strategy: Year 2 Progress Report.* Pincher Creek: Alberta Sustainable Resource Development.

Ostrom, E. 1999. "Social Capital: A Fad or a Fundamental Concept?" In *Social Capital:*

A Multifaceted Perspective, edited by P. Dasgupta, and I. Serageldin, 172–214. Washington, DC: World Bank.

Parks Canada. 2009. *Kootenay National Park of Canada: Wolf Research and Monitoring*. Accessed February 9, 2009, from http://www.pc.gc.ca/pnnp/bc/kootenay/natcul /natcul28_e.asp.

People and Carnivores. 2013. *Using Range Riders for Livestock and Wolf Monitoring*. Accessed February 10, 2013, from http://www.peopleandcarnivores.org/services /agriculture/range-riders.

Pym, W. M. 2010. *Improving Wolf Management Practices on Ranchlands in Southwestern Alberta: An Evaluation of Two Collaborative Processes*. Master's Thesis, Simon Fraser University.

Quinn, M. S., and S. M. Alexander. 2011. *Final Survey Report: Carnivores and Communities in the Waterton Biosphere Reserve*. Calgary: Miistakis Institute, University of Calgary. Accessed May 4, 2012, from http://www.watertonbiosphere .com/uploads/biosphere-resources_18_3002098588.pdf.

Robichaud, C. B., and M. S. Boyce. 2010. "Spatial and Temporal Patterns of Wolf Harvest on Registered Traplines in Alberta, Canada." *Journal of Wildlife Management* 74:635–43.

Rutherford, M. B., M. L. Gibeau, S. G. Clark, and E. C. Chamberlain. 2009. "Interdisciplinary Problem Solving Workshops for Grizzly Bear Conservation in Banff National Park, Canada." *Policy Sciences* 42:163–87.

Silvatech Consulting Ltd. 2008. *Chief Mountain Cumulative Effects Study — Executive Summary*. Cardston, Alberta: Chief Mountain Landowners Information Network. Accessed January 16, 2009, from http://www.salts-landtrust.org/cms/.

Southern Alberta Land Trust Society. 2007. *The Changing Landscape of the Southern Alberta Foothills: Report of the Southern Foothills Study Business as Usual Scenario and Public Survey*. High River, Alberta: Southern Alberta Land Trust Society. Accessed January 16, 2009, from http://www.salts-landtrust.org/sfs/sfs_reporting .html.

Steelman, T. A., and M. E. DuMond. 2009. "Serving the Common Interest in US Forest Policy: A Case Study of the Healthy Forests Restoration Act." *Environmental Management* 43:396–410.

US Department of Agriculture Foreign Agricultural Services (USDAFAS). 2008. *Global Agricultural Information Network Report: Canada Livestock and Products, Livestock Annual 2008*. Washington, DC: US Department of Agriculture. Accessed January 17, 2009, from http://www.fas.usda.gov/gainfiles/200809/146295792.pdf.

US Fish and Wildlife Service. 2012. *Gray Wolves in the Northern Rocky Mountains: News, Information and Recovery Status Reports*. Accessed May 14, 2012, from http://www .fws.gov/mountain-prairie/species/mammals/wolf/.

US Fish and Wildlife Service et al. 2012. *Northern Rocky Mountain Wolf Recovery Program 2011 Interagency Annual Report*. Edited by M. D. Jimenez and S. A. Becker. Helena, MT: US Fish and Wildlife Service. Accessed May 14, 2012, from http:// westerngraywolf.fws.gov/annualreports.htm.

Wadlow, S., C. Callaghan, and M. Musiani, eds. 2003. *World Wolf Congress 2003—Bridging Science and Community*. Banff, Alberta: World Wolf Congress. Accessed June 26, 2008, from http://www.carnivoreconservation.org/files/meetings/wolf _2004_banff.pdf.

Wearmouth, C. 2009. "Wolves on the Range." *Wild Lands Advocate* 17:1.

6 Human–Grizzly Bear Coexistence in the Blackfoot River Watershed, Montana Getting Ahead of the Conflict Curve

SETH M. WILSON, GREGORY A. NEUDECKER, AND JAMES J. JONKEL

Introduction

When large carnivore populations overlap with humans, interactions can be problematic for people, but particularly acute for wildlife. In North America large carnivores are generally not well tolerated outside of protected areas (Mattson et al. 1996). When incidents or conflicts occur, carnivores are often trapped, relocated, or removed from populations. Such is the case when grizzly bears conflict with humans at the interface of public and private lands in places like southwest Alberta, southern British Columbia, Montana, and Wyoming (Woodruff and Ginsberg 1998; Wilson et al. 2006). Conflicts or incidents include bears killing livestock, destroying beehives, foraging for garbage close to homes, or in rare cases threatening human safety. Often private lands in valley bottoms and foothills adjacent to public lands are problematic zones, especially when available bear attractants coincide with occupied grizzly bear habitat. Repeated incidents typically lead to more severe conflict, habituation, and eventually to removal of the bear through trapping, relocation, or killing.

This chapter describes how a rural community joined with wildlife agencies and conservation groups to grapple with the complex challenge of learning to live with grizzly bears in the Blackfoot River watershed of west central Montana. This effort started in 2001 when Montana Fish, Wildlife and Parks (FWP) and the Blackfoot Challenge (BC), a grassroots watershed group in the Blackfoot valley, began meeting to discuss concerns among local residents about increasing grizzly bear activity and conflicts in the region. Grizzly bears were reexpanding their range onto private lands, and there was a clear need to bring people together to determine exactly what the "problem" was and how best to address it. This chapter emphasizes how a collective decision-making process encouraged diverse local and national stakeholders to engage in a partnership where participatory efforts helped to reduce conflicts substantially for both bears and people.

Each author of this chapter has significant personal investment and professional capacity in the Blackfoot Challenge, a landowner-driven, nongovernmental organization that has worked since the 1970s to enhance, conserve, and protect the natural resources and rural lifestyle of the Blackfoot watershed. Its overarching goal is to provide a forum to support environmentally responsible resource stewardship through the cooperation of public and private interests. The BC promotes cooperative resource management of the Blackfoot River, its tributaries, and adjacent lands. Initially involved with grizzly bear management in 2000, the BC grizzly bear program was officially launched in 2002. As participants, we believed that building on the existing capacity of the BC was a sensible way to approach the problem of living with grizzly bears and that an inclusive and participatory approach to working with ranchers, landowners, conservation groups, and agencies would facilitate a more positive response to these animals. It was apparent to us that a single agency or individual would not be able to solve this challenge and that in order to make real progress, significant decision-making power would need to be in the hands of those landowners and ranchers who had to confront the daily reality of living with bears.

There were five general methodological phases in this effort. First, FWP met with the Blackfoot Challenge's executive committee to see if there was interest in creating a wildlife committee, that is, a community-driven sounding board for wildlife management. The BC agreed to form a committee with the understanding that the initial emphasis would involve grizzly bears. Second, the BC conducted a survey of thirty-five ranchers, outfitters, and small "hobby" ranch operators in 2002 and 2003 to get a better understanding of people's perspectives of grizzly bears and possible ways to coexist with them. Third, these data helped us to frame or define "problems" as perceived by residents whose livelihoods could be affected by grizzly bears. Fourth, we used geographic information systems (GIS) to map land-use practices, bear attractants, and other relevant features in the region (Wilson et al. 2005). FWP provided data on verified and reported grizzly bear conflicts and observations (1998 to 2004) that helped us to develop a GIS spatial dimension to bear ac-

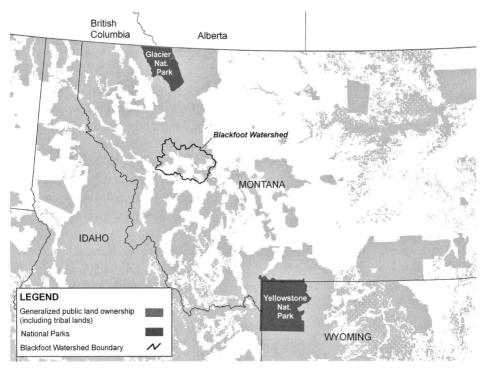

Figure 6.1. Location of the grizzly bear case study in the Blackfoot River watershed, Montana.

tivity and human-bear conflicts. From this, we determined the parts of the watershed where ranchers were experiencing the most bear problems, the areas where future problems were most likely, and the locations and types of conservation investments that should be made. Last, we brought this information back to the community and worked on the ground in diverse management activities to address problems over the ensuing years. Conflict reduction efforts have been focused on the middle portion of the Blackfoot watershed (fig. 6.1). Progress has been remarkable: a 96 percent reduction in reported and verified human–grizzly bear conflicts in the project area from 2003 to 2010 and a downward trend in known grizzly bear mortalities.

CONTEXT AND PROBLEM DEFINITION

The Blackfoot River watershed is about 1.5 million acres nestled just south of the Bob Marshall and Scapegoat Wilderness Areas and north of the Garnet Mountains (fig. 6.1). Land ownership in the watershed is approximately 49 percent federal, 5 percent state, 20 percent Plum Creek Timber Company, and 24 percent other private owners. Public lands and significant portions of Plum Creek Timber Company lands generally make up the forested, mountainous areas, whereas other private lands comprise the lower foothills and valley bottoms.

Grizzly bears were historically found throughout much of western North America, ranging from Canada to northern Mexico. Prior to European settlement, possibly more than 50,000 grizzly bears lived in the western United States (US Fish and Wildlife Service 1993). With western expansion and development, grizzly bear populations declined dramatically by the turn of the nineteenth century. By 1975 the grizzly bear was listed as a threatened species under the Endangered Species Act (ESA; 16 U.S.C. 1531–1544), and protections were put into place to begin the process of recovery. A formal recovery program was instituted in 1981 (Servheen 1998), and since then bear numbers have gradually increased in both Glacier and Yellowstone National Parks and the surrounding ecosystems. By 2007 the US Fish and Wildlife Service declared the Yellowstone population recovered and removed the region's grizzlies from the Endangered Species List. This decision was legally challenged by several conservation groups, and by 2009 the US Ninth Circuit Court of Appeals enjoined and vacated the delisting of the Yellowstone population. As of January 2013 all grizzly bear populations in the United States are considered threatened under the ESA, but population increases bode well for future viability. Unfortunately, increases in grizzly bear populations also mean increases in the potential for conflicts in places like the Blackfoot watershed.

ECOLOGICAL SETTING AND KEY STAKEHOLDERS
Ecological Setting
The landscape context of the Blackfoot watershed is unique. To the north are large wilderness areas that provide a source population of grizzly bears and to the south the Blackfoot Valley provides high-quality grizzly bear habitat. Grizzly bear activity and events associated with dispersal have been on the rise in the watershed since the early 2000s. The US Fish and Wildlife Service and Montana FWP have documented grizzly movements throughout the area (Jonkel 2002, 2006).

The Northern Continental Divide Ecosystem of Montana, an area that includes the Blackfoot watershed, shows grizzly bear population growth and expansion into formerly unoccupied habitat largely on private lands (Kendall et al. 2009). The state of Montana suggests that this grizzly population has grown at approximately 3 percent per year since population trend monitoring began in 2004. In 2004 the US Geological Survey's Northern Divide Grizzly

Bear Project identified a minimum of twenty-nine grizzlies in their sampling area that overlaps with the Blackfoot watershed. The Blackfoot Valley is functioning as habitat for a resident population of grizzlies and as a linkage or "stepping stone" habitat for grizzly bears that disperse to the south.

The geologic and hydrologic characteristics of the Blackfoot watershed have produced a rich mix of habitats, particularly wetland features like glacial lakes and ponds, bogs, fens, spring creeks, riparian swamps, and extensive cottonwood forests. This diverse mosaic of upland foothills, glacial outwash plains, and extensive creek and river bottoms is ideal habitat for a wide array of wildlife, including grizzlies. The Blackfoot has remained largely undeveloped and is sparsely populated. Located at the southern end of the Northern Continental Divide Ecosystem, the watershed has been a natural location for grizzly bears to recolonize their former habitat. The ranching community has had to confront living with bears again since the early 2000s and continues to deal with the associated costs of conflicts.

Key Stakeholders

Stakeholders in this case include ranchers, newer nonranching residents, the state and federal governments, and conservation groups. There are approximately 2,500 households in seven small communities in the watershed. The dominant land use is agriculture, primarily family-owned cow-calf ranching operations and some small-scale forestry. Cow-calf operations have cows and bulls used for breeding stock, and calves are sold annually at six to eight months of age for beef production. Landowners in the Blackfoot value their rural way of life and have worked together since at least the mid-1970s to preserve their agricultural traditions and livelihoods. The first conservation easements established in the state of Montana were in the Blackfoot. The ranching community is characterized by a desire to be independent, yet maintain strong neighborly relationships. Ranchers tend to believe in the sanctity of private property rights, and they emphasize their need to maintain economically viable operations. It was not initially easy for the ranching community to adapt to the return of grizzly bears. Bears were perceived as unwelcome visitors that threatened livelihoods and human safety, and bear conservation efforts challenged notions about private property rights and the rights of ranchers to protect their livestock.

While ranching is still a dominant land use in the Blackfoot and the cultural norms of ranchers have permeated the overall character of the valley, new residents have increasingly been moving to the region over the past two decades. In many cases these new residents are "amenity migrants," who have been drawn to the Blackfoot for its quality of life, solitude, and out-standing recreational and wildlife resources. These new residents are typically quite tolerant of grizzly bears and wolves, but may not have had much experience in actually living with these animals. However, new residents have been willing and in certain cases enthusiastic about participating in the grizzly bear projects of the BC.

Montana FWP plays the main formal role in grizzly bear management in the Blackfoot watershed. FWP is responsible for day-to-day management of grizzlies (conflict responses, monitoring, etc.) in consultation with the US Fish and Wildlife Service under the Interagency Grizzly Bear Committee guidelines (IGBC 1986). FWP has actively embraced the collaborative nature of the BC partnership concerning grizzly bear management, has shown respect for traditional ranching livelihoods, and actively supports projects that help maintain rural ranching through economic incentives and technical support.

The US Forest Service, US Bureau of Land Management, and Montana Department of Natural Resources and Conservation play minor consultative roles in bear and habitat management and have taken part in the BC efforts on an as-needed, project-by-project basis, since the bulk of the BC work on grizzly bear problems has focused on private lands.

Conservation groups have also been active participants in the efforts of the BC partnership (e.g., the Nature Conservancy, Defenders of Wildlife, Brown Bear Resources, the Great Bear Foundation, and the Living with Wildlife Foundation). The Nature Conservancy has played a major role in habitat conservation in the Blackfoot watershed by procuring conservation easements and has recently undertaken a significant project with Plum Creek Timber Company and the BC that will protect nearly 88,000 acres of land, much of which is critical grizzly bear habitat.

UNDERSTANDING CONFLICTS BETWEEN
HUMANS AND BEARS IN THE BLACKFOOT

Conflicts with grizzly bears are partly a technical biological problem, but these conflicts are also very much grounded in what people do in a given landscape. People's behaviors and practices often cause or contribute to conflicts with bears, and the level of conflict can be severe when bear populations grow, individual bears disperse to recolonize former habitats, and human communities are confronted with the challenges of living with these large carnivores.

As bear activity has increased in the Blackfoot watershed, so have human–grizzly bear conflicts and livestock losses. By the mid to late 1990s several ranches in the Blackfoot had experienced confirmed losses of cattle or sheep to grizzlies. This became a focal point for other livestock producers in the watershed, although they actually may have had few if any incidents with grizzlies on their own properties. Unfortunately, a human fatality occurred during a hunting outing in 2001, which was a tipping point that galvanized the community. This fatality and the general increase in grizzly bear observations and reported and verified conflicts on private lands stimulated strong concern among residents.

Not surprisingly, there were multiple perspectives or definitions of what exactly the "problem" was. For example, some people felt that there were simply too many bears, some celebrated the new grizzly bear activity, some defined the problem as primarily one of risk to human safety, and still others viewed the problem symbolically, linking the recolonization of bears to an erosion of personal rights and freedoms exacerbated by the regulatory burdens of the Endangered Species Act. It was clear that these competing perceptions of the problem, rooted in differing beliefs, attitudes, and values concerning grizzly bears, would make it difficult to set coherent goals. Grizzly bears presented both an ideological and material perturbation to an agrarian-based system of traditional Western values, largely centered on utilitarianism and a dominionistic view of grizzlies as predators. Nonetheless, we saw an opportunity to improve the situation for both people and bears by organizing an effective and inclusive decision-making process focused on empowering local community members, finding solutions for people, and reducing conflicts with bears.

Grizzly Bear Management Decision Process

Grizzly bear management, like other decision-making processes, is mainly about people: what we value, how we interact, how we make choices, and how we set up and carry out our day-to-day practices. Essentially, we are making decisions about how we manage ourselves as well as how we deal with bears. This includes understanding and possibly changing our behaviors and actions that can lead to problems with bears. Outcomes of this decision process affect what happens to bears, people, and the land. Ideally, a sound decision process should be open, factual, fair, and it should produce results that work.

DECISION-MAKING FUNCTIONS

Any complete decision-making process includes a set of distinct functional activities or stages. First are the activities that lead up to a decision. This includes the gathering, processing, and dissemination of information about the issue at hand (the *intelligence* activity), and the development and debate of alternatives designed to address the issue (the *promotion* activity). These pre-decision activities are followed by a decision, such as committing to a plan of action, law, regulation, policy, or program (the *prescription* activity). Prescriptions must be specific and realistic enough to work in practice and ideally should have enough support from those affected, or be backed by enough authority to be carried out and enforced. Once a decision has been made resulting in a prescription, several post-decision activities typically take place. First, the prescription must be initially instituted (the *invocation* activity) and then further interpreted and enforced (the *application* activity). Invocation and application are often referred to together as implementation, which must come to grips with all the realities of the actual situation on the ground as well as the administrative and other actions needed to put in place the selected prescription. The next step is to monitor and evaluate implementation of the decision and its outcomes (the *appraisal* activity). Good appraisals go beyond just evaluating outcomes to examine the processes through which decisions are made. If corrections are needed based on the results of appraisal, they can be made by changing course and partially or completely ending the previous policy or program (the *termination* activity). Taken together, these seven activities make up any complete decision process. Each activity, as well as the overall decision process, has standards that can be used to evaluate actual cases. In the rest of this section we examine each decision function and associated standards to evaluate the BC grizzly bear management decision-making process, with attention given to conditions before and after formalization and engagement of the BC partnership.

Gathering Relevant Information (Intelligence)

The intelligence gathering activity should be dependable, comprehensive, selective, creative, and open (Lasswell 1971; Clark 2002). The BC grizzly bear program included six main strategies to improve the intelligence activity: initial engagement between Montana FWP and the BC, which led to the formation of the BC Wildlife Committee; a survey of community members' perspectives; meetings with community members to share information and explore understandings of the problem; participatory spatial mapping of conflicts and potential attractants; regular reporting of grizzly bear activity and management actions; and working together to identify shared goals, which directed all of our initiatives and eventually provided the foundation for a formalized sub-basin management plan.

Initial Engagement and Formation of the Wildlife Committee

In 2000 the Region 2 Supervisor and bear manager for Montana FWP met with the Executive Committee of the BC to discuss the possibility of cooperating on grizzly bear management issues. Both organizations were concerned about the increase in grizzly bear conflicts in the Blackfoot Valley and were interested in exploring the possibility of a new approach to the problem. After lengthy discussions the Executive Committee agreed to work with FWP on a trial basis and then in 2002 established a formal committee of the BC to oversee issues involving grizzly bears.

Survey of Perspectives on Grizzly Bears and Grizzly Bear Management

Informal meetings in 2001 and 2002 organized by local community members and Montana FWP regarding the presence of grizzly bears in the Blackfoot watershed were characterized by strongly negative reactions to grizzlies. We saw an opportunity to improve the *dependability* and *comprehensiveness* of the intelligence activity by learning more about people's attitudes, values, and beliefs about bears and bear management. During 2002 and 2003 we conducted a survey of thirty-five active ranchers, outfitters, and small "hobby" ranch operations (appendix 6.1). Perceptions of grizzly bears were varied and complex. Of those who agreed that private landowners had a responsibility for protecting grizzly bears (42 percent), some respondents explained that grizzly bears were "part of the territory" and that it made common sense to take responsibility for running an efficient ranch that would not attract grizzly bears. Others suggested that bears played a role in the ecosystem and thus it was prudent to maintain those functions (e.g., regulation of prey populations). Of those who disagreed that private landowners had a responsibility for protecting grizzly bears (58 percent), there were three main explanations that emerged. First, some respondents felt that responsibility for grizzly bear management was clearly the role of wildlife management agencies, not landowners. Second, other respondents explained that bears should be geographically separated from humans and should not need to use private lands. Third, some respondents believed that environmentalists were a large part of the problem since the protective and legalistic actions of environmentalists on behalf of grizzly bears had contributed to population increases and had led to unnecessary problems for ranchers.

When respondents were asked about whether the Blackfoot would be a better place to live without grizzly bears on private lands, 52 percent agreed, and two themes emerged: (1) fear for personal and family safety and (2) risk to livelihoods and property. Of the 48 percent who disagreed with the statement that the Blackfoot would be a better place to live if there were no grizzly bears on private lands, explanations fell into three themes: (1) the biodiversity theme was consistent with earlier sentiments suggesting that bears play a role in the system that should be valued; (2) the place and lifestyle theme was characterized by the notion that grizzly bears help make the Blackfoot a special and unique place and that grizzly bears help to define the rural lifestyle; and (3) the biocentric theme involved the recognition that humans are one species among many, all of which have value, including grizzly bears.

Fifty-eight percent of respondents agreed with the statement that people shouldn't have to change their habits to accommodate grizzly bears that use private lands. Three overarching themes emerged: (1) the loss of freedom theme was characterized by the concern that use of private lands by grizzly bears was limiting personal management authority and the ability to run an efficient ranching operation; (2) this was accompanied by fear and anxiety about potential injury by bears, particularly in situations where ranchers might have to contend with grizzlies during fence repair, irrigation set-up, livestock monitoring in remote settings, or during the calving season (particularly during night work); and (3) the perception that grizzlies could impact livelihoods and destroy private property was reiterated here in the context of ranchers not wanting to change their habits to accommodate bears because they perceived this as meaning that they would have to change their traditional rural lifestyles and practices.

One finding from the survey that was particularly important for the development of the BC grizzly bear program was that 90 percent of respondents agreed that private landowners should take precautions to reduce conflicts with grizzly bears. At the same time 90 percent of respondents disagreed with the proposition that grizzly bears should remain off limits to hunting. Finally, respondents were split over the statement that "grizzly bears are a serious threat to my livestock," with 45 percent in agreement and 55 percent in disagreement.

The survey concluded by asking about the most important issues that needed to be addressed so that grizzly bears and people might coexist in the area. Responses fell into two broad categories representing solutions that emphasized changing human behaviors versus those that emphasized changing grizzly bear numbers and bear behaviors. This breakdown of responses was a simple yet effective way to gain insights about how local residents thought about the problem and what they offered as possible solutions (appendix 6.2).

Meetings with Community Members

The survey responses and other discussions with community members revealed a need for a more *open* and *creative* intelligence gathering process for decision making about grizzly bears in the Blackfoot. It was apparent that local residents sought to develop strategies, understand the system, and receive valid information from wildlife managers. It was also clear that many locals recognized that human-based changes could occur to reduce conflicts with grizzlies. These types of solutions were more likely to be identified and brought into practice if there was an organized venue in which to share information, discuss creative alternatives, and make decisions.

To meet this need, the BC partnership organized a series of meetings and discussions with private ranchers, residents, Montana FWP, the US Fish and Wildlife Service, conservation groups, and Allied Waste Services (western Montana's largest waste hauler). These early meetings allowed the BC group to further explore the ways people defined the problem and to identify shared perspectives. The BC agreed to develop a formal Wildlife Committee to serve as a forum to continue to bring participants together for regular communication, information sharing, and decision making.

A key task at this point was to make sense of the problem. Based on the survey results and discussions in group meetings, we could see that there were multiple and complex perceptions of just what the problem was as local residents confronted the reality of an expanding grizzly bear presence. It was important for all of us to understand that competing definitions of the problem fell into ideological and symbolic realms. For example, the notions that grizzly bears should be completely geographically separated from human activities, or that there were simply too many bears, or that environmentalists were largely to blame for the situation, posed serious barriers to constructive discussions about how best to respond, since there were no feasible solutions available if the problem was characterized in these ways. There were other problem definitions apparent, however, that were quite practical in nature. These reflected concerns about human safety, property, and livelihood interests—all areas that theoretically could be addressed by understanding the problem as one of risk management. This provided an avenue for productive discussion during initial goal setting.

Participatory Spatial Mapping

The fourth initiative to improve the intelligence activity was to bring a spatial dimension to the collective understanding of the problem of grizzly bears on private lands in the Blackfoot. We organized mapping workshops to demonstrate how Geographic Information Systems (GIS) could help to understand the problem of human–grizzly bear conflicts from a spatial perspective. Using GIS, it was possible to display grizzly bear activity and conflict locations visually and demonstrate that human-bear interactions were occurring across ownerships comprising more than 650,000 acres. This illustrated that potential solutions might require a community-level response to match the scale of grizzly bear home ranges and conflict locations.

From experience in previous research conducted in Montana's Rocky Mountain Front it was evident that asking ranchers to talk about their ranching operations and to map their livestock pasture arrangements digitally was a powerful way to collect meaningful data in a nonthreatening manner that also helped invest ranchers in the process of data generation (Wilson et al. 2006). The mapping workshops in the Blackfoot demonstrated the participatory nature of GIS mapping and showed how locations of agricultural and other human-related attractants could be compared with known conflict locations and grizzly bear observations collected by Montana FWP. The hope was that by *opening* up the intelligence process and *selectively* focusing on conflict locations, attractants, and preventive techniques, we could diminish the polarization and symbolic rhetoric often associated with the presence of grizzly bears in ranching country and work on tractable and feasible solutions to reduce conflicts. The objective was to get a better understanding of where regular conflicts were occurring (current conditions), share this with the community, then jointly discuss trends and projections, and offer some possible solutions. At these early workshops, local FWP bear managers discussed preventive techniques, including the electric fencing of calving areas and other proven tools that they had used over several decades of experience. Collectively, ranchers and residents showed an interest in building on the existing proactive efforts of FWP and using GIS data to guide and augment these efforts.

The participatory and interactive nature of mapping as proposed to the ranching community was important for stimulating people's interest and asking for their direct involvement in solving the problem. Mapping reverses the traditional flow of information in practical problem solving, so that rather than flowing from the "experts" to the local people, information flows from the local people to the experts. Further, this approach created the opportunity for mutual learning. This is distinctly different from traditional approaches to land and wildlife management, where, for example, rural people have endured heavy-handed regulation or have resented being told what to do (Brick and Cawley 1996).

With the aid of maps participants were able to explain their management practices readily. When ranchers located calving or lambing areas they often explained why they had chosen specific pastures for their livestock or why they had selected a portion of their ranch for livestock carcass disposal. For example, during the calving season, ranchers lose some calves to natural causes. Since the calving process is labor intensive and requires constant management, it is typical for "boneyards" or "dead piles" to be located near calving and lambing areas. This saves labor in disposing of dead animals, but also increases the risk that scavenging grizzly bears could come into conflict situations. Mapping produced spatial data sets that included the underlying explanations that ranchers helped generate, led to other insights regarding management practices, and, perhaps most importantly, developed trust and credibility. The integration of livestock management practices and known locations of grizzly bear observations and conflicts allowed us to produce maps of conflict hotspots using a variety of geospatial analysis techniques (Wilson et al. 2005; Wilson et al. 2006). This enabled us to prioritize specific locations in the landscape to focus on in possible management and conservation actions with landowners.

Reporting of Grizzly Bear Activity and Management Actions
The fifth intelligence-oriented initiative of the BC project was to make *dependable* information on grizzly bear behavior and management actions more available. In addition to the GIS mapping and risk assessment work, the formation of the Wildlife Committee offered a powerful means for the Montana FWP grizzly bear management specialist to provide a factual accounting of grizzly bear activity and management actions on a regular basis. This was a critically important step that allowed landowners to have regular contact with a wildlife manager and to learn about grizzly bear foraging activity, when to expect emergence of grizzly bears from their dens, what travel routes were preferred by bears, and other insights on bear behavior and life history needs.

Regular communication with FWP also provided landowners with detailed information about grizzly bear management activities. For example, seasonal updates were provided on the causes of human-bear conflicts, the number of grizzlies outfitted with radio telemetry, the numbers of grizzlies trapped because of conflicts, and the locations to which these bears were moved. These data from FWP helped to minimize the likelihood of false information being disseminated into the community and maximized opportunities for FWP to provide factual and comprehensive information about grizzly bear activity. This information included FWP's finding that in the period from 1998 to 2005 about 59 percent of conflicts in the Blackfoot area (known as Region 2) resulted from poorly contained residential and agricultural attractants and that more than a third of known human-caused grizzly bear mortality in the region arose from attractant-related incidents and repeated livestock depredations (Jonkel 2006).

Setting Goals

By integrating expert and local knowledge through regular meetings about the social and ecological factors contributing to human–grizzly bear conflicts, the BC program *opened* up the intelligence activity and provided an atmosphere that was more conducive to generating shared goals. The group eventually settled on the following goals: to formalize community-supported management of human-wildlife interactions in the Blackfoot and to address human–grizzly bear conflicts and grizzly bear-livestock conflicts by emphasizing preventive techniques, protecting human safety, and helping to maintain rural livelihoods by reducing risk of livestock losses. These goals generated within the community complement national goals of grizzly bear recovery by focusing on reducing bear mortalities by reducing conflicts. Moreover, these goals were powerful in that their development and linguistic framing supported rather than threatened the personal identities of local people involved in understanding the problem and ways to confront it. The process of clarifying a shared sense of the problem and developing community-supported goals was time consuming, costly, and required a great deal of patience. However, this process helped to develop trusting relationships for sharing information and a shared sense of responsibility for working toward viable, common sense solutions that benefit both people and bears.

One ongoing difficulty with the intelligence activity for the BC grizzly bear program is that, despite concerted efforts to have open and inclusive meetings, a small group of Blackfoot ranchers have elected not to take part actively.

Developing Support for Action (Promotion)

In addition to its involvement in the intelligence activity, the Wildlife Committee of the BC has been actively engaged in the promotion activity. Ideally, promotion should be rational, integrative, and comprehensive (Lasswell 1971; Clark 2002). The Wildlife Committee was structured to encourage the involvement of a wide range of values and interests in its decision-making activities. Afternoon meetings of the full group are held quarterly and are supplemented by quarterly meetings of an evening work group called the Landowner Advisory Group. Together, the Wildlife Committee and the Landowner Advisory Group represent both communities of interest and communities of place, thereby contributing to more *comprehensive* and *integrative* development and debate of alternative strategies for action. For example, meetings of the full Wildlife Committee are heavily attended by both state and federal agency personnel and several prominent, national-level conservation groups like Defenders of Wildlife. One factor that contributes to attendance by these groups at these daytime meetings is that many agency interests or conservation groups have the ability to meet during the day and are, in effect, paid to do so. These agencies and organizations broadly represent larger, national-level communities of interest regarding threatened and endangered species conservation, and they hold values that may not be shared by many local residents. Conversely, the Landowner Advisory Group allows for broad local geographic representation from key opinion leaders, business owners, and respected ranchers throughout the Blackfoot watershed, who represent a community of place. Coordinating, balancing, and including the sometimes competing value demands that participants bring to discussions about grizzly bears through these multiple decision-making forums has been challenging, but it is critical for mutual learning, developing shared goals, and promoting strategies that are *rational*—in that they advance shared goals—and effective in achieving on-the-ground successes.

Our experience in the BC provides further evidence that inclusiveness is absolutely fundamental to progress in contested conservation settings. The Wildlife Committee and the Landowner Advisory Work Group offer multiple opportunities for broad representation of different value systems and perspectives. In an ideal world there might be one overall committee coordinating all the various interests (all of the other BC committees operate in this way). However, if certain individuals such as ranchers prefer a small group setting and can only attend meetings during the evening because they are working during the day, then it is important to attend to these details. Open and frequent communication through multiple means for all participants is a key factor for managing competing value demands and finding ways to arrive at solutions that serve the common interest.

Although having these two sets of meetings has been critical to ensure broad participation in the BC grizzly bear program, there are also some draw-backs. The process is time consuming and delicate in terms of maintaining a neutral stance. Ideally, one forum would simplify the task of managing and integrating competing values and would bring more efficiency to the decision-making process. However, it appears that members of the Wildlife Committee are satisfied with the current structure and appreciate the role of the Landowner Advisory Group. Regular dissemination of information to all stakeholders and annual field tours highlighting different projects have been critical for integrating multiple values.

The Wildlife Committee was also involved in a community-wide, sub-regional planning process led by the BC and the Nature Conservancy to develop an overall conservation plan for the watershed. The Wildlife Committee focused on developing a set of conservation targets for grizzly bear habitat conditions, conflict, and mortality thresholds along with a strategy for achieving those targets. This effort brought additional community members into the decision-making process and helped to document formally the Wildlife Committee's plans and strategy for long-term grizzly bear conservation under the larger, sub-regional plan for the watershed.

Making Decisions (Prescriptions)

The Wildlife Committee and the Landowner Advisory Work Group fostered civil discourse and rational discussion of various programs (prescriptions) for ameliorating human–grizzly bear conflicts in the Blackfoot. The two groups looked for prescriptions that were pragmatic, nonthreatening, and participatory. They selected four: electric fencing, livestock carcass removal, a neighbor-to-neighbor communication network, and a waste management program. These initiatives used tools that had been proven to work in other settings, provided ranchers with a central role in making decisions, and involved cost sharing among ranchers, beekeepers, and residents.

Standards for the prescription activity include effectiveness, rationality, and comprehensiveness (Lasswell 1971; Clark 2002). The BC partnership intentionally focused on modifying management practices and human behaviors rather than killing bears or attempting to change value systems. As a result, decisions have been highly *effective*, that is, they have met the expectations of all partners and have been supported by all partners. For example, small changes in ranching practices like shifting from traditional barbed wire to electric fences around calving areas, or eliminating dead cows and calves from a ranch during the calving season, are concrete ways to prevent and reduce conflicts with bears and avoid livestock depredations. Although the *rationality* of recognizing the livelihood interests of ranchers and honey producers may seem obvious, close attention was paid in the BC to making programs and practices more *comprehensive* by tailoring them so that, in addition to the direct benefits of capital investments in nonlethal deterrence of bears and other predators, there were indirect economic benefits to ranchers. For example, several of the fences built around calving areas also keep elk away from haystacks, thus helping ranchers to maintain adequate supplies of winter feed for their cattle. Many of the electric fences installed in the Blackfoot also protect grain storage facilities, household garbage, pet foods, and other attractants associated with ranch operations.

Some of the risks of ranching, such as disease, fire, land-use change, weather-related calf mortality, or global market fluctuations in cattle prices, are largely beyond the control of livestock producers. However, providing incentives to modify existing land-use practices is a nonthreatening way to help ranchers manage risk better, without demanding that they give up control over their operations. Initially, some ranchers in the Blackfoot were reluctant to adopt new practices like electric fencing or livestock carcass removal because these innovations challenged traditional norms. However, the results of our initial trials encouraged broader participation. For example, once some of the first fencing projects successfully deterred grizzly bears from calving areas and greatly reduced bear activity, many ranchers began to openly acknowledge the benefits of the fencing.

Implementation (Invoking and Applying Decisions)
Implementation consists of invoking (initially instituting) and applying (further interpreting and enforcing) decisions. Implementation should be timely, dependable, rational, nonprovocative (nonthreatening), realistic, and uniform (Lasswell 1971; Clark 2002). As just discussed, the programs implemented by the BC partnership included electric fencing, livestock carcass removal, a neighbor-to-neighbor communication network, and a waste management program. All of these efforts were directly related to the original goals of addressing human–grizzly bear conflicts through preventive techniques, protecting human safety, and reducing impacts to rural ranching livelihoods.

First, during 2003 to 2010, 61,000 linear feet of electric fences were constructed around fourteen calving areas, a rural livestock transfer site, a composting facility for dead livestock and deer, and sixteen apiaries. Electric fences nonlethally deter grizzly bears from attractants like calves, garbage, or beehives. All projects were paid for using funds from public and private foundations, which provided ranchers and beekeepers with substantial cost savings on the capital investments. Ranchers and beekeepers helped share the total costs by providing in-kind donations of labor to prepare sites and remove old fences.

Our early experience with the electric fencing project serves as a good illustration of how the implementation of a new program can be perceived as threatening. At first, some ranchers were concerned that electric fences would require excessive maintenance or would be susceptible to ungulate damage. In some cases, ranchers were unfamiliar with the technical aspects of electric fencing, and the adoption of this new technology challenged norms such as their pride in self-reliance regarding routine work like fixing barbed wire fences. We have worked closely with ranchers to monitor and maintain electric fences so that they are functional and inexpensive to ranchers.

Second, the livestock carcass pick-up program was designed to remove the cows, calves, ewes, and other livestock that naturally die during the calving and lambing season (mid-February through mid-May), so that carcasses would not be found by foraging grizzly bears and other predators. Cow-calf ranches in this area are characterized by winter feeding, centralized and spatially fixed operations, irrigated hay production, and docile breeds of cattle (Dale 1960; Jordan 1993). The calving season typically overlaps with the emergence of grizzly bears from their dens in the early spring. Bears routinely visit calving areas, and the traditional practice of dumping dead livestock into spatially fixed boneyards (carcass dumps) can lead to chronic livestock-grizzly bear conflicts. Grizzlies that are drawn to a ranch because of livestock carcasses may be tempted to kill live calves or lambs or find other foods like grain, protein licks, pet foods, or bird seed. The BC grizzly bear program has been responsible for the removal of more than 1,700 livestock carcasses from our project area since 2003.

Our initial efforts to remove livestock carcasses generated considerable concern as ranchers did not want to have the numbers of livestock deaths on their ranches disclosed to neighbors for fear of being stigmatized for poor animal husbandry. This concern was addressed by establishing centralized drop-off locations where ranchers could bring carcasses for pick up. Since 2007, composting of livestock carcasses has proven to be a highly effective disposal method and has been widely applauded by the ranching community as a more appealing method of disposal than past practices of depositing carcasses at landfills on their properties.

Third, the "neighbor network" initiative of the BC partnership connects local residents so they can help each other to reduce and prevent human-bear conflicts. The network consists of over 120 residents who work together: (1) to minimize the availability of human-related attractants, (2) to communicate among neighbors about grizzly bear and wolf activity using phone trees and e-mail alerts, and (3) to report to a designated area coordinator any incidents or observations of bear or wolf behavior that may pose problems. The goal of this program is to improve communication among neighbors and with Montana FWP in order to prevent conflicts with carnivores before they occur. Nine networks have been set up in the project area, each with a coordinator, to help facilitate communication among neighbors and to FWP when there is grizzly bear or wolf activity. A free check-out program allows residents to borrow bear-resistant trash cans, portable electric fencing, electrified bird feeders, and other nonlethal deterrent tools that help residents reduce conflicts.

Fourth, the BC program also focuses on common sense management of waste and household garbage for all residents of the Blackfoot. Waste haulers and residents use bear-resistant garbage cans or take simple precautions to keep garbage secure from scavengers. These efforts are integrated with the neighbor network.

Participants in the BC have worked extremely hard to deliver all of these projects in an efficient, *dependable*, and *timely* manner. Demonstrations and one-to-one discussions with ranchers have helped to overcome fears and suspicions about innovative projects, making implementation less *provocative*. Moreover, our efforts have been open and available to anyone in the Blackfoot, including ranchers, landowners, and new residents, thus providing *uniformity*. One key to effective implementation has been to have a representative of the BC program available on short notice to respond and deal with questions or problems as they arise. We have made extensive efforts to be highly responsive to landowners and their needs. Often a telephone call or personal visit is used to assess the situation and take necessary steps to alleviate any issues. As a result of all these factors, participants appear to be satisfied that program delivery is *rational* in that it serves the broader common interest in the Blackfoot watershed and nationally.

Although the BC partnership has made good progress with implementation of its grizzly bear projects, we anticipate future challenges with funding. Initial funding was provided for virtually all projects from public and private foundations, which provided a strong economic incentive for ranchers to participate. To date, there has been fairly limited direct financial contribution to ongoing projects by ranchers, although they have made in-kind contributions of labor. Programs like livestock carcass removal are clearly valued by ranchers as an important service, but the long-term sustainability of these benefits is not secure, considering the small, annual contributions of actual funds (less than $1,000 in total) by ranchers to a program that costs approximately $14,000 per year.

Monitoring and Evaluating Decision-Making
Processes and Outcomes (Appraisal)

Criteria for the appraisal activity include dependability, rationality, comprehensiveness, selectiveness, independence, and continuity (Lasswell 1971; Clark 2002). Thus far, evaluation of the BC grizzly bear effort has been informal and internal, without the involvement of an *independent* third party. Although the present chapter *rationally* evaluates the decision-making processes of the BC partnership using explicit criteria (summarized in appendix 6.3), this is still an internal evaluation because the authors are all involved in the program. Nonetheless, the Wildlife Committee and the Landowner Advisory Group of the BC were structured to be as inclusive as possible and to provide opportunities for broad community representation, so the self-appraisal that has taken place incorporates a variety of perspectives (contributing to *dependability* and *rationality*). Although appraisal has not been *continuous*, the BC program does regularly assess its projects and look for opportunities for improvement. These informal appraisals typically focus *selectively* on specific projects, in contrast to this chapter, which is a more *comprehensive* evaluation of the program as a whole.

Participants are justifiably proud of the outcomes of the BC grizzly bear program. From 2003 to 2010 there was a 96 percent reduction in reported and verified human–grizzly bear conflicts in the BC project area. There has also been a downward trend in known grizzly bear mortalities in the Blackfoot watershed. In 2003 there were five grizzly mortalities in the watershed, and in 2004 there were three known mortalities, one resulting from a hunter-related incident, one from an illegal kill, and one from a vehicle collision. In 2005 there was one road-kill mortality of a sub-adult grizzly of unknown sex. In the same year, no grizzly bears were trapped in the Blackfoot for conflict management purposes, nor were there any management-related conflicts or mortalities in 2006. In 2007 human-bear conflicts continued to decline with fewer than five conflicts reported. There were two hunter-caused mortalities (one adult male, one adult female) in backcountry settings outside the project area and one adult male grizzly killed by a vehicle collision outside the project area. In 2008 there were no grizzly bear mortalities and twelve minor, attractant-related conflicts. In 2009 only five reported conflicts occurred and there were no grizzly bear mortalities. In 2010 there were only three conflicts in the project area but several grizzly bear mortalities on the outside edges of the project area as a result of garbage-related attractants.

Recent population analysis by the US Geological Survey, which used DNA hair-snare methods, reported approximately twenty-nine individual grizzlies in the Blackfoot area in 2004, and population estimates by FWP suggest that the Northern Continental Divide Ecosystem population is growing at about 3 percent per year. We recognize that the downward trend in conflicts may partly be a result of there being fewer "problem" bears, or changes in bear foraging behaviors, or the heightened awareness by local residents, who may report conflicts less often. However, the combination of targeted prevention efforts—electric fencing, carcass removal, improved decision making, and citizen-based monitoring—are clearly helping to reduce conflicts between humans and grizzly bears. Ongoing FWP monitoring of grizzly bear activity, conflicts, and bear mortality will help to evaluate the long-term outcomes of this initiative.

Ending or Changing Programmatic Direction (Termination)

The grizzly bear conflict work of the BC grizzly bear program is now entering a monitoring and maintenance phase. The past five years have resulted in the development of several successful initiatives and dozens of on-the-ground projects. Monitoring and maintaining the investment in a coexistence infrastructure in the Blackfoot watershed is the next task ahead. While the grizzly bear work is certainly not completed, new projects will be initiated as needed and overall direction will be guided by the grizzly bear strategy in the subregional plan.

The BC partnership is now shifting energy, skills, and resources to the challenging issue of wolf-livestock conflicts, with the hope that the decision-making systems put in place for grizzly bear management will enable a proactive response to the emerging and contentious issue of wolves. Initial efforts suggest that the BC will again have to contend with widely differing definitions of the problem, including differing perceptions regarding wolf numbers, management authority, direct and indirect impacts of wolves on livestock, impacts on ungulate numbers and hunter opportunity, and levels of lethal control. The BC has elected not to take an advocacy position on wolf management or the status of wolves under the Endangered Species Act, and a small number of individuals have formally withdrawn from the BC partnership over this issue.

Moving Forward: Lessons and Recommendations

The success of the BC grizzly bear program shows that a community-based initiative that attends to principles of good decision making can be effective in building social capacity and reducing conflicts with carnivores. By opening up the intelligence and promotion activities, developing more reliable information about important social and biological trends, and selectively focusing on practical ways of managing attractants and reducing conflicts, the BC has helped to promote coexistence between grizzlies and humans in the Blackfoot Valley. In this section we identify field-based lessons for managers from our experiences and then discuss general recommendations for designing and improving decision-making processes in this and other complex conservation programs.

LESSONS FOR MANAGERS

Our field-based lessons for managers are not meant to be an exhaustive inventory, but merely an attempt to share some of the most important and useful lessons reflecting our collective experience with the BC. These lessons are summarized in the following sub-sections.

Build on Existing Institutional Capacity

When this effort started it was clear that it would be foolish to ignore the existing institutional capacity of the BC. The long-term commitment that ranchers, residents, and other landowners in the Blackfoot have made to preserve rural ways of life has resulted in strong working relationships within the community and across multiple natural resource agencies and other government entities. This history of collaboration has generated and maintained a stock of social capital, which has been drawn on successfully from time to time to confront complex issues like drought, invasive plants, and landscape-scale habitat conservation (Putnam 2000). It made sense that if the BC was willing to work on the grizzly bear issue we should build on and expand this existing institutional capacity.

Take a Proactive Approach

While it is not always possible to take a proactive and precautionary approach to natural resource issues, we did so in this case. A few human–grizzly bear conflicts and livestock losses to grizzlies had already occurred when the BC grizzly bear program began in the early 2000s, but it is likely that without a collective preventive response at that point the intensity and frequency of grizzly bear conflicts would have increased substantially. The use of GIS analysis helped us to predict and target areas that were at greatest risk for conflicts in the future, thereby avoiding the need to "bear-proof" the entire landscape. If the expansion of grizzly bear activity and their population continues on private lands in this area, it is hoped that these proactive efforts will enable the community to stay well ahead of the problem.

Develop a Broad Understanding of "the Problem"
Considerable time and research was invested in understanding how the problem was perceived by key stakeholders and how these perceptions were grounded in the local context. The spatial and biological dimensions of

grizzly bear activities and the opportunities for managing attractants were important considerations that the group learned about collectively. Care was taken to be holistic and to understand that a long-term solution to the problems associated with grizzly bears would depend on finding common ground across multiple stakeholders.

Be Patient and Listen

While it may appear obvious that patience is an important skill to bring to collaborative efforts, this skill is critical in rural, agrarian contexts. Participatory efforts are time intensive and costly, but developing inclusive and meaningful decision-making processes greatly increases the likelihood of achieving desired outcomes. The process can be as important as the end product. Well-developed listening skills are also essential. We have found that people involved in our efforts have different ways of communicating important information. Some are direct, others more subtle. For example, it is common for a rancher to refer to himself in the third person (e.g., "a guy could do such and such" to describe what he himself might be willing to do). This may serve as a way to deliver positions in a thoughtful and nonthreatening manner and may have origins in agrarian needs for neighborly communications. Indeed, the word "neighbor" is sometimes used as a verb, as in "he knows how *to neighbor* well." There are unique complexities in the ways people in different settings communicate, and being attuned to this by being a good listener is essential to comprehending the situation and being effective.

Share Power

Power-sharing arrangements can be difficult to create and manage. Nonetheless, our experience with the BC partnership demonstrates that power sharing among stakeholders, specifically in making decisions on both a tactical level (project-based) and strategic level (management and conservation approach) are beneficial. The willingness of Montana FWP and other agencies to invest in the BC effort and actually give and take direction from landowners, ranchers, and residents shows that the BC developed a high degree of trust among stakeholders. This has resulted in stakeholders who are willing to participate on a long-term basis and who know that their values are being incorporated into decisions. The BC partnership will need to work to maintain this level of trust as it tackles the highly divisive issue of wolf management.

Scale Matters

One likely explanation for the downward trend in conflicts between bears and humans is that our programs were designed to match the biological scale of grizzly bear activity. For example, the carcass pick-up and removal program covers ranches across nearly 650,000 acres, a scale commensurate with grizzly bear home ranges and their twenty-four- to forty-eight-hour foraging bouts. By having dozens of ranchers participating in this program, there is a collective benefit realized from individual participation. If only a handful of ranchers had taken part in the program, grizzlies would likely have found and exploited carcasses on the ranches of nonparticipants, eventually creating spillover conflicts across our project area.

RECOMMENDATIONS

In addition to these lessons for managers, there are three main recommendations that emerge from this case study for designing and improving community-based approaches to large carnivore conservation and other complex natural resource cases: build and maintain partnerships, communicate effectively, and strive for long-term sustainable outcomes.

Build Partnerships

Beginning with the intelligence activity and continuing through all the decision-making activities of the BC program, at the core of our success has been a strong partnership. This has enabled us to bring a diverse group of stakeholders together to capitalize on our collective knowledge, skills, and financial resources. Integrating local and expert knowledge in the intelligence and promotion activities was a powerful means to develop a holistic understanding of the problem of coexisting with grizzly bears. The people involved in our efforts have multiple skill sets––from bear monitoring to interpersonal communication, fund-raising, facilitating meetings, and ranch management. These skills, and many more, constitute a "skill network" that can be readily tapped on a project-by-project or issue-by-issue basis to advance the goals of the partnership. In addition, funding institutions (both public and private) are often interested in supporting partnerships, which has helped the BC partnership to garner substantial financial resources to implement key programs.

Communicate Effectively

Effective communication is critical to carnivore conservation programs in both formal and informal settings. The formal communication structure that we put in place using the Wildlife Committee and the Landowner Advisory Group gave all invested stakeholders a voice to share information and exchange ideas. This opened up the intelligence and promotion activities, giving participants access to better data about social as well as biological trends and helping them to settle on a tractable definition of the problem. Both communities of place and of national interest were given representation and encouraged to communicate with each other to make decisions. The ability for national-level conservation groups to understand the local reality of living with bears has been particularly helpful in crafting approaches that are locally supported and practical but also benefit the national goal of conserving grizzly bears.

At the informal level, the network of partners that organized under the BC has helped to provide options for communication in decision making based on trusting relationships. For example, certain individuals in this partnership have known some of the ranchers in the Blackfoot for more than three decades, which makes it possible to present new ideas or grapple with sensitive subjects initially in one-on-one informal settings. Being able to broach sensitive issues at the individual level prior to group settings can be an important way to probe or test out new ideas in a nonthreatening manner. In these types of cases, testing or prototyping a new innovation, such as fencing, is possible and presents a low-risk way to try new approaches.

Additionally, we recommend that communication be based on nonthreatening language choices and a nonadvocacy approach to discussing the issues, and that, at least in the Blackfoot watershed, the communication flow be largely upward from the grassroots. Being highly attuned to the values and needs of landowners and ranchers through regular communication and being willing to "listen from the ground up" have been important ways to frame communication strategies.

Strive for Long-Term Sustainable Outcomes

Sustainable coexistence with large carnivores like grizzly bears in the long term will ultimately depend on local people's tolerance for these animals and their willingness to exert a degree of ownership, in the sense that there is shared responsibility for wildlife management rather than the traditional model in which people rely wholly on agency action. Ideally, living with carnivores will become part of cultural expectations in these regions. Over the past ten years of our work, people in the Blackfoot have increasingly developed a sense of responsibility for their actions with respect to coexisting with grizzly bears. Our neighbor network is a prime example of residents taking an active and participatory role in peer education and reversing the flow of information about grizzly bear activity to Montana FWP. Ranchers are developing their own portable electric fencing systems to deter bears and wolves nonlethally, and across the community the general level of "bear awareness" has increased dramatically.

Appendix 6.4 depicts a continuum of local involvement in grizzly bear conservation and indicates the implications for sustainable outcomes. We recognize that in the BC program technical assistance and financial resources were provided to ranchers for many of our programs and that we have not yet reached the far end of the continuum in terms of local ownership of grizzly bear conservation. However, the improved decision-making process developed in the Blackfoot is a critical step in reaching this long-term and most sustainable outcome. We also believe that the long-term support that this partnership has enjoyed from the ranching community is largely a result of having developed communication forums where local values, needs, and concerns were regularly articulated and met in a meaningful way. The future outlook of the effort appears to be sustainable. The state and federal agencies involved recognize that sharing power in decision making has invested local people in the long-term success of this effort. However, long-term financial sustainability to maintain core programs may require more cost sharing by ranchers, more investment by local landowners who can act as donors, and continued and diversified support from private foundations. It may also be desirable to develop new legislation that could direct congressional appropriations to support proven tools and techniques as described in this effort.

Conclusions

Much of the success of the BC grizzly bear program can be attributed to understanding and communicating effectively with people. In too many cases of natural resource management, undue attention and emphasis are placed on observing, studying, and analyzing biological processes or problems. We have found that people's interactions, communications, and decision making are the key areas on which to focus attention. We did not ignore important biological data relevant to reducing grizzly bear conflicts, but we found that focusing our efforts on building nonthreatening forums for understanding how the problem was perceived and cooperatively generating shared goals and solutions has been well worth the investment.

At the beginning of this effort it was expected that the issue of grizzly bears would be perceived differently among interested stakeholders, that there would be multiple and competing value demands made by stakeholders, and that an organized approach was needed to work through this complexity. Doing nothing would clearly have been a poor choice for people and grizzly bears. The BC partnership focused on changing people's practices and behaviors, not changing their value systems, and it gave people who had to live with grizzly bears the means to develop coherent goals of reducing and preventing human–grizzly bear conflicts. While not all ranchers and landowners have been involved in the effort, there has been enough core support to make the programs a success. The inclusive and multiple forums to guide decision making relied on participatory projects and had the support of wildlife managers who were willing to invest in this collaborative effort. This has produced a strong partnership and enhanced communication and trust among participants, thereby opening up opportunities for innovation and on-the-ground success.

Appendix 6.1
Likert-Scaled Statements Regarding Perceptions of Grizzly Bear Activity and Appropriate Landowner/Resident Behaviors in the Blackfoot Watershed, Montana

1. Grizzly bears that use private land are a threat to human safety.
    71 percent agree
    29 percent disagree
2. I do not feel safe when I am outside on my property because of grizzly bears.
    45 percent agree
    55 percent disagree
3. There are too many grizzly bears using private lands in this area (Blackfoot).
    71 percent agree
    29 percent disagree
4. I am comfortable with the current level of grizzly bear activity in this area (Blackfoot).
    32 percent agree
    68 percent disagree
5. Private landowners have a responsibility for protecting grizzly bears.
    42 percent agree
    58 percent disagree

6. This would be a better place to live if there were no grizzly bears on private lands.

   52 percent agree

   48 percent disagree
7. People shouldn't have to change their habits to accommodate grizzly bears that use their private land.

   58 percent agree

   42 percent disagree
8. Private landowners should take precautions to reduce conflicts with grizzly bears.

   90 percent agree

   10 percent disagree
9. Grizzly bears should remain off-limits to hunting.

   10 percent agree

   90 percent disagree
10. Grizzly bears are a serious threat to my livestock.

    45 percent agree

    55 percent disagree

Appendix 6.2
The Relative Difficulty of Important Issues for Improving Human–Grizzly Bear Coexistence

Aggregated responses of respondents when asked about the most important issues that should be addressed to improve human–grizzly bear coexistence and the relative difficulty of addressing issues.

ISSUES CHARACTERIZED BY MANAGING HUMAN BEHAVIORS
Less Difficult

Develop a strategy.

Focus on protecting garbage, grain, bird feeders, etc.

"Keep a clean camp."

Better communication.

Truthful communication from wildlife managers.

Better education on how to live with bears.

Protect human safety.

Improve cooperation among landowners and managers.

More Difficult
Stop developments in bear habitat.

ISSUES CHARACTERIZED BY MANAGING
BEAR BEHAVIORS AND NUMBERS
Less Difficult
Manage "problem" bears more aggressively.
Improve bear monitoring.
More Difficult
Know where "problem" bears are.
Know what the bear population is doing.
Reduce the bear population.
Institute a hunting season on bears.
Have a legal right to protect livestock from bears.
Increase bear relocation distances.

Appendix 6.3
Decision Activities, Standards, and Effectiveness for Grizzly Bear Management in the Blackfoot River Watershed, Montana

An overview and appraisal of the grizzly bear management decision process in the Blackfoot River watershed, Montana (standards adapted from Lasswell 1971; Clark 2002).

INTELLIGENCE
Recognizing the problem and gathering information.
Dependable
Comprehensive
Selective
Creative
Open

Pre-engagement (1997–2001)
FWP data on bear numbers and activity were perceived with some skepticism by the community and viewed as *undependable*. Bears were perceived as great threats to human safety and livelihoods. Data were needed to integrate land-use practices, attractants, and conflict location data to create a *comprehensive* and integrated understanding of the problem while *select-*

ing or targeting key practices (e.g., calving areas, livestock carcass management) that would influence conflict probability. Information gathering was traditional and limited in scope (emphasis on monitoring movements of a small number of collared bears) or *uncreative* and was not widely available or *open* to the community.

Engagement (2002–Present)

Analysis and clarification of the "problem" of having grizzly bears return to private lands helped to show that ideological and symbolic problem definitions would be difficult to solve, and they led to general support by the community and a focus on understanding what human behaviors and practices would reduce the risk of conflict. A spatially explicit and more *comprehensive* understanding of conflict was generated with the support of FWP and ranchers through data sharing and valuing local knowledge. The process of data collection, *open* dissemination of information, and regular communication with FWP and ranchers helped to create trust in data (*dependability*) and opened opportunities to find solutions that respected ranchers' exclusive interest in property while attending to inclusive national interests of reducing grizzly bear-human conflicts and bear mortality.

Key Challenges

The process of clarifying problem definitions is time consuming and costly. Power sharing and generating the trust and support of FWP was a time consuming and delicate process that required patience. A small vocal minority of ranchers have refused to take part in any information generation or sharing activities since 2002.

PROMOTION

Open debate, in which various groups advocate for their interests or preferred policies.

Rational

Integrative

Comprehensive

Pre-engagement (1997–2001)

Forums for *rational* and open discussion and debate were limited. Disparate values were simultaneously promoted and dismissed by various stakeholders with little or no *integration* or synthesis. Special interests were largely dominating informal discussions at the community level.

Engagement (2002–Present)

The BC provided structure to elevate the discussion and debate to a more *rational* and *comprehensive* level that *integrated* place-based and interest-based values through multiple communication forums.

Key Challenges

Multiple communication forums require intensive coordination (time-consuming). Ideally, the BC would have a single committee to debate and discuss grizzly bear management.

PRESCRIPTION

Setting the policy, rules, or guidelines.
Effective (stable expectations)
Rational
Comprehensive

Pre-engagement (1997–2001)

Ranchers and landowners were concerned that state and federal management actions would trump or ignore local decision making and limit property rights. FWP was understaffed and underbudgeted.

Engagement (2002–Present)

Committing to specific decisions was carried out in a collaborative manner with cogeneration of voluntary plans deemed *rational* or balanced by ranchers. The *expectations* of ranchers that they should have a central role in decision making were met, helped build support of existing FWP management plans, and led to new voluntary projects (characterized by economic incentives and risk-reducing practices), furthering a *comprehensive* and proactive approach to the problem. Funding and personnel were acquired to expedite project delivery.

Key Challenges

Initially, ranchers were skeptical about the efficacy of certain proposed actions such as electric fencing and livestock carcass removal since adoption of these practices challenged traditional cultural norms.

INVOCATION AND APPLICATION

Implementation, enforcement, and dispute resolution.
Timely
Dependable

Rational
Nonprovocative (nonthreatening)
Realistic
Uniform

Pre-engagement (1997–2001)

Efforts by FWP existed to manage grizzly bears efficiently, but funding and personnel shortages made progress slow. Efforts were not systematically organized.

Engagement (2002–Present)

The existing institutional capacity and ability of the BC were used to catalyze new partnerships and synchronize state, federal, and NGO involvement, leverage significant funds, and deliver results on projects efficiently (*timely*), *dependably*, and *nonprovocatively*. Widespread support of the ranching community for the project ensued. Efficient and reliable project implementation may have resonated with the work ethic of ranchers, whose survival is dependent on a problem-solving and solution-oriented practice. The institutional capacity of the BC to implement projects efficiently has created efficiency and rationality in program efforts that are perceived as a favorable private-sector solution that served common interests (*rational*).

Key Challenges

The use of economic incentives that have resulted in relatively low-to-no-cost projects may have created expectations of conservation subsidies and could jeopardize the ability to sustain core programs in the long term if some proportions of costs are not borne by the ranching community.

APPRAISAL

Review and evaluation of the activities so far.
Dependable
Rational
Comprehensive
Selective
Continuing
Independent

Pre-engagement (1997–2001)

Initial efforts by FWP to address the grizzly bear issue were largely based on traditional, expert-driven wildlife management and self-appraisal with limited engagement of the local community in decision making.

Engagement (2002–Present)

Although the appraisal process presented here is informal (no external third party has yet evaluated this effort), systematic, regular, and coordinated communication among all stakeholders has helped create a *dependable* and *rational* process that appears to be generally supported by those invested in the issue. The *comprehensive* approach to identifying and properly *selecting* and removing/securing attractants has helped reduce conflicts with grizzly bears by 93 percent from 2003 to 2009. A downward trend in grizzly bear mortality has been observed in the project area and no known livestock kills have been attributed to grizzly bears since 2004 while FWP reports a slowly expanding (1 to 3 percent per year) grizzly bear population in the Northern Continental Divide Ecosystem. FWP monitoring of grizzly bear activity, conflicts, and bear mortality will help evaluate the long-term success or failure of programs. *Independent* appraisal might generate new ideas for improvements.

Key Challenges

Recolonization of wolves in the Blackfoot watershed beginning in 2007 has generated widespread concern among the community, not unlike the negative perceptions of grizzly bears described earlier in this chapter that were common in the late 1990s and early 2000s. However, the widely differing problem definitions among stakeholders in the Blackfoot watershed regarding wolf numbers, management authority, direct and indirect impacts to livestock, purported impacts to ungulate numbers and hunter opportunity, levels of lethal control, and the BC's nonadvocacy position on wolf management or status under the Endangered Species Act have led a small number of individuals to withdraw formally from BC activities.

TERMINATION/SUCCESSION

Ending or moving on.

Timely

Comprehensive

Dependable

Balanced
Ameliorative

Pre-engagement (1997–2001)
Previous land-use practices (e.g., livestock carcass management, unpro-
tected calving areas) led to increased conflicts.

Engagement (2002–Present):
Previous practices that led to conflicts were abandoned in favor of new
practices that helped *ameliorate* conflicts in a *timely* manner. The process
of stopping specific practices and shifting to alternatives was collaborative and sought out *balanced* and *comprehensive* solutions. With a strong
downward trend in conflicts and bear mortalities observed in the project
area, specific programs like electric fencing of calving areas and beehives
have been terminated since the bulk of the high-risk areas are now secure.
Livestock carcass removal requires long-term maintenance. Overall, the
grizzly bear management approach is now characterized by monitoring
and maintenance, and additional projects will be implemented as needed.

Key Challenges
Maintaining funding to sustain the annual costs of livestock carcass removal may be difficult.

OVERALL STANDARDS
Honest
Economical
Technically efficient
Loyal and skilled personnel
Complementary and effective impacts
Differentiated structures
Flexible and realistic in adjusting to change
Deliberate
Responsible

Throughout all decision making, *honesty* and trust have been hallmarks of
this effort. Regular communication among wildlife managers, researchers, landowners, and ranchers has generated trusting relationships among
key stakeholders. The use of *economic* incentives facilitated adoption of

new land-use practices, yet the long-term financial sustainability of core programs and willingness of ranchers to defray future costs will remain a challenge. The partnership that was created by the BC relies on diverse *skill* sets of *technically* trained personnel who can rely on one another to address diverse problems as they arise. This group of stakeholders has come to a place where *loyalty* and *responsibility* are largely invested in the decision process itself, not situated at individual agency or institutional levels, which has helped to prevent any one individual or organization from attempting to appropriate success of the effort.

Appendix 6.4
A Continuum of Local Involvement in Grizzly Conservation

A continuum of local involvement in grizzly conservation and implications for sustainability of carnivore populations and conservation programs (adapted from Wilson, Primm, and Dood 2007).

Least Sustainable
"You're on your own."
   Outcome—People eliminate large carnivores.
"Tell them how."
   Outcome—People slowly eliminate large carnivores.
"Do it for them" (expert dependency model).
   Outcome—Program may be successful in the short term but costly and likely unsustainable.
"Show them how" (technology transfer).
   Outcome—More sustainable.
Ownership: Peer-educators
   Outcome—Coexistence becomes part of local culture.
Most Sustainable

NOTES
We would like to thank sincerely the livestock producers, landowners, and residents of the Blackfoot Valley for their time, support, and involvement with all aspects of this effort. It would not have been possible without your support and interest. Special thanks to the Blackfoot Challenge and all members of the Wildlife Committee, the Landowner Advisory Group, the Waste Management and Sanitation Work Group, and the Neighbor Network.
Special thanks to Montana Department of Fish, Wildlife and Parks and the Mon-

tana Department of Transportation. Thanks to all the many individuals, too numerous to name.

Support came from Allied Waste Services (formerly BFI), Blackfoot Challenge, Brown Bear Resources, Bunting Family Foundation, Chutney Foundation, Defenders of Wildlife, Great Bear Foundation, Keystone Conservation, Living with Wildlife Foundation, Montana Department of Fish, Wildlife and Parks, Montana Department of Transportation, Montana Department of Natural Resources and Conservation, Nature Conservancy, Natural Resources Conservation Service (NRCS), Northern Rockies Conservation Cooperative, Pumpkin Hill Foundation, Powell County Extension, private landowners, University of Montana, College of Forestry and Conservation, US Fish and Wildlife Service, US Forest Service, US Geological Survey, Y2Y/Wilburforce Foundation, and the Yale University School of Forestry and Environmental Studies. D. Casey, M. Rutherford, S. Clark, D. Mattson, J. Ellis, M. Wilson, and P. Wilson provided critical review.

REFERENCES

Brick, P. D., and R. M. Cawley. 1996. *A Wolf in the Garden: The Land Rights Movement and the New Environmental Debate*. London and Maryland: Rowman and Littlefield.

Clark, S. G. 2002. *The Policy Process: A Practical Guide for Natural Resource Professionals*. New Haven, CT: Yale University Press.

Dale, E. E. 1960. *The Range Cattle Industry: Ranching on the Great Plains from 1865 to 1925*. Norman: University of Oklahoma Press.

Interagency Grizzly Bear Committee (IGBC). 1986. "Interagency Grizzly Bear Guidelines." Available from http://www.igbconline.org/IGBC_Guidelines.pdf.

Jonkel, J. J. 2002. *Living with Black Bears, Grizzly Bears, and Lions: Project Update*. Missoula: Montana Department of Fish, Wildlife and Parks.

———. 2006. *Living with Predators Project: Preliminary Overview of Grizzly Bear Management and Mortality 1998–2005. Region 2, Montana Fish, Wildlife and Parks*. Missoula: Montana Department of Fish, Wildlife and Parks.

Jordan, T. G. 1993. *North American Cattle-Ranching Frontiers: Origins, Diffusion, and Differentiation*. Albuquerque: University of New Mexico Press.

Kendall, K. C., J. B. Stetz, J. Boulanger, A. C. Macleod, D. Paetkau, and G. C. White. 2009. "Demography and Genetic Structure of a Recovering Brown Bear Population." *Journal of Wildlife Management* 73:3–17.

Lasswell, H. D. 1971. *A Pre-View of Policy Sciences*. New York: American Elsevier.

Mattson, D. J., S. Herrero, R. G. Wright, and C. M. Pease. 1996. "Designing and Managing Protected Areas for Grizzly Bears: How Much Is Enough?" In *National Parks and Protected Areas: Their Role in Environmental Protection*, edited by R. G. Wright, 133–64. Cambridge: Blackwell Science.

Putnam, R. D. 2000. *Bowling Alone: The Collapse and Revival of American Community*. New York: Simon and Schuster.

Servheen, C. 1998. "The Grizzly Bear Recovery Program: Current Status and Future Considerations." *Ursus* 10:591–96.

US Fish and Wildlife Service. 1993. *Grizzly Bear Recovery Plan*. Missoula, MT: US Department of Interior, Fish and Wildlife Service.

Wilson, S. M., J. A. Graham, D. J. Mattson, and M. J. Madel. 2006. "Landscape Conditions Predisposing Grizzly Bears to Conflict on Private Agricultural Lands in the Western USA." *Biological Conservation* 130:47–59.

Wilson, S. M., M. J. Madel, D. J. Mattson, J. M. Graham, J. A. Burchfield, and J. M. Belsky. 2005. "Natural Landscape Features, Human-Related Attractants, and Conflict Hotspots: A Spatial Analysis of Human-Grizzly Bear Conflicts." *Ursus* 16:117–29.

Wilson, S. M., S. A. Primm, and A. Dood. 2007. *Summary Report: Beyond Boundaries: Challenges and Opportunities for Grizzly Bear Management and Conservation in Montana*. Helena: Montana Department of Fish, Wildlife and Parks.

Woodruff, R., and J. R. Ginsberg. 1998. "Edge Effects and the Extinction of Populations inside Protected Areas." *Science* 280:2126–28.

7

Collaborative Grizzly Bear Management in Banff National Park Learning from a Prototype

J. DANIEL OPPENHEIMER AND LAUREN RICHIE

Introduction

In 1885, two years after construction workers for the Canadian Pacific Railway stumbled on a cave of hot springs in Alberta's Rocky Mountains, the federal government set aside the area as a reserve. It would become Canada's first national park and the world's third. Today Banff National Park is a world heritage site and icon for Canadians, attracting more than three million visitors annually and supporting 8,000 permanent residents (Jamal and Eyre 2003; Parks Canada 2008, 2010; Dickmeyer 2009). Parks Canada, the federal agency that manages the national parks, strives to "protect and present nationally significant examples of Canada's natural and cultural heritage and foster public understanding, appreciation, and enjoyment in ways that ensure their ecological and commemorative integrity for present and future generations" (Parks Canada 2010, 2). Embedded in this mandate is a delicate balance of objectives to protect, educate, and provide a quality visitor experience. With increasing frequency, the discourse and divisiveness surrounding park and grizzly bear management decisions boil down to those who support use and enjoyment of the park versus those who favor preservation and ecological integrity. Most of the disagreement has played out in the media and public processes of consultation, leaving stakeholders and agency personnel beating their heads against the wall and pointing their fingers at each other. In recent years, alternative arenas and processes have been created in Banff National Park. One in particular allowed Parks Canada officials and stakeholders to sit down on the same side of the table to resolve conflicts and work on grizzly bear management problems together, rather than seeing each other as the problem.

Here we provide a case study of the Grizzly Bear Dialogue Group (GBDG). Parks Canada initiated this stakeholder process in 2005 as a prototype to resolve conflict and make collaborative decisions to manage and conserve grizzly bears in Banff National Park. As a prototype, the GBDG represented a small-scale, practice-based, trial intervention, one with guiding goals yet enough flexibility to identify and solve problems (Brunner and Clark 1997, 54). As with other prototypes, the GBDG trial intervention had the underlying goal of learning about a system and improving outcomes (Clark 2002, 143). This dialogue group met between 2005 and 2009, during which members learned new skills for problem solving and then applied this new knowledge to bear management issues. In 2009, after successfully developing several prescriptions to manage bears, the GBDG fell apart. Here we describe what the GBDG did in terms of learning, collaborating, and making recommendations to Parks Canada. We also appraise the functions of the joint decision-making process. Finally, we provide lessons learned and recommendations for prototypical elements that might be useful to others. While we hope that our recommendations may be helpful in other collaborative processes and contexts, we do not suggest they are a cookbook or regimen for collaboration (Conley and Moote 2003).

We lived in Banff from June through August of 2010, during which we conducted about forty interviews. Most of the interviewees were participants in the GBDG. But we also spoke to journalists, community members, and experts knowledgeable in bear management and stakeholder engagement processes. We met interviewees in coffee and bagel shops, living rooms, conference rooms, and sometimes on the phone. Following each interview, we cross-checked and discussed transcripts, analyzed the contents of the interviews in a social matrix, and built on interviews to flesh out or clarify new information. Toward the end of the summer we began to hear renditions of the same story, indicating a comprehensive set of data about the collaborative process. Our goal in this project was not to take sides, find fault, or place blame on individuals or organizations. Rather we sought to determine what worked well, what did not, and to identify the causal mechanisms for why the collaborative process failed to sustain the common interest and ultimately fell apart. To do so, we immersed ourselves in Banff, reading local newspapers, meeting minutes, and management plans, attending community events, and conducting informal interviews throughout the summer. As we grounded ourselves in the realities of grizzly bear management and the controversies surrounding it, we used a number of concepts and theories to find meaning and understand some of the things we saw and heard. Additionally, we conducted research in a range of disciplines, familiarizing ourselves with other collaborative conservation processes and prototypes.

At that time both of us were graduate students from the United States with similar interests in problem solving, conflict resolution, and wildlife conservation. We have worked professionally and in academia on an array of biodiversity conservation issues, from biological, land-use planning, and natural resource policy perspectives. With this experience and shared interests, we both sought to understand, appraise, and learn from this particularly complex case of collaborative grizzly bear management. However, our individual experiences with, connections to, and understanding of wildlife, as well as broader worldviews, are different. One of us is a hunter from Texas, the other an East Coast animal lover. As such, there were some interesting parallels in the case we were studying and how we went about studying it in terms of bridging and building on different perspectives.

Context and Problem Definition

In this section, we consider the human and environmental context that shaped the Grizzly Bear Dialogue Group. We examine the impetus and preceding conditions of acrimony, the competing perceptions for how to advance beyond the rancor, and the goals sought and problems encountered by the GBDG.

BEARS, PEOPLE, AND THE PARK

The Banff-Bow Valley, situated in the Central Rockies Ecosystem in western Alberta, includes Banff National Park and the Bow River watershed and constitutes a stunning and accessible landscape that has attracted significant development (Kölhi 2010; Rutherford et al. 2009; Gibeau 2000). It encompasses montane, subalpine, and alpine ecosystem types. The Canadian Pacific Railway, Trans-Canada Highway, and a network of town roads and secondary highways transect the Banff-Bow Valley, which supports tourism, residential development, transportation, and logging. An hour's drive from Banff to the east is Calgary, which in 2000 passed the one-million population mark and has since continued to expand (City of Calgary 2007). Between Calgary and Banff lies one of the fastest growing areas in the province.

Historically, grizzly bears roamed a large portion of North America, although they have been extirpated from much of their historical range, including in Alberta (Alberta Grizzly Bear Recovery Team 2005). Today Banff National Park is home to a population of about sixty grizzly bears affected by human development, habitat fragmentation, slow reproductive rates, and human-caused mortality, mostly near highways and railroads. In fact, the park is nestled in one of the most heavily developed and used areas where grizzly bears still exist (Garshelis, Gibeau, and Herrero 2005; Gibeau 2000; Gibeau and Stevens 2005; Proctor 2005). Especially given the poor status of the species in Alberta, the remaining bears in Banff "have added symbolic and political significance simply because they exist in one of Canada's premier national parks" (Rutherford et al. 2009, 165). One cannot step into the Banff-Bow Valley without seeing the grizzly bear promoted in some shape or form. Images of grizzlies are posted on billboards along the Trans-Canada with claims such as "Best Grizzly Bear Viewing and Scenery in the Rockies!" In the local newspapers, bears have found celebrity status somewhere between outlaws and part of the community. Local headlines from the *Banff Crag and Canyon* and *Rocky Mountain Outlook* vary from the cautionary (e.g., "Grizzly Bluff Charges Runner" or "Berries Ripening, Be Bear Aware"), to the dismayed ("Bear, Elk Killed on TCH"), to the admiring ("Year of the Bear"). Seasoned bartenders and first-time tourists alike have their bear stories. Bears are a part of the life and lore of this place and are celebrated in historical accounts, myths, and art (Primm and Murray 2005).

Given their political, ecological, and symbolic importance, grizzlies in Banff National Park are often found at the center of debate and acrimony. The participants in the grizzly bear debate comprise a number of individuals and organizations, including Parks Canada, conservation groups, tourism industries, First Nations, commercial developers, recreation and user groups, transportation agencies, extractive agencies, wildlife scientists, local government, provincial agencies, and interested citizens. While some stakeholders discuss bears in terms of pride and reverence, some associate them with fear (especially following rare but highly publicized incidences like a mauling or killing), and others are displeased about restricting human behavior and use of the park. According to their differing identities and personal experiences, these participants have had conflicting perspectives, value demands, problem definitions, and preferred solutions (Rutherford et al. 2009; Chamberlain 2006). Specifically, participants have held opposing views on the state of the grizzly population and the question of whether limitations on human access are warranted in order to protect grizzlies (Richie and Oppenheimer 2012).

Parks Canada has the ultimate authority to manage bears and human access in national parks. With changing demographics and declines in visitation over the past decades, Parks Canada has faced a number of obstacles on both the national and park levels. Struggles to fulfill mandates of resource protection, education, and visitor experience are intertwined with struggles to remain relevant to a transitioning (e.g., older, more urban, increasingly foreign-born) constituency (Jager and Sanche 2010; Statistics Canada 2006; Dearden 2008; Chartier 2004). As lifestyles and values change, Parks Canada has reoriented to connecting Canadians to their natural and cultural heritage (Jager and Sanche 2010). This shift in focus, from ecological integrity to visitor experience, however, has stirred a number of concerns, especially among those oriented toward preservation of wildlife and wild places. Caught in the middle, grizzly bears are used as a symbol to promote various interests and claims, including preserving ecological integrity and fostering use and enjoyment of the park.

Often these debates and promotion of special interests emerge during stakeholder engagement processes. In accordance with the Canada National Parks Act (Canada 2000, C. 32, 6), "opportunities for public participation" are provided for a number of situations, including "the development of parks policy and regulations, the establishment of parks, the formulation of management plans, land use planning and development in relation to park communities." In recent decades opportunities for public participation have noticeably increased (Gunton, Day, and Williams 2003; Spyke 1999; Beierle and Cayford 2002). Typically, participation has occurred under top-down consultation, whereby Parks Canada has taken input from stakeholders and then used its authority to make decisions (e.g., Clark and Rutherford 2005). With limited access to decision making and authority in public consultation, stakeholders have frequently claimed that such decisions have been made behind closed doors, that their public input is a sort of "window dressing," taken so that Parks Canada could check off its mandated consultation box.

Given the highly politicized debate about managing bears and the lack of suitable arenas to sit down and discuss their views, many stakeholders in the region went to the media to promote their conflicting views and voice their discontent. Years of intense acrimony characterized the debate, with blame placed on Parks Canada, scientists, environmentalists, business interests, and others. The outcome was that many people were deprived of respect, affection, and power—and by some accounts, the bears were not doing so well either. With high mortality rates (e.g., Alberta Grizzly Bear Recovery Team 2008) and intense rancor, something new and different needed to be done.

SETTING THE STAGE

Although there was recognition by many that business-as-usual was not working, there was great variability in stakeholders' framing of "the problem" of grizzly bear management. Recognizing the problematic trends in bear mortality and acrimony in the Banff-Bow Valley community, a Parks Canada biologist, Mike Gibeau, reached out to the superintendent of his field unit and several academics with experience in the human dimensions of large carnivore conservation. After several interactions, they decided to explore the perspectives of stakeholders involved with or interested in grizzly bear management (Rutherford et al. 2009). In 2004 Emily Chamberlain, a graduate student from Simon Fraser University, conducted a study that would set the stage for the GBDG workshops (see Chamberlain 2006; Chamberlain and Rutherford 2005; Rutherford et al. 2009). Examining stakeholders' perspectives on bear management, she conducted a series of individual interviews and then convened a group meeting with stakeholders to review and discuss preliminary results. Chamberlain found four distinct sets of views about the source of the problem. The first set emphasized deficient directives. Stakeholders with this view believed that, with a lack of overall conservation strategies, goals, visions, and targets for success, bear management was reactive and not grounded in science. The second set asserted that the problem was, in fact, that management problems were exaggerated or overstated, overlooking the many successes. The third set pointed to problematic institutions and the fourth politicized management. With such different understandings of the problem, participants subsequently had different views on the best corresponding solutions for grizzly bear management. Without an arena to share and meld these perspectives, participants would presumably remain discontented and fragmented in their views. In conducting this study, Chamberlain developed a body of research and an arena that, first, set the focus on perspectives and attitudes and, second, identified and established willingness among participants to talk. With a prevalent sense of bitterness and crisis, a critical mass willing to engage in an open and honest dialogue, and a partnership with several academics from Simon Fraser University and Yale University, Parks Canada began to develop a small-scale trial intervention.

GOALS OF THE GRIZZLY BEAR DIALOGUE GROUP

In 2005 Parks Canada invited about thirty stakeholders to join them in a series of three problem-solving workshops. Known as the interdisciplinary problem-solving (or IPS) workshops, these skill-building sessions served as the inception of the GBDG, "to promote more effective and efficient collaborative thinking about grizzly bears in Banff National Park and the Bow Valley region" (2005 IPS Invitation).[1] Murray Rutherford from Simon Fraser University and Susan Clark from Yale University introduced participants to an interdisciplinary approach to understanding and solving problems. Between eighteen and twenty-two participants, including the park's superintendents and stakeholders with diverse interests, attended these three workshops. With Rutherford's and Clark's guidance, participants examined individual perspectives and potential biases, considered how to build more comprehensive and constructive understandings of management problems, and studied the human dynamics of grizzly bear management, that is, how people interact and how decisions are made (Rutherford et al. 2009).

Commencing in May 2005, these three workshops set up an ongoing decision seminar that continued to meet until 2009. A concept that dates back several decades, the decision seminar is "concerned with knowledge of the policy process and with the evaluation of knowledge for policy" (Lasswell 1971, 44). As a setting for learning, the decision seminar provides an open system highly responsive to information demands of participants and an effective problem-solving culture that is contextual, problem-oriented, and egalitarian (Lasswell 1971; Burgess and Slonaker 1978). Gibeau, Clark, Rutherford, and a facilitator from the CSE Group, Felicity Edwards, designed the flexible and responsive system of the decision seminar in stark contrast to traditional, top-down, scientific and bureaucratic management. Instead, stakeholders built social capital and practiced adaptive governance. Applying the principles of adaptive governance, they produced policy recommendations based on real people (i.e., not caricatures), pragmatism (accounting for on-the-ground context), and the notion of the common interest (as opposed to special interests only held by a few) (Brunner et al. 2005, 19). Unlike conventional consultation, where stakeholders typically participate as advocates in what many consider a predetermined process, members of the GBDG participated as decision makers seeking a single voice and common ground (Edwards and Gibeau n.d.).

THE PROBLEM

Creating such an innovative prototype for people to learn and interact in new ways requires a great deal of creativity, patience, and dedication. In the case of the GBDG, the architects of the process worked within the confines of institutional cultures and agency regulations while promoting a new way of doing business. Add to the mix people from different backgrounds with conflicting perspectives on how parks should be managed and for whom, and community dynamics in which participants have shared past experiences (some good, some bad) with others at the deliberation table, and finding ways to engage people in a way that fosters common ground can be quite difficult. Indeed, effectively implementing such a community process "necessitates a careful and difficult blending of local, national, and sometimes international interests and institutions. . . . The complexity of goals, interests, and organizational features of [community natural resource management] renders its implementation exceedingly difficult" (Kellert et al. 2000, 707). To overcome these and other challenges, the GBDG would need to bring people to the table in a manner that clarified and secured common ground, avoided polarization, and fostered meaningful dialogue.

In the Banff case, the GBDG deliberations led to increases in social capital, amelioration of conflict, and effective management prescriptions. However, the dialogue group encountered a number of concurrent and intertwining obstacles that placed strains on the decision-making process. The underlying shift and goal substitution in national park management, as described earlier, fed into the local Banff intervention, effectively raising the stakes and creating uncertainties about the implications of these shifts in Parks Canada management. Meanwhile, the prototype encountered substantial turnover of participants: some moved away, others changed jobs, and still others defected because of competing demands or frustrations with the process. Many of these frustrations manifested in misaligned expectations. GBDG participants were not accustomed to the open, flexible nature of the decision seminar and sought more deliberate, constrained terms of reference. Further, as participants at the table changed through the years, newcomers were not adequately inculcated to the interdisciplinary problem-solving approach, and skills were not maintained or regularly refreshed. In the final years, with declining skills and waning cohesion, the dialogue group experienced two important events (see appendix 7.1). First, the two park superintendents engaged in the process since its inception were replaced in 2008 by new leaders who had not grown up with the process and whose actions were interpreted by many GBDG members as requisitioning the dialogue group's power. Second, the dialogue group took on a thorny topic with a long history of controversy: management of the Bow Valley Parkway. During deliberations on the parkway, a series of events occurred that brought the curtain down on the joint decision-making process. Ultimately, the GBDG encountered a reversion to special interests and was unable to sustain the goal clarity and social capital needed for effective collaboration toward the common interest.

Decision Process of the Grizzly Bear Dialogue Group

Here we break down the functions of the joint decision-making process to identify how each of these functions was designed and played out as well as to appraise each function. We then consider some of the underlying constitutive issues that may have played a role in these functions (appendix 7.2).

The Grizzly Bear Dialogue Group may be understood in a number of ways. On one level it created an arena for people to come together and interact in new ways (Cherney, Bond, and Clark 2009). Stakeholders from a variety of sectors, including provincial and national park agencies, a horse outfitter, hotel-motel associations, environmental and conservation groups, the ski industry, and Canadian Pacific Railway brought their expectations, demands, and perspectives to the table. Of note, superintendents from both field units in the park (the Banff field unit and the Lake Louise, Yoho, and Kootenay field unit) participated in the workshops and meetings and did so as two of many voices at the table. Rather than seeing each other as the problem, participants engaged in civil discussions and learned new ways to analyze problems and the policy process. Collectively, they offered a strong suite of skills and expertise and effectively represented the community's interests and demands. However, as the deliberations evolved and years went by, people came and went, making it difficult to maintain cohesion and problem-solving skills.

On another level the GBDG created a new process for making decisions. In conventional consultation, stakeholders provide input via written suggestions or spoken comments during structured meetings; input is given, but little dialogue and learning occurs. In this conventional form of decision making, the process is dominated by promotion, as many stakeholders do not have meaningful involvement in developing or processing information, or in crafting, implementing, or adjusting prescriptions. But in the GBDG stakeholders inherited additional power and responsibility to collect and share information, discuss findings and policy alternatives, and reach consensus for management recommendations. Pimbert and Pretty (1997, 309) provide a typology for participation, which can take passive and active forms. The GBDG may be understood as one of the more active forms, "interactive participation," whereby participants conducted joint analysis, employed an interdisciplinary methodology, and sought multiple perspectives for making local decisions. Using this form of participation, many GBDG members felt they were making real contributions to making decisions, that the closed door of consultation had been opened to collaboration.

INTELLIGENCE (PLANNING)

Intelligence, the act of collecting, processing, and disseminating information, may also be understood as the planning stage (Clark 2002, 59). In the GBDG members worked together to collect and share information in a variety of ways. During the initial skill-building workshops, participants reflected on themselves, drew mind-map diagrams, and shared fundamental information about their values, identities, and perspectives. One participant, for example, noted that he or she was "a person of rectitude," motivated by certain ethical standards. Others at the table recalled this observation years later for its insight into that particular individual's views and actions. Such social intelligence was shared with all workshop participants and led to noticeable increases in respect, empathy, and social capital.

In the years of deliberations that followed the skill-building workshops, the GBDG used several avenues for gathering and processing information. First, participants contributed their own expertise. Two biologists, for example, presented spatial maps and other data pertaining to habitat security while a representative from an organization supporting sustainable tourism provided an overview of market strategies and components of the visitor experience. Second, the facilitator and Parks Canada worked strenuously to provide members with intelligence, from briefings and meeting notes to social and natural science, requested by the dialogue group. Third, the group invited external experts to share information from a variety of fields. One bear biologist from British Columbia presented on several occasions about monitoring, mortality targets, and road/trail access issues. Many participants, some of whom perceived Parks Canada scientists as biased, echoed the perspective that having this "independent" scientist was of great help and appreciated this scientist's clear articulation of what from the scientific literature is suspect and what is not. Other presenters included an expert on human-wildlife relations and social scientists who presented results of a social survey commissioned for the dialogue group and a human-use simulation tool illustrating how various scenarios might impact grizzly bears.

As a stakeholder group constituted to apply an interdisciplinary framework, the GBDG moved beyond conventional disagreement about the number of grizzly bears in the park to broader, value-based discussions that considered social as well as ecological information. It was, as one participant described, "scientifically informed" rather than "scientifically driven" management, based on information that was evidence-based, interdisciplinary, and sanctioned by the group. As another interviewee described, a collaborative process such as the GBDG is not like playing poker: rather than keeping straight faces and concealing their cards, participants shared as much information as possible with others in the dialogue group so that they could say "yes" to each other. One subject in particular prompted a number of intelligence demands and captured the competing value demands that stood between the participants—managing the Bow Valley Parkway. Many locals and visitors treasure the parkway for its scenic and wildlife viewing opportunities, while some are concerned about the road's impact on wildlife dispersal and habitat quality. Parks Canada's previous efforts to manage the parkway with voluntary closures had proven both ineffective and contentious. In 2007, when GBDG participants began to clash and entangle during discussion of managing the parkway, Parks Canada commissioned for the group an online survey of park visitors to gain insight into social information such as usage patterns of the parkway, reasons for driving on the parkway, and expectations about crowding. Following the study, the GBDG created a subgroup to examine further the details of managing the parkway.

The intelligence function played a number of roles and varying importance in the GBDG. It shaped who was invited to the table, and equally important, who was not. While some participants felt that the group included all relevant participants, others claimed that the GBDG failed to bring the right people to the table in 2007 when the group started to address the Bow Valley Parkway. This management issue encompassed broader ecological and social factors, including several family businesses located along the parkway. Whether and how these family businesses were represented by a member-based organization that supports the tourism industry was called into question. Ultimately, these businesses were invited to join the stakeholder process, but not until 2008, when the dialogue group had already begun to discuss solutions. As such, the timing and manner in which these family businesses were brought in led to a great deal of hostility at and beyond the deliberation table. Second, intelligence was used as a fallback to making decisions. One participant noted that some issues were "debated, rehashed, and reconstituted umpteen times," and right as they were on the brink of making a decision, there was a retreat by participants to the scientific management model, with demands for more technical information and certainty (Brunner et al. 2005). Third, intelligence laid the foundation for making several decisions. Compared to other stakeholder engagement processes, participants described GBDG deliberations as more fact-based, emphasizing rational intelligence over emotionalism. Equal exposure and access to information provided tremendous gains in enlightenment, as group members learned to think in new ways and increased their understanding of various management issues. One participant noted that by having this intelligence and using the IPS framework, they were able to begin to understand the problem rather than just jumping to solutions. However, there were exceptions where information did not inform decisions. One such case was the discussion about bear habitat security. Multiple participants averred that, after requesting information for several months and during several meetings, the group ignored the data and instead retreated to their foxholes and personal preferences.

Using the expertise represented at the table, inviting experts from several fields to offer presentations, and making use of Parks Canada's resources and willingness to find information, the GBDG was both creative and comprehensive in finding information on which to craft prescriptions. Within the private deliberations of the GBDG, information was freely shared across the table and through meeting minutes. However, beyond this, the intelligence function was not particularly open for participants who joined the process after the initial skill-building workshops. While they did receive briefings and meeting minutes, new members neither shared nor had access to the social data (e.g., values and worldviews) of their colleagues, nor received hands-on training in skills (such as occurred in the three initial workshops) on entering the process. The social data were critical not only in getting people on the same page (i.e., recognizing that decisions were value based), but also building relations and trust critical to making future decisions. Furthermore, in terms of public access, information was not actively shared with the media, agencies, or community members. As a result, community members tended to be largely unaware of the process or, to a lesser extent, perceived it as secretive and elitist. The group strove to address these sentiments with several outreach efforts, but with mixed results. In part this stemmed from the difficulties and esoteric aspects of the IPS method and the dialogue group's work. Additionally, the group's reticence about going public as well as a lack of participants' focus on their broker role (presenting information from the deliberations table to the constituencies, and from the constituencies to the table) limited where and how information was shared.

PROMOTION (DEBATE)

Promotion may be understood as recommending and mobilizing support for policy alternatives (Clark 2002, 61). It is through promotion that mass support and enthusiasm are developed to drive collective action, to push people to move to act. Traditionally, promotion might be seen in campaigns, letters to the editor, protests, and advertisements, urging for different courses of action. In the case of the GBDG, the types of promotional activities, by design, were limited. Participants were asked to leave their baggage at the door so as not to promote specific prescriptions that advanced only their special interests. It was, as one participant stated, not about winning, bantering, or badgering, but rather identifying the problem and then figuring out solutions. With shared responsibility and new accountability, participants could no longer simply "throw rocks" during deliberations. Rather, they had to work together. Certainly there were times when some individuals deviated from the group's norms and either dug their heels in or reverted to their foxholes. One admitted to taking a defensive position to make sure that he or she did not "get screwed." Another recalled a participant bringing in positional pamphlets supporting a particular special interest. But compared to other processes, the GBDG created a place for civil discourse, rather than for media wars and emotionalism, where people "got down to the facts . . . by stripping their tribalism." Central to maintaining this civil discourse, the facilitator and the ground rules steered participants away from circling the wagons in order to find common ground.

Seeking common ground, given the contention of several issues, the reticence of some participants, and the impetuousness of others, necessitated mechanisms for propelling collective action. These mechanisms took form in guiding voices at the table, effective facilitation, application of the IPS framework to understand and solve problems, and use of social capital to navigate through difficult value dynamics. And in one instance, an invited speaker called the group out to make a decision, which they did. However, in other instances, the dialogue group was unable to mobilize in the face of contentious issues. Especially for participants representing organizations and not just their own views, balancing dual roles as individuals at the table and as representatives of organizations with positions was very trying at times. Indeed, there were instances of people "going back to their camps to wave their flags." Some would go back to their positions because, as one person put it, that was their job. This reverting from common ground to personal or organizational preferences played out in discussions of bear habitat security. From November 2006 to June 2007, the group and an ad hoc subgroup considered the issues (e.g., adopting habitat security targets and making improvements to security), but never reached consensus. Some interviewees attributed this to the fact that this discussion would impact town sites and ski areas, others stated that commercial interests and hikers were unwilling to close certain trails, and still others pointed to the difficulty of the technical and nuanced science. That was, as one person noted, when they "began to step on each other's toes." In the end, the group failed to make a decision

regarding habitat security. Instead, the outcome was frustration and enerva-
tion around the table.

Often promotion took form not in urging for particular policy alternatives but in pushing for the IPS methodology. Several participants referred to
an incident when two participants from purportedly disparate sectors stood
up together to promote the GBDG process and chastised a newcomer who
was acting counter to the IPS method. Other means of promotion might be
less noticeable, possibly manifested in respectful listening or questions to
gain greater understanding rather than fall back to conventional positions
and argument. Beyond the deliberation table, participants promoted the IPS
methodology in their communication efforts. In March 2006, after the group
acknowledged the need for more external public buy-in and internal pride in
what the group had accomplished, they took several outreach steps, including convening a series of briefings for interested Parks Canada staff, developing a communication subcommittee, and selecting a general spokesperson.
However, tentativeness about "going public" prevailed and the group failed
to promote its gains and successes. Consequently, the Parks Canada agency
and some of the organizations represented at the table were not invested in
or wholly knowledgeable of the outcomes of the GBDG prototype.

Initially, promotion met the standards set by the ground rules and norms
of the group. Participants effectively used the wealth of intelligence to consider and promote various alternatives, resulting in several prescriptions.
But as the years went on, as the issues became more difficult, and the faces
at the table changed, some grew weary and frustrated. Conversations stalled
and others derailed. Many interviewees echoed a lack of direction and clear
objectives. It was also at this point that new superintendents came into the
process. Like some of the new participants, they had not gone through the
three skill-building workshops. Further, they had different leadership styles
that did not fully align with the expectations of the group. When the process
started to flag and people felt as though they had reached a dead end, cham-
pioning leadership to spur action was needed the most, but it was not to
be found. Without collective promotion for the process at this point, things
fizzled out. Agreement could not be reached in terms of what to work on and
how to work through it.

PRESCRIPTION (RULE MAKING)

Prescription is the activity that establishes the rules, policies, or guidelines
for actions (Clark 2002, 63). In the GBDG process participants held equal
stalemating power and worked together through joint decision making to

formulate prescriptions (Zartman 1977). These prescriptions took the form of recommendations to Parks Canada. Provided that these recommendations were deemed reasonable and consistent with the park management plan and that they met other standards, the superintendents would take the recommendations to the CEO of Parks Canada to be adopted.

There are a number of ways to make decisions in stakeholder processes. Groups may use, for example, a majority vote in which 51 percent approval is needed to reach agreement. Others may use some form of consensus. In the case of the GBDG a form of consensus defined as overwhelming agreement and based on a "heartburn scale" was used. After discussions were completed, the group members sat around the table and the facilitator synthesized and summarized what the group was saying. When heads were nodding, the facilitator asked, "Can you live with this?" With a 1-to-5 heartburn rating scale, if everyone nodded in agreement, then consensus was considered reached. If someone suggested they could not live with a decision (i.e., they had a 5 on the heartburn scale), then additional discussions followed.

The group reached consensus on several prescriptions, all of which Parks Canada adopted. The first of these, in 2006, represented a trial-run application of the IPS framework in areas with a history of human-bear conflict: the Bryant Creek-Allenby Pass and Aylmer Pass areas. With the summer and berry season around the corner, there was a need to develop a new interim management plan for these popular hiking and recreation destinations. Bryant Creek, recognized as a core reproductive area for grizzly bears and common feeding area for females with cubs, had experienced several bear attacks in recent years (Rutherford et al. 2009). Using the IPS method of problem orientation, which considers management goals, past trends and conditions, and alternatives, a GBDG subgroup selected a policy option that placed a seasonal restriction on the Aylmer Pass trail and a nearby campground from late summer through the early fall berry season (when bears and people are most likely to cross paths). The interim management plan also restricted access on certain trails in the Bryant Creek-Allenby Pass for several months during the 2006 and 2007 hiking seasons, limiting access to horse users and certified guided hiking groups (Rutherford et al. 2009). Lessons learned included the recognition that it was a "value-based discussion," that there was an agreed-on goal, and that there was readiness to deal with the issue because there was a perception of a real problem. Later that year the group made its second recommendation, putting forth new and more realistic mortality targets that were more specific and consistent with the scientific literature (Rutherford et al. 2009).

Some deliberations, however, did not result in prescriptions. The first major hurdle came in the form of developing prescriptions for bear habitat security. In 2006, using studies and models, the group was to put forth a recommendation for the revised park management plan. Over the duration of several meetings, participants reviewed technical information while clarifying their standpoints and asking problem-oriented questions about the goals for habitat security. After requesting additional maps, hearing presentations from several biologists, and forming a subgroup to deliberate further, participants were unwilling to make concessions and the group failed to reach a decision on this issue. Following discussions of habitat security, the group moved on to what became the most intractable and volatile issue, the Bow Valley Parkway. This was the final issue the group deliberated and it became the bomb that sent some flailing beyond the table, looking for alternatives other than the GBDG. While a list of recommendations was proposed for managing the parkway, including a mandatory seasonal restriction from dawn to dusk, paid access by buses for wildlife viewing during these restricted times, and several educational components, there was internal disagreement—what some referred to as second thoughts—about the recommendations and whether the group had in fact reached a consensus. What is clear is that things really fell apart at this point: social capital in the GBDG deteriorated and Parks Canada sought and created another public forum for addressing the parkway.

Using an interdisciplinary framework, the group formed prescriptions that were both rational (reflecting a diverse and comprehensive body of information) as well as inclusive (reflecting the voices represented at the table through the heartburn scale). Many echoed the sentiment that the group's prescriptions were informed decisions based on facts, not just passion and emotion. While most considered the initial topics of mortality targets and Bryant Creek to be fairly simple, "low-hanging fruit," some saw these as substantial issues and representative of real achievement. Indeed, the group faced hurdles of entrenched positions, differing expectations and demands, and baggage from past and other ongoing efforts to overcome, all of which resulted frequently in lengthy and sometimes exhausting discussions.

INVOCATION AND APPLICATION (IMPLEMENTATION)

The GBDG was constituted to provide recommendations to Parks Canada for managing grizzly bears in Banff National Park. As such, its role was not to implement or enforce any of these management recommendations (though each member had the responsibility to uphold the dialogue group's ground rules).

While the group did not have any official role in invocation or application, it may have affected these functions in indirect and direct ways. One participant noted that with the joint decision-making process of the GBDG, the bulk of time is spent before and leading up to the decision (e.g., in intelligence, promotion, and prescription functions), resulting in more easily implementable prescriptions. By contrast, with top-down decisions the bulk of the time is spent after the decision is made, when the stakeholders are up in arms. Additionally, having the superintendents at the table had notable implications for implementation. They were invested in the process and able to provide information and details from the process to the Chief Executive Officer of Parks Canada. However, this stage of the process, when recommendations were taken from the deliberation table up the Parks Canada chain of command, was a persistent source of confusion. While the superintendents explicitly discussed the role and authority of the group during its inception, as time went on, as new participants replaced some from the core group, and particularly as new superintendents entered the process, uncertainty grew about whether the group was writing public policy or providing advice subject to others' discretion.

APPRAISAL (EVALUATION)

Appraisal may be understood as an evaluation of a decision process as well as how successful prescriptions are at achieving respective goals. In conservation programs appraisals are seldom comprehensive, in-depth, or external, yet they provide opportunities for learning, course correction, and improved program performance (Clark 2002; Kleiman et al. 2000). The GBDG included a provision in its ground rules for appraisal after the first year of the process. Following the third skill-building workshop in March 2006, participants were asked to respond to the following question: How well is the decision process working now? Positive observations included openness of information, the variety of issues being addressed, and inclusivity of participants. Looking forward, concerns were expressed about the authority of the group, differing expectations about timeliness, and delays and decreasing momentum.

But beyond the first year the GBDG had no clear provisions for appraisal, in terms of if and when it would be appropriate to conduct, who would lead such efforts, and how recommendations from the appraisal would feed into the GBDG process for improved course of action. As years went on, participants began to ask questions nearing the appraisal function. For example, in a December 2007 meeting participants began to articulate concerns about the need to clarify the scope of the group, to assess how the group was functioning, and to address concerns about individual viewpoints prevailing over the notion of common ground. One person expressed concerns about how Parks Canada's sharing of leadership and decision making impacted the functioning of the GBDG and other stakeholder processes. Concerns about impediments to seeking common ground, unclear expectations, and grinding lengthy discussions continued throughout 2008. However, these discussions occurred amid a flurry of other events, including the replacement of superintendents and the burgeoning controversy surrounding the Bow Valley Parkway. Although facilitated discussions about the direction of the group did lead to an amendment of the ground rules, the major struggles about scope, positioning against common ground, and unclear expectations persisted in the final years of deliberations.

While the group did not engage in the appraisal process in a comprehensive manner, participants did engage in two other efforts that might be considered partial appraisals. The first was a thesis conducted by Jutta Kölhi (2010), a master's student from Simon Fraser University. In the summer of 2008 she examined how perspectives on grizzly management in the Banff-Bow Valley had changed since Chamberlain's 2006 study. Kölhi's study included an assessment of the GBDG: "In spite of the criticisms and shortcomings of the [GBDG] process, most participants had either a positive view or mixed feelings of the group and felt that it was beneficial and worthwhile" (129). Another partial appraisal was a workshop held to upgrade skills in September 2009. During this two-day workshop participants who were not involved in the initial three workshops received their first in-depth training in applying the IPS framework, while original members received a refresher on topics ranging from standpoint clarification to problem orientation to social and decision process mapping. The timing of this workshop was such that it was considered by some to be post-mortem, when the process had already fallen apart and frustrations had peaked. These dire circumstances may explain in part why participants were not particularly interested in a full-fledged appraisal and reaching a good termination. Without provisions for continued evaluation or collective and concerted group focus on improving the process, the appraisal function was not dependable or comprehensive.

TERMINATION (REPEAL AND ADJUSTMENT)

Termination may be understood as the repeal or large-scale adjustment of a prescription. An example might be planning for and implementing delisting of an endangered species. This function, like appraisal, is often overlooked and underappreciated in policy processes (Clark 2002, 69). The GBDG was designed to be a flexible learning environment for improving thinking and fostering collaboration for grizzly bear management. As such, it was inherently designed not to be tightly bound. When invitations were sent out in 2005 Parks Canada invited stakeholders to "participate in a series of three workshops being planned over the next year on grizzly bear conservation and management." While the invitation noted the working assumption that "there is a willingness to devote time to this effort in order to move away from the way things were done in the past," there was no clear indication then or later just how much time this would be.

As of August 2010 the GBDG process was at best in hibernation. Participants had not met in any official capacity in over a year and Parks Canada had initiated a new stakeholder engagement process, the Bow Valley Parkway Advisory Group, which had picked up approximately where the GBDG had left off, addressing the management of the parkway.[2] Nearly all interviewees agreed that the GBDG process was, for all intents and purposes, dead. In fact this was one of the few points where we heard near consensus. However, Parks Canada has not officially ended the process. Thank you letters have not been sent out nor have participants been told how their work will be used. With a shift in Parks Canada officials' responsibilities, there is no one to carry the GBDG from start to finish. As one interviewee stated, "They haven't closed the loop on [the GBDG]. It's hanging out in space." While several participants expressed uncertainty about the status of the GBDG, either because they had defected years before and not kept up, or saw it as being officially on hiatus, nearly all considered it dead in the water.

Our analyses of interviews identified several distinct narratives on the problems and factors in the GBDG that led to the termination of the process. They may be grouped as: (1) leadership and governance, (2) time and momentum, (3) people and process-design, and (4) Bow Valley Parkway. Some participants placed great emphasis on the role of the new superintendents and, to a lesser extent, on paradigm shifts in the Parks Canada agency. Within Parks Canada and among individual stakeholders, there are very strong and differing expectations of and demands for superintendents in stakeholder engagement processes. One salient belief is that these leaders should attend processes like the GBDG and that their active attendance is what gives the process legitimacy, an authoritative signature (i.e., the public stamp of approval), and control intent (i.e., the clear intent of authorities to implement the group's prescriptions). These interviewees claimed that a promising prototype was terminated by new Parks Canada superintendents who joined the process in 2008 and who did not grow up with nor appreciate the virtues of the process. Consequently, Parks Canada took back the shared power that the group was accustomed to, shifting from interactive participation and power sharing to a more traditional authoritative style of leadership common to many bureaucracies. Some of these participants also claimed that forces in Ottawa, including a shift in the agency focus from ecological integrity to visitor experience, as well as consolidations of power formerly vested in superintendents to higher authorities, may also have been at play.

Others aver that, after several years, the process simply lost steam and fell on top of itself. For them this was not particularly problematic as stakeholder processes are intended to come and go. Still others pointed to deficient terms of reference that failed to provide clear timelines, scope, goals, and planning to address situations when new people joined the group. One participant, for example, noted that no one knew when the work would end; as a consequence, people started to lose focus, the process started to dissolve, and they started to look for and define problems to address that went beyond how the group was constituted. A final narrative focused on the Bow Valley Parkway, a symbolically laden topic involving sensitive montane habitat, several wildlife species, family businesses not adequately represented at the deliberation table, and a history of contested management. During deliberations about the parkway, one participant deviated from the ground rules by sharing confidential information with the media, causing a steep decline in social capital. Another participant took actions that called into question the very mechanism for reaching consensus. And the family businesses were brought to the table in an untimely manner that led to a great deal of hostility. The tone of civility and respect that characterized the prior years of collaboration dissipated and hostility quickly took its place during discussions of the parkway. While there was overlap in these four narratives about termination, it is clear that people had different perspectives and placed varying emphasis on the multiple causal mechanisms at play. It is the very nature of these differing perspectives that makes collaborative, consensus-based processes so challenging.

A shift in agency attention from the GBDG to the new Bow Valley Parkway Advisory Group seemed to be the most official signal of termination. The shift was neither timely nor comprehensive. After Parks Canada initiated the new stakeholder process and shifted personnel responsibilities, the GBDG lost opportunities to appraise the group's work, determine how to carry components of the prototype forward, and look back and celebrate years of collaboration. Certainly, some participants were fatigued and frustrated at this point. However, the reasons for frustration and lessons learned from the prototype were never realized. Instead, many participants were left disillusioned and wondering.

CONSTITUTIVE DECISIONS

Earlier we described the functions of the GBDG decision process, focusing on ordinary, day-to-day policy choices that occurred within the structure of the stakeholder process. It is worth noting, however, the higher scale of constitutive decision making, which consists of "deliberations and choices about how policy should be made and, by implication, who ought to be involved in the decision process" (Clark 2002, 71). These "rules for making rules" are foundational yet not always tangible (Clark 2008, 24).

In the GBDG, constitutive issues came into play in three major ways. The first of these was the ground rules, which dictated how participants were to engage and make decisions. These were developed jointly during the first meetings of the process and via e-mail exchanges between the facilitator and group members. The facilitator carried a copy at all times, but after being adopted, the rules were not explicitly discussed for multiple years. Some interviewees expressed confusion about acceptable behavior (e.g., what they were permitted to tell their constituents and when) under the ground rules, while several either noted that there were no repercussions for abrogation of the rules or that they were frustrated that the rules were dismissed and lost meaning. Participants very seldom broke the ground rules, but these rare instances led to significant and adverse impacts, decreasing trust and social capital.

A second major constitutive issue was the change from explicitly bear-specific issues to deliberations on the Bow Valley Parkway. While the parkway was long tied to the issue of bear habitat security, it represented a shift in scope, taking the group beyond grizzlies to an issue with a new suite of ecological and social considerations. However, planning did not reflect this shift and the right people were not brought to the table in a suitable way. While some participants felt that bringing the family businesses to the table represented an effort to "stack the deck" in favor of business interests, most believed their initial absence was an oversight that needed rectifying. An owner of one of the family businesses located along the parkway felt that the GBDG had already made its decision by the time the businesses were invited: "We were invited for their output, not our input." The timing and mismatch of expectations resulted in a sharp spike in acrimony at the table and in the community. Reacting to the fallout, Parks Canada superintendents convened a public forum that led to the creation of a new stakeholder process, the Bow Valley Parkway Advisory Group.

The final and greatest constitutive issue took form in the authority vested in Parks Canada. Parks Canada is the land agency with mandated power for managing national parks. How the agency and its designated leaders share or wield this power impacts the authority and day-to-day operations of stakeholder processes. While the GBDG did have shared power, there was chronic confusion about the role and power of the group, whether they were simply providing advice, writing public policy, or something in between. Ultimately, superintendents are mandated under the Canada National Parks Act to fulfill certain regulatory and enforcement responsibilities. However, there appeared to be different and subtle manners in which some superintendents divested their power to support this joint decision-making process. Participants described the first two superintendents who engaged in the process from its inception in 2005 to 2008 as fully engaged in and supportive of the process. Their leadership styles were such that they gave up some of their power by participating as two voices among many at the table. In contrast, the leaders who replaced them in 2008 took actions that, accurately or not, were perceived by many participants as authoritarian and as an appropriation of power. These different leadership styles created and exacerbated existing confusions about the dialogue group's authority.

Moving Forward: Lessons and Recommendations

Much like everything else, stakeholder processes come and go. But how they come, how they go, and what happens in between matters dearly. The nature of the termination of the Grizzly Bear Dialogue Group was particularly problematic. It occurred without an appraisal and without retaining any potentially effective prototypical elements of the process. After stakeholders worked together for five years, there was never an opportunity to evaluate and understand where and how to take elements of their hard work and experience forward in other park management efforts. Further, termination in the wake of the Bow Valley Parkway controversy left a sour taste that many still carry with them. A fascinating and complex interaction of people, organizations, and institutions, the GBDG case provides a number of lessons learned in terms of how to identify, create, and sustain a process to resolve conflicts and manage resources in the common interest. Here, we provide several recommendations for creating, maintaining, and learning from collaborative processes, based on lessons learned in Banff.

GETTING OFF ON THE RIGHT FOOT

The GBDG did a number of great things to start off on the right foot, especially in setting the stage. When graduate student Emily Chamberlain initially interviewed individual stakeholders in Banff on their perspectives

Box 7.1

Lessons for Managers from the Grizzly Bear Dialogue Group

- In settings of acrimony, conventional forms of public consultation may be inadequate for engaging stakeholders and developing effective management prescriptions. Arenas based on civil and respectful discussion, adaptive governance, and joint decision making may be suitable if not critical alternatives.
- Initiation of collaborative processes should include clear ground rules and expectations of participants' roles and representation.
- In highly controversial contexts where science is debated, intelligence should include social as well as biological information, and it should be jointly found, processed, and sanctioned by a given stakeholder group.
- Promotion should be geared not only to reaching consensus at the table, but also to promoting the prototype to apposite organizations so that they are familiar with and invested in the prototype.
- Termination and appraisal, frequently overlooked functions in conservation policy processes, should be performed in a timely and deliberate manner. From the inception, ground rules should be clear about who should perform these functions, when, and under what circumstances.
- Collaborative processes provide opportunities to learn and adapt management techniques. Knowing how to build capacity and adapt means knowing how to learn and respond as individuals and as organizations. Leaders should recognize where and how to create opportunities to learn, particularly following the termination of stakeholder processes.

and then brought them together to discuss the preliminary results, she provided them with a forum to discuss their perspectives in a relaxed, informal setting (i.e., it was a student project, not a forum for the park management plan). The discussion set the focus on perspectives and attitudes and affirmed stakeholders' willingness to discuss these issues moving forward. Our personal experience reinforced this notion that stakeholders tend to feel more comfortable speaking to students, who are outsiders, sitting on the academic sidelines, with little on-the-ground impact. While we lack the authority to impact policy, we do provide a number of services for these situations, including bringing people together and asking questions that trigger new thinking. As one interviewee noted, students "add to and reinvigorate the process."

Identifying the right stakeholders, their concerns, and willingness to engage in negotiations is a critical step to determining whether and how to construct a collaborative process. Organizations like the Consensus Building Institute will conduct conflict assessments "to help stakeholders and assessors increase understanding of interests and issues, clarify options, determine if a consensus building process is appropriate, and if so generate a process design" (Ferguson 2010, 1). Based on individual interviews, these assessments help the convening party determine whether a collaborative process is appropriate in a particular case. For those considering whether to join such a process, they should recognize that conservation and management of imperiled species is "an involved process requiring the long-term commitment of individuals and organizations and substantial monetary resources. The process can become contentious and highly politicized and is not for the weak of heart" (Clark and Gillesberg 2001, 141).

With preceding conditions ripe for building consensus, Parks Canada set forth with the GBDG. However, initial planning did not fully clarify what constituencies the stakeholders represented. The issue of roles and representation became problematic, especially when the group began to address the Bow Valley Parkway. At this point some had assumed that the umbrella organizations were representing family businesses along the parkway. This assumption became troublesome when the family businesses were finally brought to the table and, as one person noted, the process went off the train tracks. Roles and representation also played out in less dramatic forms. Some individuals did not appear to be attuned to their dual roles of representing their constituencies at the table and the table to their constituencies. Clarifying roles and representation is paramount to identifying what information is acceptable for sharing and what promotion roles are expected of participants.

Getting off on the right foot also includes looking forward to consider the final steps. Many conservation and recovery programs tend to overlook the termination function early in the process. Early attention to the termination function can provide for opportunities to modify or stop ineffective practices as well as reduce potential conflict and distress associated with ending an emotionally and time-intensive process (Clark and Gillesberg 2001).

MAINTAINING EFFECTIVE DECISION-MAKING PROCESSES

For partnerships and collaborations to be effective, "considerable attention must be given to the decision-making process" (Clark and Brunner 2002, 4). When creating or participating in a local collaborative process, little can be done about national demographic shifts and agency goal substitutions. However, facilitators, local leaders, or stakeholders engaged in such a process can take steps to enhance the day-to-day decision process applied in a given stakeholder process. Clark and Brunner (2002, 3) note that, "by knowing how the decision process works, or does not work, partners in endangered species recovery can maintain good practices or correct a poorly functioning process." Decision making, as outlined in this chapter, is not an event but a process distinguished by several functions. Designing or requesting an intelligence function that is open to new participants (e.g., providing continued training in skills and opportunities to share perspectives and social values) will better inculcate newcomers and refresh others in effective, contextual thinking toward building common interest. Although time- and resource-intensive, skill-upgrading workshops can not only provide opportunities for learning and applying the skills in trial cases, but can also invigorate and foster enthusiasm for the process. Between the initial workshops and the 2009 skill refresher, GBDG participants lagged in internal pride and had varying understandings of and commitments to the IPS method. However, we heard resounding enthusiasm for the 2009 workshop, which introduced newcomers and brought old-timers back to the IPS framework. With greater understanding and internal pride, processes may be able to sustain the momentum required to navigate difficult management subjects.

Having periodic appraisals to identify what is and is not working well is particularly important for creating innovative processes, assessing how goals are being achieved, considering whether goals should be reconfigured, improving performance, refining components of the process, and potentially changing approaches entirely (Kleiman et al. 2000). In the case of the recovery program for the Australian eastern barred bandicoot, for example, after encountering a number of problems, including diverging goals and lack of learning, several program participants conducted a comprehensive and systematic "crisis intervention" appraisal, and the reorganization that resulted from this appraisal streamlined and upgraded all decision processes (Clark and Brunner 2002, 2). The Banff case illustrates a number of participants coming into a process with expectations and demands that did not align with the GBDG's open flexible system for learning and interacting in new ways. Being clear from the inception about the goals, scope of work, and differences of prototypes from other processes will help manage these expectations. But more than that, constituting prototypes with provisions for continued appraisal will provide additional time and space for discussion of expectations and potential course correction of the process.

LEARNING FROM STAKEHOLDER PROCESSES

As a small-scale, hands-on, and practice-based approach, the GBDG prototype represented something new and different. One of the greatest virtues of the GBDG was the learning that took place. Participants learned from each other and from experts who were brought in to provide presentations. However, one of the greatest opportunities for learning was never realized. When the GBDG was replaced by another stakeholder group, the Bow Valley Parkway Advisory Group, it was terminated without an appraisal (until now). Consequently, the reservoir of insights and lessons learned were not captured or integrated in capacity-building or organizational learning. Given the amount of time, resources, and energy put forth by the dialogue group, this is particularly unfortunate. Constituting stakeholder processes to conduct detailed and timely appraisals will certainly help. But the individual and organizational capacity to learn from experience is also needed to perform effective adaptive management (Fazey, Fazey, and Fazey 2005; Kleiman et al. 2000; Salafsky et al. 2002). Fazey et al. (2005) outline the basic factors for individual and organizational learning. Organizational learning requires not only individual members willing to engage in learning, but a "learning culture" based on openness and freedom of expression. Key to fostering this learning culture is leadership. While the role of managers in formal organizations is typically conceived as performing tasks such as defining goals, distributing resources, and influencing processes and procedures, managers are also facilitators of learning (though most do not view themselves as such). Whether the superintendent of a national park or a municipal leader, authoritative figures may create opportunities to learn from collaborative processes, to turn experience and hardship into effective learning and improved conservation practices.

Conclusion

For years grizzly bear management in Banff National Park was dominated by bitterness, finger pointing, and defensive measures by Parks Canada. Stakeholders had different and fragmented understandings of problems and pursued special interests with varying success. As an intervention to build social capital, resolve conflicts, and collaboratively manage bears in order to serve common interests, Parks Canada created a prototype, the Grizzly Bear Dialogue Group. The prototype effectively shifted decision making from scientific management to adaptive governance and created a place for civil and meaningful dialogue. But as a flexible process open both to information and to identifying emerging problems to address, the GBDG clashed with many participants' expectations and demands for a finish line, clear timelines, targets, and outcomes. This, along with participants' varying exposure to the IPS framework and skill set, as well as changes in group membership and Parks Canada leadership, led to mounting frustrations and declines in social capital. Ultimately, the GBDG failed to sustain goal clarity and maintain social capital needed for continuing collaboration.

Knowing if, when, and how to construct a collaborative process, given the context of the particular situation, is critical and can lead to gains in human dignity and on-the-ground, ameliorative management measures. The GBDG provides an example of an intervention that clarified and secured, but was unable to sustain, the common interest. As a prototype, it provides important lessons, including how to initiate, maintain, and learn from a collaborative decision process in wildlife management.

Appendix 7.1
Overview of Grizzly Bear Management in Banff National Park, Alberta, Canada

"Pre-intervention (2000–2005)" encapsulates the acrimonious years preceding and leading up to the GBDG process; "GBDG (2005–2009)" includes the five-year period during which the GBDG was active; and "Bow Valley Parkway Advisory Group (2009–2010)" represents the stakeholder process that superseded the GBDG. (Source: Based on Chamberlain 2006; Rutherford et al. 2009; and personal interviews.)

GRIZZLY BEARS
Pre-intervention (2000–2005)
Unsustainable mortality.
GBDG Process (2005–2009)
Unsustainable mortality; new mortality targets agreed on.
Bow Valley Parkway Advisory Group (2009–2010)
Unsustainable mortality.

SOCIAL PROCESS

Pre-intervention (2000–2005)

Acrimony and debate over conflicting perspectives, largely in the media.

GBDG Process (2005–2009)

Arena for civil dialogue and building of trust and social capital.

Bow Valley Parkway Advisory Group (2009–2010)

New arena and advisory group created, with considerable disillusion and initial setbacks.

VALUE TRENDS

Pre-intervention (2000–2005)

Deprivation of respect and unbalanced power dynamics.

GBDG Process (2005–2009)

Increases in respect and more balanced power dynamics.

Bow Valley Parkway Advisory Group (2009–2010)

(To be determined.)

DECISION PROCESS

Pre-intervention (2000–2005)

(1) Bureaucratic and scientific management.

(2) *Consultation*: low public access to decision making and authority.

(3) Criticisms of grizzly bear science and park management, contests among special interests, growing use of stakeholder groups.

GBDG Process (2005–2009)

(1) Adaptive governance.

(2) *Collaboration*: high public access to decision making and authority.

(3) Social and natural science jointly gathered and sanctioned by group in pursuit of common ground.

Bow Valley Parkway Advisory Group (2009–2010)

(1) "Future best picture" design and planning.

(2) *Consultative-collaboration*: high access to planning but less access to authority and decision making.

(3) Joint visioning and taking steps toward future-best goals.

PROBLEM

Pre-intervention (2000–2005)

Acrimonious social process and competing special interests resulting from lack of public access to decision making.

GBDG Process (2005–2009)
  Goal clarity and social capital were not sustained as new participants
    and park leaders joined with incongruent expectations.
Bow Valley Parkway Advisory Group (2009–2010)
  The new advisory group fired its first facilitator and has had difficulty
    finding a suitable replacement; acrimony and frustration have
    characterized discussions.

COMMON INTEREST
Pre-intervention (2000–2005)
  Not clarified or secured.
GBDG Process (2005–2009)
  Clarified and secured, but not sustained.
Bow Valley Parkway Advisory Group (2009–2010)
  (To be determined.)

Appendix 7.2
Decision Activities, Standards, and Effectiveness in the Grizzly Bear Dialogue Group

The source of these assessments is personal interviews, meeting minutes, and newspaper articles.

INTELLIGENCE
  Recognizing the problem and gathering information.
  Dependable
  Comprehensive
  Selective
  Creative
  Open

The information was dependable and factual, placing greater emphasis on evidence-based and group-sanctioned intelligence over emotions and positions. Using an interdisciplinary approach, the group moved beyond conventional disagreements about bear numbers and biology to consider a comprehensive body of social and biological data. The group used a number of creative methods, bringing in experts, requesting information from Parks Canada, and sharing their own expertise and social values.

Within the private deliberations of the GBDG, information was freely shared across the table and through meeting minutes. However, beyond this, the GBDG process was not particularly open for participants who joined the process after the initial skill-building workshops. While they did receive meeting minutes and briefings about the distinctiveness of the process, they neither shared nor had access to the social data (e.g., values and worldviews) of their colleagues, nor received formal training in and hands-on experience with IPS skills on entering the dialogue group. Furthermore, in terms of public access, information was not actively shared with the media, agencies, or community members, leading to a perception by some that the group was secretive and elitist.

PROMOTION

Open debate, in which various groups advocate for their interests or preferred policy.

Rational

Integrative

Comprehensive

Participants were asked to leave their baggage and special interests at the door. While there was great initial success in abandoning bantering and badgering for identifying problems and seeking common interest solutions, in later years some participants dug in their heels, reverted to individual positions, and deviated from the ground rules by going to the media. The GBDG was inherently an integrative process, bridging different views and values during lunchtime presentations, briefings, and group and subgroup committee meetings. While the GBDG may not have been comprehensive in terms of promoting its work beyond the table to achieve community buy-in, it was comprehensive in deliberating on all appropriate and relevant policy options for managing bears.

PRESCRIPTION

Setting the policy, rules, or guidelines.

Effective (stable expectations)

Rational

Comprehensive

As prescriptions were consensus-based, there were certain expectations of the group, the facilitator, and the process for reaching consensus. For the most part, prescriptions were effective, meeting expectations laid out in the ground rules and taking promising measures to solve the problems at hand. Ground rules, effective facilitation, and the interdisciplinary process fostered balanced discussions and rational approaches to finding agreement. This function was comprehensive in terms of the interests and values represented at the table.

INVOCATION

Implementation.
Timely
Dependable
Rational
Nonprovocative

While the group was not constituted to implement its prescriptions for Parks Canada's management of grizzly bears, it was self-charged with implementing its own ground rules. These constitutive decisions were introduced in a timely manner, at the beginning of the process. However, they fell to the back burner (in part because there was not initial breaking of the rules) and were not talked about explicitly for several years. With the change in participants, especially park superintendents, there was added confusion regarding the GBDG's authority and if and how group recommendations would be implemented.

APPLICATION

Dispute resolution and enforcement of the prescription.
Rational and realistic
Uniform

Both the facilitator and, at times, group members would take on the role of enforcing ground rules in a manner that was timely and prompt to certain discussions. However, as time went on and new participants entered the process and frustrations increased, several members deviated, either intentionally or inadvertently, from the ground rules and norms, suggesting that invocation was not dependable or uniform. The outcome was a loss of trust and greater difficulty to reach common interest.

APPRAISAL

Review and evaluation of the activities so far.
Dependable
Rational

Comprehensive
Selective
Continuing
Independent

Appraisal occurred after the first three skill-building workshops and partially during the final years of the GBDG, as participants began asking questions about the group's purpose, successes, and challenges. However, this review occurred amid changes in leadership, problem definitions, and other significant shifts. While there was great interest in learning about each other and bear management, learning about and improving the process was not prominent. As such, appraisal was to some extent continuing and selective toward issues broached by a select group of concerned and vocal participants, but did not meet any of the other listed standards as it was lost beneath other tasks and commitments.

TERMINATION/SUCCESSION
Ending or moving on.
Timely
Comprehensive
Dependable
Balanced
Ameliorative

As termination was not something actively planned for or thought about, it was not timely. Many participants felt as though the process had run its course long before the process was unofficially terminated, but others felt it was prematurely restricted and terminated. While the process had not convened in over a year at the time of this study and nearly all participants considered it dead, as of August 2010 Parks Canada has not formally ended the GBDG stakeholder engagement process, making termination neither comprehensive nor dependable. A new stakeholder process looking at issues addressed by the GBDG is seen by many as the alternative that followed the now defunct GBDG process.

OVERALL STANDARDS
Honest
Economical
Technically efficient
Loyal and skilled personnel

Complementary and effective impacts
Differentiated structures
Flexible and realistic in adjusting to change
Deliberate
Responsible

As an intervention, the GBDG created an arena for civil respectful dialogue where honesty and respect were fostered. As a prototype it was flexible and highly responsive to participants' request for information, but less responsive to their expectations for clear timelines and targets. Consensus has many strengths in terms of building social capital and building prescriptions with community buy-in; however, it tends to be the most costly in terms of time and resources needed to reach that form of agreement. While some participants dropped out of the process for a variety of reasons, there was a core group of respected community members that was loyal to the GBDG process and way of doing things.

NOTES

Above all we would like to thank Mike Gibeau for his courage, patience, and support. Without him there would be no innovation to appraise or learn from. Susan Clark and Dave Mattson provided invaluable support and insight, keeping us grounded, on our toes, and thinking. The Williams Internships Fund and Berkley Conservation Scholars Fund provided the means for traveling to and working in Alberta for three months. We would also like to thank the Northern Rockies Conservation Cooperative, Parks Canada, Murray Rutherford, Felicity Edwards, Doug Clark, Emily Chamberlain, Dave Cherney, Colleen Campbell, Peter Otis, Catherine Picard, Ona Ferguson, Jutta Kölhi, and Kirsten Leong for their support; Denise Casey for her discernment and review of our work; our respective families for their love and encouragement; and the GBDG interviewees for their kindness, patience, time, thoughts, and experience.

1. The Grizzly Bear Dialogue Group was also referred to as the "IPS Group," after the interdisciplinary problem-solving method it employed. Here, we use the former term, acknowledging that some participants were accustomed more to one name than the other.

2. The stakeholder process that superseded the GBDG has been referred to by several names, including the Bow Valley Parkway Area Planning Committee, Bow Valley Parkway Area Advisory Group, and the Bow Valley Parkway Area Advisory Committee. Here, we use the term Bow Valley Parkway Advisory Group.

REFERENCES

Alberta Grizzly Bear Recovery Team. 2005. *Draft Alberta Grizzly Bear Recovery Plan 2005–2010. Alberta Species at Risk Recovery Plan No. X.* Edmonton: Alberta Sustainable Resource Development, Fish and Wildlife Division.

————. 2008. *Alberta Grizzly Bear Recovery Plan 2008–2013. Alberta Species at Risk Recovery Plan No. 15*. Edmonton: Alberta Sustainable Resource Development, Fish and Wildlife Division.

Beierle, T., and J. Cayford. 2002. *Democracy in Practice: Public Participation in Environmental Decisions*. Washington, DC: Resources for the Future.

Brunner, R. D., and T. W. Clark. 1997. "A Practice-Based Approach to Ecosystem Management." *Conservation Biology* 11:48–58.

Brunner, R. D., T. A. Steelman, L. Coe-Juell, C. M. Cromley, C. M. Edwards, and D. W. Tucker. 2005. *Adaptive Governance: Integrating Science, Policy, and Decision Making*. New York: Columbia University Press.

Burgess, P. M., and L. L. Slonaker. 1978. *The Decision Seminar: A Strategy for Problem-Solving*. Columbus: Mershon Center of the Ohio State University.

Canada. 2000. *Canada National Parks Act. Statutes of Canada, C. 32*. Canada National Parks Act.

Chamberlain, E. C. 2006. "Perspectives on Grizzly Bear Management in Banff National Park and the Bow River Watershed, Alberta: A Q Methodology Study." MRM Planning Project No. 394. Burnaby, BC: School of Resource and Environmental Management, Simon Fraser University.

Chamberlain, E. C., and M. B. Rutherford. 2005. "Perspectives on Grizzly Bear Conservation in the Banff-Bow Valley: Views of Problems and Solutions." Unpublished Report. Burnaby, BC: School of Resource and Environmental Management, Simon Fraser University (on file with authors).

Chartier, A. 2004. *Parks Canada—Corporate Intelligence Bulletin 2004*. Gatineau, Quebec: Parks Canada.

Cherney, D. N., A. C. Bond, and S. G. Clark. 2009. "Understanding Patterns of Human Interactions and Decision Making: An Initial Map of Podocarpus National Park, Ecuador." *Journal of Sustainable Forestry* 28:694–711.

City of Calgary. 2007. "Civic Census Overview." Accessed December 28, 2010, available from http://www.calgary.ca/DocGallery/BU/cityclerks/city.pdf.

Clark, S. G. 2002. *The Policy Process: A Practical Guide for Natural Resource Professionals*. New Haven, CT: Yale University Press.

————. 2008. *Ensuring Greater Yellowstone's Future: Choices for Leaders and Citizens*. New Haven, CT: Yale University Press.

Clark, T. W., and R. D. Brunner. 2002. "Making Partnerships Work in Endangered Species Conservation: An Introduction to the Decision Process." *Endangered Species Update* 19:74.

Clark, T. W., and A. Gillesberg. 2001. "Lessons from Wolf Restoration in Greater Yellowstone." In *Wolves and Human Communities: Biology, Politics, and Ethics*, edited by V. A. Sharpe et al., 135–50. Washington, DC: Island Press.

Clark, T. W., and M. B. Rutherford. 2005. "The Institutional System of Wildlife Management: Making It More Effective." In *Coexisting with Large Carnivores: Lessons from Greater Yellowstone*, edited by T. W. Clark, M. B. Rutherford, and D. Casey, 211–53. Washington, DC: Island Press.

Conley, A., and M. A. Moote. 2003. "Evaluating Collaborative Natural Resource Management." *Society and Natural Resources* 16:371–86.

Dearden, P. 2008. "Progress and Problems in Canada's Protected Areas: Overview of Progress, Chronic Issues and Emerging Challenges in the Early 21st Century." Paper commissioned for Canadian Parks for Tomorrow: 40th Anniversary Conference, May 8–11. Calgary: University of Calgary.

Dickmeyer, L. 2009. "The Banff-Bow Valley: Environmental Conflict, Wildlife Management and Movement." Master's Thesis. University of British Columbia.

Edwards, F., and M. Gibeau. n.d. "I Wouldn't Start from Here: Some Lessons from the Real World of Wildlife Conservation." Unpublished Presentation (on file with authors).

Fazey, I., J. A. Fazey, and D. M. A. Fazey. 2005. "Learning More Effectively from Experience." *Ecology and Society* 10:4.

Ferguson, O. 2010. "Situation, Stakeholder or Conflict Assessment." Unpublished Article. Overview presented October 18, at Yale School of Forestry and Environmental Studies, Negotiation and Conflict Resolution course (on file with authors).

Garshelis, D. L., M. L. Gibeau, and S. Herrero. 2005. "Grizzly Bear Demographics in and around Banff National Park and Kananaskis Country, Alberta." *Journal of Wildlife Management* 69:277–97.

Gibeau, M. L. 2000. "A Conservation Biology Approach to Management of Grizzly Bears in Banff National Park, Alberta." PhD diss. University of Calgary. Available from http://www.canadianrockies.net/grizzly/mikes_thesis.html.

Gibeau, M. L., and S. Stevens. 2005. "Study Areas." In *Biology, Demography, Ecology, and Management of Grizzly Bears in and around Banff National Park and Kananaskis Country: The Final Report of the Eastern Slopes Grizzly Bear Project*, edited by S. Herrero, 11–16. Calgary: University of Calgary.

Gunton, T. I., J. C. Day, and P. W. Williams. 2003. "The Role of Collaborative Planning in Environmental Management: The North American Experience." *Environments* 31:1–4.

Jager, E., and A. Sanche. 2010. "Setting the Stage for Visitor Experiences in Canada's National Heritage Places." *George Wright Forum* 27:180–90.

Jamal, T., and M. Eyre. 2003. "Legitimation Struggles in National Park Spaces: The Banff Bow Valley Round Table." *Journal of Environmental Planning and Management* 46:417–41.

Kellert, S. R., J. N. Mehta, S. A. Ebbin, and L. L. Lichtenfeld. 2000. "Community Natural Resource Management: Promise, Rhetoric, and Reality." *Society and Natural Resources* 13:705–15.

Kleiman, D. G., R. P. Reading, B. J. Miller, T. W. Clark, J. M. Scott, J. Robinson, R. L. Wallace, R. Cabin, and F. Felleman. 2000. "The Importance of Improving Evaluation in Conservation." *Conservation Biology* 14:356–65.

Kölhi, J. K. 2010. "Stakeholder Views on Grizzly Bear Management in the Banff-Bow Valley: A Before-After Q-Methodology Study." Master's Thesis. Simon Fraser University.

Lasswell, H. D. 1971. "The Continuing Decision Seminar as a Technique of Instruction." *Policy Sciences* 2:43–57.

Parks Canada. 2008. *Banff National Park of Canada State of the Park Report*. Accessed August 19, 2010, available from http://www.pc.gc.ca/~/media/pn-np/ab/banff /plan/pdfs/REP_SPR_e.ashx.

———. 2010. *Banff National Park of Canada: Management Plan 2010*. Resource Document. Accessed August 10, 2010, available from http://www.pc.gc.ca/pn-np/ab /banff/~/media/pn-np/ab/banff/pdfs/2010/Banff-Management-Plan-EN-2010 .ashx.

Pimbert, M. P., and J. N. Pretty. 1997. "Parks, People, and Professionals: Putting "Participation" into Protected-Area Management." In *Social Change and Conservation: Environmental Politics and Impacts of National Parks and Protected Areas*, edited by K. B. Ghimire and M. P. Pimbert, 297–330. London: Earthscan Publications.

Primm, S., and K. Murray. 2005. "Grizzly Bear Recovery: Living with Success?" In *Coexisting with Large Carnivores: Lessons from Greater Yellowstone*, edited by T. W. Clark, M. B. Rutherford, and D. Casey, 99–137. Washington, DC: Island Press.

Proctor, M. 2005. "East Slopes Grizzly Bear Fragmentation Based on Genetic Analyses." In *Biology, Demography, Ecology, and Management of Grizzly Bears in and around Banff National Park and Kananaskis Country: The Final Report of the Eastern Slopes Grizzly Bear Project*, edited by S. Herrero, 126–32. Calgary: University of Calgary. Available from http://www.canadianrockies.net/wp-content /uploads/2009/03/Complete_ESGBP_FinalReport2005.pdf.

Richie, L., and J. D. Oppenheimer. 2012. "Social Process in Grizzly Bear Management: Lessons for Collaborative Governance and Natural Resource Policy." *Policy Sciences* 45:265–91.

Rutherford, M. B., M. L. Gibeau, S. G. Clark, and E. C. Chamberlain. 2009. "Interdisciplinary Problem-Solving Workshops for Grizzly Bear Conservation in Banff National Park, Canada." *Policy Sciences* 42:163–87.

Salafsky, N., R. Margoluis, K. H. Redford, and J. G. Robinson. 2002. "Improving the Practice of Conservation: A Conceptual Framework and Research Agenda for Conservation Science." *Conservation Biology* 16:1469–79.

Spyke, N. P. 1999. "Public Participation in Environmental Decision Making at the New Millennium: Structuring New Spheres of Public Influence." *Boston College Environmental Affairs Law Review* 26:263–68.

Statistics Canada. Census Snapshot—Immigration in Canada: A Portrait of the Foreign-Born Population, 2006 Census." Available from http://www.statcan.gc.ca /pub/11-008-x/2008001/article/10556-eng.htm.

Zartman, W. 1977. "Negotiation as a Joint Decision-Making Process." *Journal of Conflict Resolution* 21:619–38.

8

Large Carnivore Conservation
A Perspective on Constitutive
Decision Making and Options

SUSAN G. CLARK, DAVID N. CHERNEY, AND DOUGLAS CLARK

Introduction

The overall goal of large carnivore conservation is to reverse fragmentation and declines in populations and secure viable, sustainable populations and habitats in ways that enjoy enduring public support. Various forces and factors in society militate against this goal, evident in the behavior of individuals (e.g., poaching, indifference), institutions (e.g., forestry policies that encourage habitat destruction), and society (e.g., materialism and consumerism). Carnivores mean different things, both positive and negative, to people, often issuing from deep feelings, and they are put to the service of many worldviews, agendas, and practices, of which conservation is just one (Primm 2000; Charon 2007). If anything is clear, it is that large carnivore conservation is truly a human endeavor. This leads to conflict, not only between people and carnivores, but also among people and their institutions. Beyond what to do with these animals, the conflict is largely about whose interests should be considered and whose interests will dominate. Consequently, as Boitani, Asa, and Moehrenschlager (2004, 159) noted, large carnivore conservation "needs to be viewed as a complex system of decision making that requires an interdisciplinary approach." Despite this good advice, conservation is rarely viewed as such and interdisciplinarity is rarely used (e.g., Smith 2011, who looks structurally and conventionally, not interdisciplinarily and functionally, at wildlife conservation).

In this chapter we examine the constitutive decision-making process that is the functional and cultural foun-

[251

dation for the complex system for carnivore conservation. We elaborate this important concept, which may be new to some readers, distinguish it from ordinary decision making, and examine large carnivore conservation using a constitutive lens—focused on cases from Canada, the United States, and elsewhere—to diagnose widespread, recurring weaknesses, especially at the institutional level. Finally, we draw lessons and offer recommendations for more successful conservation worldwide.

Our data for this analysis are drawn from our collective field experience with large carnivores, the diverse wildlife conservation projects in which we have participated, and our observation of the work of others. We also draw on a broad base of literature (e.g., T. Clark, Paquet, and Curlee 1996; Gittleman et al. 2001; Ray et al. 2005; T. Clark, Rutherford, and Casey 2005; MacDonald and Sillero-Zubiri 2004; MacDonald and Loveridge 2010; Reading et al. 2010; Hornocker and Negri 2010). Although most of the data we discuss in this chapter come from the North American West, there are similar dynamics at play in other settings in the world and for other species. We use a functional analysis that looks at people's values, worldviews, and interactions, based on Lasswell and McDougal (1992), who detail a genuinely interdisciplinary analytic framework for guiding problem-oriented, contextual, and multimethod inquiry. We apply widely recognized standards of "meta" analysis for sound decision-making processes (see Lasswell 1971; chapter 1, this volume). We also use procedural, substantive, and pragmatic tests of the common interest, as described by Brunner et al. (2002, 2005) and Steelman and DuMond (2009).

Our own roles in large carnivore conservation over forty-five years have been diverse. We have carried out fieldwork, led conservation teams, conducted appraisals, written management plans, been first responder to acute carnivore-human conflicts, trapped, relocated, and even killed "problem" bears in national parks, consulted, researched, and published papers and books, and taught workshops, courses, and seminars on carnivores, their ecosystems, and the professional, organizational, and policy dimensions involved. Our collective wildlife experience encompasses dozens of species and cases, including sensitive, rare, threatened, and endangered species, and a wide variety of carnivores, including polar bears, grizzly bears, black bears, wolves, cougars, and meso- and small carnivores. We have worked in more than ten countries. We seek to enhance wildlife conservation for the common good via open, inclusive, comprehensive, targeted, fair, cooperative, and effective means.

Context and Problem Definition

To introduce some of the issues associated with decision making in carnivore conservation, we begin by describing grizzly bear conservation in Alberta. This leads to a discussion of context, the social and ecological conditions in which decision processes operate. Most conservation work focuses on the ecological context, while largely ignoring social factors that are critical to decision-making processes and outcomes. We highlight major variables of the social context, including the concepts of society, culture, myth, and symbols.

THE ALBERTA GRIZZLY BEAR CASE

Grizzly bear conservation in Alberta illustrates the importance of context and the constitutive process at play behind the daily operation of ordinary decision-making processes that many people focus on. Despite an enormous investment of resources, numerous researchers have shown that both the constitutive and ordinary decision-making processes in this case are weak and have failed to produce conservation outcomes as promised (e.g., Stenhouse, Boyce, and Boulanger 2003; Boulanger and Stenhouse 2009; Clark and Slocombe 2010; Festa-Bianchet 2010; Gailus 2010). Consequently, the bear's future in the province is genuinely in doubt outside of Alberta's national parks (Nielsen, Stenhouse, and Boyce 2006).

An analysis of this case was presented by D. Clark, Gailus, and Gibeau (2010) at the 2010 annual meeting of the International Society for Conservation Biology (ISCB) in Edmonton, Alberta. The authors appraised current provincial policies in a contextual and multimethod way. They found both the constitutive and ordinary decision processes wanting and provided evidence that many aspects of the constitutive process actually work against conserving grizzlies over the long term. Since it is this flawed constitutive process that establishes and governs ordinary, everyday decision processes, government actions in the realm of ordinary decision making are unlikely to affect the prevailing and dominant constitutive trends and conditions. Yet there is little doubt that the public in Alberta wants to conserve grizzly bears. In a large-scale mail survey, Stumpf-Allen, McFarlane, and Watson (2004) documented widespread public pro-ecological views and support for grizzly conservation across all geographic, demographic, and economic sectors. There is a clear disconnect here.

A recent report by Festa-Bianchet (2010) provides an estimate of the current status of Alberta grizzly bears and a review of relevant research. The small population of fewer than 700 individuals is being fragmented in places into smaller, isolated units. Some population segments are declining as well. Mortality remains high, especially near roads. This report follows up on a formal, authoritative 2002 recommendation by Alberta's Endangered Species Conservation Committee that Alberta grizzlies be listed as threatened. Although the province made no decision about listing at that time, a provincial recovery team was organized in 2003 and in 2004 submitted its recommended recovery plan. The province approved a revised recovery plan in 2007, which contains a comprehensive set of recommendations, many of which approximate standards of sound decision making (see chapter 1, this volume).

Unfortunately, implementation of the approved plan has been slow. D. Clark, Gailus, and Gibeau (2010) noted that the few recommendations that have been implemented are far from the most critical ones. They appear, rather, to be actions that are the least costly in political terms to government and business. One example is the decision in June 2010 to list grizzlies as a provincially threatened species. Occurring eight years after the original recommendation to list, this is largely a symbolic gesture since Alberta's non-legislated species-at-risk policy does not compel any action that has not already been taken. Consequently, as Clark and Slocombe (2010, 9) concluded in another review of the Alberta case: "There is no reason to expect that the decision to list grizzlies will materially change the decision process for conserving this species in the province. The challenge of grizzly bear conservation is not so much a lack of scientific information, formal agreements, or even authoritative policy prescriptions: it is their effective implementation on the ground."

D. Clark, Gailus, and Gibeau (2010) argued that the Alberta government is committed, first and foremost, to goals and programs that favor natural resource extraction over conservation. These goals are deeply institutionalized today through a constitutive process (e.g., reflecting worldviews, political systems) geared toward extraction of the province's abundant nonrenewable resources, including coal, oil, and natural gas. The scale of this industry is noteworthy: the mining and oil and gas extraction sectors employ more than 136,000 people (15 percent of the province's workforce), and in 2008 and 2009 oil and gas royalties yielded $12.2 billion, more than 30 percent of the provincial government's revenues (Alberta Energy 2010). D. Clark et al. also noted that Alberta's political system is dominated by a well-established political elite and a set of institutions that favor natural resource extraction over other concerns and support minimal governmental interference with industry. This ideology and its political elite perpetuate a "power and wealth outlook" that has long indulged people who support reelection of politically elected elites (see Leadbetter 1984; Harrison and Laxer 1995; Freeden 2003). The authors also present data on voting, financing, and other factors that support their conclusions.

The authors of the ISCB paper concluded that the constitutive decision-making process responsible for grizzly bear conservation in Alberta: (1) serves the special interests of the resource extraction industries and political views of the established elite, (2) is unresponsive to widespread public support for conservation, (3) is unresponsive to scientific information that calls for new policy and programs that are at odds with the dominant policy that favors natural resource extraction industries, and (4) fosters apathy and distrust of government in an otherwise trusting public (see Hawley 2007). In sum, Alberta's constitutive decision process, supported by ordinary decision process, produces outcomes and effects that favor special interests over common interests, expedient interests over principled interests, and exclusive interests over inclusive interests.

One major weakness that contributes to these constitutive failings is an inadequate appraisal function. The quality of the present constitutive process has never been adequately taken on as an object of inquiry, at least not with any enduring effect on the process itself. For example, there is no arena available for citizens publicly to challenge governmental policy decisions on substantive grounds, and there is no assurance that meritorious challenges will lead to better decision making or better outcomes. In addition, the other decision-making functions, namely, intelligence, debate, planning, implementation, monitoring, and termination, also show weaknesses, according to these authors. Not only is the process failing to meet recognized standards for sound natural resource management and policy, it is not operating in the common interest (see Brunner et al. 2002, 2005; Steelman and DuMond 2009). Thus, in this case, as in many others, the constitutive decision-making process is not farsighted, open, or effective enough to conserve large carnivores over the long term in ways that serve common interests (e.g., MacDonald and Sillero-Zubiri 2004; MacDonald and Loveridge 2010; Ascher 2009). These weaknesses in the constitutive process reflect the context of bear conservation, natural resource extraction, and many other cultural and institutional forces and factors.

CONTEXTUAL CONCERNS

The Alberta case demonstrates that large carnivore conservation is a societal activity that determines not only how animals will be treated, but also which groups will get their way and how decision making itself will proceed. The central questions are: How should decision making be organized and carried out in large carnivore management? Who should be included, when, how, and why? Functionally, whose interests should be served? These questions inspire others, as Parsons (1995, 56–57) notes: How is knowledge defined? Whose knowledge is to be used? What kind of knowledge does it claim to be? How does knowledge come to be produced, propagated and used, abused, or ignored (see S. Clark 2002; Ascher, Steelman, and Healy 2010)? At heart, these questions are about how power and other values will and should be allocated, used, enjoyed, and justified in society. Different people holding different worldviews or ideologies offer different answers (see Freeden 2003).

Despite its importance, social context can be an invisible or underappreciated set of variables, especially for science-oriented, technical experts and advocates who focus on biology and technical matters (Torgerson 1985; Honadale 1999; Wilkinson, Clark, and Burch 2007). Many of the challenges in carnivore conservation, especially among conflicting people and interests, stem from a lack of realistic attention to contextual variables. Often people have only an intuitive, dim grasp of social context, even though they may be aware of social codes, symbols, and conventions that are at play in society at large and in themselves, variables that may become more apparent in times of crisis.

SOCIETY AND CULTURE AS CONTEXT

Society and culture orient us to our place in the world, providing rules for how to live with one another and how to solve problems. They allow us to make meaning in our lives, tell us what life is all about, and structure our lives and our scope of thought and action (Eagleton 2007). The broad societal and cultural context shapes our perspectives and actions and pervades all aspects of our individual and collective lives, whether we are conscious of it or not (see Bell 2011). Thus in order to understand decision making, in large carnivore conservation as in any other arena, it is important to know something about society and culture, what they are, how they operate, and how individuals reflect them in their perspectives and practices (see Berger 1963). We can learn about these variables through, for example, sociological studies of people's attitudes toward carnivores, nature, and other subjects (e.g., Kellert 1997; Bath 2000). Many scholars have discussed these subjects (e.g., Berger 1963; Robinson 2005; Nie 2003), but we need much deeper knowledge than what has been provided to date by the "human dimensions" school of research on wildlife and natural resource management (Manfredo, Decker, and Duda 1998; Bonar 2007). We need to delve into the foundations of culture, its embedded worldviews, and how these are institutionalized in society and play out in both ordinary and constitutive decision making.

The development of culture helps groups of people ensure their own survival over time. Individuals take on culture as they grow up, through socialization, education, and daily living. Each individual typically belongs to a dominant culture and at the same time many subcultures. Conservation groups, government entities, local associations, and other collectives all constitute such subcultures, each of which produces its own outlook and outward expressions of who the members are, what they value, and how they view situations. According to Charon (2007, 162), "culture means the 'consensus' of the group, the agreements, goals, knowledge, understandings, shared language and values that emerge together." According to Freeden (2003, 32), the way of comprehending culture, or the "set of ideas, beliefs, opinions, and values" that dominate, is to observe the recurring patterns held by significant groups (e.g., elites, the most powerful) and how the status quo or proposed changes are justified by them. Culture determines our view of reality, and through it we come to a shared or "inter-subjective" reality that provides coherency for coordinating society and making and implementing policy (Berger and Luckmann 1987). Culture is also an essential source of the rules that individuals use to control themselves, interact with others, and make individual and collective decisions. These rules largely play out symbolically in decision making. Growing up inside a culture makes it very difficult to see it in oneself, but easy to see in others, especially in those with dramatically different worldviews.

In the end, no matter where we grew up or how we feel about large carnivores, we are all part of an "ongoing social interaction that is characterized by cooperation among actors and that creates a shared culture" (Charon 2007, 167). Much of the conflict over large carnivores is among people with different worldviews (i.e., subcultural perspectives). As such, culture should be an important object of study, especially for those interested in problems of public policy and decision making. When people are thrown together in the arena of carnivore conservation, there are bound to be differences, conflicts, and disagreements. Yet sound decision making can often transcend these differences to find shared common interests, and in fact good decision process is the means through which people harmonize different subcultural outlooks and manage political conflict.

MYTH AS CONTEXT

Many claims and counterclaims circulate about carnivores. The dispute over the effect of wolves on elk in the northern Rockies is a prime example. Jackson Hole outfitter B. J. Hill, for instance, argues that "if we don't do something with this wolf in the next year or two or three, sport hunting is going to be gone in the West" (Neary 2010, 6). In contrast, Jackson Hole resident Jim Stanford (2007) claims that "elk are getting along just fine despite the presence of wolves in northwest Wyoming." With such drastically opposing claims, most people assume that one side must be right and that the other perspective is unsubstantiated by fact. But the scientific answer is much more complex. Between 1984 and 2009 elk numbers increased by 5 percent in Idaho, by 35 percent in Wyoming, and by 66 percent in Montana (Rocky Mountain Elk Foundation 2009). Thus, there is little room for debate that overall elk numbers have increased in the presence of wolves, which were reintroduced to the Yellowstone ecosystem in 1995. At the same time, several specific herds have declined substantially since the reintroduction of wolves. Between 1995 and 2008 the Northern Yellowstone elk herd dropped by 68 percent, the Gallatin Canyon elk herd by 67 percent, and the Madison Firehole elk herd by 78 percent (Allen 2010). While a number of factors are likely responsible for these declines, including weather and climate, wolves have certainly played a significant role (Middleton 2012).

We can interpret this information to mean that while overall elk numbers are up, wolves have perhaps hit certain herds hard. However, neither side in the debate takes this reasonable position. Both see their opponents' beliefs as lies or fictions and their own beliefs as truth based on empiricism and scientific evidence. The reality is that both sides adhere to "social facts" that are partially constructed in experience and in science and partially created in worldviews, a phenomenon that Sarewitz (2004) calls the "excess of objectivity." He argues that so much scientific information is available today that it is easy for contrasting political perspectives to justify their worldviews by selectively choosing legitimate facts to support their claims. Often people do not even realize they are engaged in "cherry picking" their facts. In fact, political facts come from the social construction of reality; they are an expression of an individual's worldview, society, and subculture. The facts are real and undeniable to those who hold them, even though counterclaimants dismiss them as misinformed, misguided, or malevolent. Melding such conflicting myths and subcultures can be difficult. However, having noted this, the world is not just a subjective construction.

The concept of myth or ideology is helpful in understanding people's worldviews, paradigms, or belief systems (see Freeden 2003). Myth is used here not as a traditional, legendary, or imaginary story, but in its cultural anthropological and political sense as a mechanism for creating meaning out of our experiences and relationships (Flores and Clark 2001). A myth is a belief system that may or may not be true, that is, it may or may not comport with empirical, verifiable observations about the world. Because myths are a tool that people use to make meaning, understanding myths is essential to understanding why people behave as they do. Myths are powerful. They function as worldviews and animate people to action. They also lay out a blueprint about how power is to be used and justified. For example, when Europeans arrived in North America, they brought with them many practices that changed the face of the continent, such as grazing, logging, and other technologies, all supported by a myth of human domination over nature. Directly and indirectly, many of these beliefs and practices affected grizzly bears, mountain lions, and wolves. In the last few centuries, North Americans and their practices have irrevocably altered much of the continent's biology, and in the West large carnivores have been greatly reduced or extirpated. Thus it is people's belief systems that drive carnivore management because, just as people cannot be separated from the land, they cannot be separated from their myths of themselves and the land. Consequently, decision making is about addressing both grizzly bear conservation and people's beliefs, worldviews, and ideology. Creating a sustainable future depends on changing or adjusting currently unsustainable perspectives and damaging practices to be more realistic and adaptive. This is one function of sound decision making.

The job of public policy making then is to meld people's views to find common ground, not for each side to try to vanquish the other as though it were an epic struggle of right over wrong or good over evil. Melding perspectives to develop a shared view of goals, problems, and solutions is a practical problem of democracy and governance, to be addressed rationally and responsibly through sound decision-making processes. It also requires sound leaders who understand these dynamics and are skilled in managing them constructively.

SYMBOLS AS CONTEXT

"What does happen to a species that has the mixed fortune of becoming a potent political symbol?" asks Primm (2000, 6). Sometimes the good of the species takes a back seat to control of the symbol by some interest group. Primm (6) says that "grandstanding politicians use grizzlies as a potent symbol to rally conservative voters: they equate grizzlies with loss of resource-extraction jobs, reduced recreational access, and interference from 'outside interests.'" For some people, depending on their outlook and subculture, "grizzlies symbolize a loss of power, a loss of opportunity to practice skills (e.g., logging), and a loss of esteemed position (e.g., bread-winners, providers) in our culture" (6). Used symbolically this way, large carnivores become a target for people's discontent and anxieties stemming from other issues in their lives (see Yalom 1980). On the other hand, for some people grizzlies symbolize "many positive things—wilderness, freedom, and America's bounteous natural heritage" (6).

Symbols, especially symbolic words, but also symbolic deeds, are central to human existence and figure prominently in all constitutive and ordinary decision-making processes. Social scientists have studied how symbols are used in society. A symbol can represent or stand in for a myth. It can serve as a kind of shorthand for the whole belief system of a culture. To make this relationship clearer, all myths rest on a *doctrine* or ideology that is largely taken as a matter of faith and is essentially invisible to the people who hold it. How the doctrine is applied in everyday life is guided by a *formula*. For example, in the United States the doctrine about how people will live, treat each other, and go about the governance of community life is expressed in part in the preamble to the Constitution (which uses broad terms such as justice, tranquility, welfare, and liberty), while the Constitution itself functions as the formula for American democracy. In turn, the doctrine and formula are symbolized and reinforced in daily life by the American flag, folk heroes like Lewis and Clark, the pledge of allegiance, or the uniforms of the armed forces. Our lives are symbolically very rich even though few people fully appreciate the pervasiveness or the social and political functions of the symbols (Charon 2007; Freeden 2003). The underlying myths in any situation can be partly discerned by looking for the symbols people use. To find symbols, we have to analyze ordinary objects and the words and behavior of people for signs of hidden cultural interest. Many social scientists, such as Lasswell and Kaplan (1950), offer principles for analyzing symbolic politics, which provides great insight into people's behavior in decision making.

Large carnivores are powerful cultural and political symbols, with the feelings they elicit varying contextually with the animals, circumstances, individuals, societies, and subcultures involved. Scientists, environmentalists, wise-use advocates, conservation-minded people, and popular culture all interpret nature according to their own mythologies, which in turn embody their value outlooks and interests. It is the job of decision makers to sift through the multitude of cultural constructs to find a community's common interests (and to recognize that they themselves also have myths and subjectivities and may decide in favor of their own myths and the groups that they serve) (Primm and Clark 1996).

Not everyone sees the symbolic content of carnivores, and rarely does anyone treat symbols or myths, much less society and culture, as socially constructed objects for formal observation and study. As a result, these factors are almost never taken into account in conservation. Nevertheless, having knowledge of symbols and myths can bring about better understanding, cooperation, and decision making. Symbols are the building blocks of society, and that is the reason we need to be cognizant of them and their potency and not be taken in by their careless, shortsighted, or strategic use.

In sum, context is a critical variable in all human activity, and consequently, it is important to appreciate the social matrix and its effects on conservation, science, management, and policy, on the individuals and groups involved (including the public, agencies, and positional leaders), and on the constitutive process through which we organize and make decisions. Coexistence with large carnivores can come about only by changing decision making, but change ultimately rests with adjusting society, culture, myths, and symbol systems.

The Decision-Making Process for Large Carnivores

In looking at the decision-making process in some detail, we again delve into the social sciences, including jurisprudence (the philosophy of law) for guidance. We examine both the constitutive process that sets broad guidelines about how decisions will be made in society and who will be involved, as well as the ordinary process that is carried out in myriad everyday practices at individual, family, and community levels. We end by revisiting and expanding on our definition of the problem of large carnivore conservation.

Conserving large carnivores depends on making and carrying out sound decisions. But this is not simply one person making one decision at a particular time and place. Rather, it is a whole process of numerous small and large activities that together produce the processes and outcomes we see in our lives. The decision process for any one issue may take years and may blend into other decision processes. For example, modern grizzly bear conservation in the Greater Yellowstone Ecosystem has been underway since at least 1959, when John and Frank Craighead first conducted field studies in the region (Craighead 1979). Knowing about this entire process, its context, its constituent activities, how it is structured and how it functions, and how to assess it and what standards apply opens up ways to work more effectively toward practical improvements in the science, formulation, debate, planning, choosing, executing, monitoring, and succession of decisions.

ORDINARY DECISION MAKING

We distinguish between constitutive and ordinary decision-making processes, although they are interrelated (see McDougal, Lasswell, and Reisman 1981; Lasswell and McDougal 1992; fig. 8.1). *Ordinary decision making* in large carnivore conservation can most easily be conceived as conventional, everyday science and management, typically focusing on biophysical matters, such as determining hunting seasons, estimating population size, or removing problem animals. Ordinary decision making is about the more or less routine tasks of management, often structured in the form of governmental (and nongovernmental) programs (see Rosen and Bath 2009; GYE Interagency Grizzly Bear Committee, www.IGBConline.org). Established, stable rules, generally accepted by participants, exist to carry out this kind of decision making. Lasswell (1971, 77, 98–111) notes that ordinary decision making is the deliberation about substantive policy choices (e.g., should a problem grizzly bear be removed) within a given structure or context (e.g., established agency procedures and operations as in the Alberta bear case). Ordinary process is about what people want and how policies are viewed as legitimate. In the Alberta example, people evidently want both economic prosperity and wild grizzly bears, and bear managers have considerable faith that scientific management will allow them to have both (Clark and Slocombe 2010). Chapters 2 through 7 in this book mostly focus on ordinary decision making.

Three cases, all from Wyoming from the summer of 2010, exemplify ordinary decision making. First, the presence of a small grizzly bear in an outfitter camp in the Upper Green River on the Bridger-Teton National Forest led to closure of the area to overnight camping (Hatch 2010a). Second, in Grand Teton National Park a grizzly was hit and killed by a driver, who said, "it happened very quickly." According to the report, "the bear darted out onto the road. . . . It was one of those unfortunate situations where they had no time to respond" (Hatch 2010b, 2). Officials said the incident serves to warn drivers to watch for wildlife on the roads. Third, a horse was euthanized because of injuries sustained from a wolf attack (or perhaps merely because the horse was frightened by the wolf), a management action consistent with established rules and procedures. US Forest Service officials met with the US Fish and Wildlife Service to determine this course of action (Staff Reporters/ *Jackson Hole Daily* 2010a). These three examples are about more or less rou-

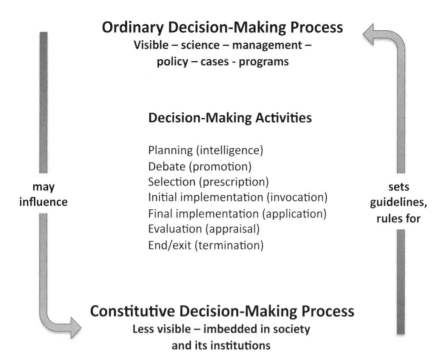

Ordinary Decision-Making Process
Visible – science – management –
policy – cases - programs

Decision-Making Activities

Planning (intelligence)
Debate (promotion)
Selection (prescription)
Initial implementation (invocation)
Final implementation (application)
Evaluation (appraisal)
End/exit (termination)

may influence

sets guidelines, rules for

Constitutive Decision-Making Process
Less visible – imbedded in society
and its institutions

Figure 8.1. The constitutive decision process and its relationship to ordinary process in society. Both processes have the same decision-making activities.

tine, everyday kinds of problems. The patterns of authority and control, the allocation of competence, and the structure of decision making were well established in advance and supported by the public. There was little ambiguity or "decision space" available for government to respond.

CONSTITUTIVE DECISION MAKING

The *constitutive decision-making process*, on the other hand, determines how ordinary decision making is conducted. It is a higher-order process that comprises deliberations and choices regarding how policy and other decisions should be made and, by implication, who ought to be involved in choosing. In other words, it is deciding about how we will make decisions. McDougal, Lasswell, and Reisman (1981) give a detailed, functional description of the constitutive process, reminding us that, "in many cases, it is difficult to pinpoint a deliberate and conscious effort to decide about decision making" (99). The outcomes of constitutive process lay out how the common interest will be clarified and secured, including whether it will be enforced or not (McDougal et al. 1981). Constitutive policy making goes well beyond the ordinary decision process and everyday management operations to focus on how institutions function, how they are structured, the analytic techniques used and those overlooked, who and what is ignored or discounted, and what procedures and people ought to be involved and what they should be doing. These issues have real-world, hard consequences for the people involved and for large carnivores and their habitats, as the Alberta case amply illustrates. Constitutive process is about institutions in society—organizations, such as game and fish departments, but also other entrenched social practices and relationships (see Clark and Rutherford 2005; fig. 8.1).

Constitutive decision making sets the rules for ordinary decision making, which are institutionalized in society. Deeply rooted in culture, it is a manifestation of how we view ourselves as a society (flowing from our myths), how we see our relationship to nature, and how we think it best to organize and make decisions for ourselves. It typically focuses on preserving the dominant mythic view in the body politic, the status quo, and the structure and legitimacy of government and its programs, defending against all challengers. It allocates and regulates competency (i.e., some people are deemed competent and valid participants, while others are excluded from the process for various reasons), and it determines how decisions will be made over resources that are seen as economically important or otherwise socially relevant at any particular time. The constitutive process determines what analytic techniques, procedures, and participants ought to be used in decision making, and it establishes and maintains authority and control within society (Lasswell 1971, 77, 98–111).

The constitutive process is implicitly recognized as playing a fundamental role in the dynamics of carnivore conservation, as noted, for example, by Karanth and Chellam (2009, 2): "In our rapidly changing world the conservation of carnivores now stands at a crossroads. How effectively scientists, conservationists, governments and society at large will study, understand, collaborate and move forward to meet the ecological needs of these mammals will determine where and how many species and populations will survive." Even though the constitutive process may be only dimly acknowledged in the world of carnivore conservation, it has been well conceptualized in other policy areas and used in the study of natural resource management (e.g., McDougal, Lasswell, and Vlasic 1963; Arsanjani 1981; Johnston 1987; Sahurie 1992) and other arenas (e.g., McDougal, Lasswell, and Chen 1980; McDougal, Lasswell, and Reisman 1981).

The workings of constitutive process are much harder to discern than those of ordinary decision making. For example, although it is relatively easy to go to Alberta and find people, offices, and activities that make up the ordinary process of managing grizzly bears, it is not possible to locate the constitutive process in the same way. Again, three recent Wyoming news items illustrate the process. First, Wyoming Governor Dave Freudenthal was reported to be "steamed over how federal courts have handled wildlife questions affecting his state" (Staff Reporters/*Jackson Hole Daily* 2010b, 7). In this case of federalism versus states' rights, the governor challenged the existing constitutive rule about authority and control (vested in the federal courts) and called for a change in favor of the state, or more technically, he demanded a new arrangement in the "application" phase of the constitutive process and a different "allocation of competence." In another example, a federal judge in Helena, Montana, held hearings to decide how the federal Endangered Species Act would be interpreted and whether the federal government could use political considerations in making decisions about how and where a species could be listed under the act. The decision was contested. At issue, substantively at least, was the US Fish and Wildlife Service's decision in April 2009 to consider the northern Rocky Mountain gray wolf as a distinct population segment (Staff Reporters/*Jackson Hole Daily* 2010b, 1). Again, claimants were calling for changes in the relationship of authority and control and the allocation of competence. Third, Cromley (2000, 174) examined a highly contentious incident in which a grizzly in Grand Teton National Park was killed by management authorities. She reported that the incident "indicates that expectations about whose activities—grizzlies' or humans—should take precedence in which case and about who should resolve such conflicts remains unclear." In other words, the constitutive rules were not clear. Answers in this last case are still being worked out in practice more than ten years later.

STANDARDS AND WEAKNESSES

Generally speaking, decision making, both ordinary and constitutive, has seven interrelated functions: (1) intelligence gathering and planning; (2) debate and promotion about the nature and status of the problems and potential solutions; (3) deciding on a plan to solve the problems, in other words, setting new rules; (4) invoking the new rules in specific cases; (5) applying, that is, interpreting and enforcing, the rules through administrative and judicial activities; (6) appraising progress or lack of it; and finally (7) terminating the rules when they are no longer appropriate or necessary (S. Clark 2002). These functions do not necessarily occur in any particular order; several functions may be ongoing at any given time in any single program. The participants involved often differ in the different functions, and they tend to focus selectively and incompletely only on parts of the overall process. Knowing about these functions allows us to see them in action in carnivore conservation.

Both ordinary and constitutive decision processes should strive to meet widely accepted standards of quality and effectiveness. These standards can guide appraisal of the adequacy of decision making, learning, and improvements. Criticisms, challenges, debates, and other public or scientific "appraisals" that appear in the newspapers and elsewhere commonly speak to one or more of these standards (e.g., Mattson and Chambers 2009). Among the overall standards for good decision making are factuality, comprehensiveness, timeliness, economy and technical efficiency, honesty as well as a reputation for honesty, skill of official personnel involved, effectiveness of impact, flexibility, and responsibility (see S. Clark 2002; a series of questions about the adequacy of decision making as it pertains to large carnivores appears in T. Clark et al. 2001). Close examination of any large carnivore case shows how hard it is to meet these standards. Nevertheless, programs that are perceived as successful do in fact meet many or most of these standards. Since large carnivore conservation everywhere is fraught with destructive conflict, it is clear that these standards are often not being met to the satisfaction of the majority of the participants. Improving conservation, management, and policy in effect means upgrading performance to approximate these standards in practice.

Constitutive decision process should meet five additional standards (Lasswell 1971). First, the process should ensure that common interests prevail over special interests. The goal is to discover and work toward common interest outcomes, and the procedural solution is to include and involve all people, or legitimate representatives of all people, throughout the process. This is democracy. Second, the process should give precedence to high-priority rather than low-priority common interests. Third, the process should protect both inclusive and exclusive common interests. Common interests are inclusive if they genuinely and significantly affect all people. If decisions affect one or a few participants or only part of a community, then they are exclusive. Fourth, when there are conflicting exclusive interests, preference should be given to the participants whose value position is most substantially involved. And fifth, values of sufficient magnitude (e.g., power, money, knowledge, skill, and so on) should be allocated to provide an "authoritative signature" to the process and to enable authority to be controlling. Ordinary and constitutive decision-making processes are the means by which society distinguishes common interests from special interests and strives to achieve the former.

Common weaknesses also exist in both kinds of decision making. In planning (intelligence), for example, simplistic problem definitions with a single objective may dominate. Sometimes single organizations take over the process. Erroneous assumptions may be made that the program will benefit everyone. Long-term consequences may be discounted. Some hard-won advice exists to avoid these and other common pitfalls. Analysts recommend that decision processes bring in diverse experts and citizens early on and minimize their biases. Recommendations include not letting expert opinions overrule the preferences of those affected by decisions, not exaggerating the expected benefits or underestimating or ignoring potential costs, and not focusing on easily measured or easily understood data, but instead asking the hard questions. T. Clark (1997, 171) offers additional recommendations. These standards, along with ways of avoiding common pitfalls, provide a practical guide to appraise decision making and make it as effective as possible.

Defining the Problem

Looking at large carnivore conservation from the vantage point of decision making in the context of society, myth, and symbol, we see a complex intermixed set of three problems, concerning first the constitutive decision-making process, second the ordinary process, and third the individuals involved in decision making. There are other problems, but this grouping is a good way to begin a systematic conversation about the challenges in large carnivore conservation.

CONSTITUTIVE PROBLEMS

The existing constitutive process is not a sufficient vehicle for successful, sustainable coexistence between large carnivores and people (e.g., see Craighead 1979; Primm 1996; Primm and Clark 1996; T. Wilkinson 1998; Gailus 2010; Clark and Rutherford 2005; MacDonald and Sillero-Zubiri 2004; MacDonald and Loveridge 2010). One aspect of this problem is that the goals of large carnivore conservation—and nature conservation and sustainability in general—are typically subordinated to other societal goals, such as natural resource exploitation, national security, energy production, the economy, or solving various social ills (see Bell 2011). Conservation will likely remain a relatively low priority goal in western North America and in most countries with growing resource scarcity. These circumstances make it hard to gain broad public and elite support for carnivore conservation, gain the resources required for success, create a process of inclusive competency that serves common interests, and find the skilled constitutive leadership required.

Another issue is that the constitutive decision process in large carnivores conservation does not measure up to widely recognized standards for sound decision making. Too often, processes fall short of being inclusive, timely, open, constructive (i.e., helping the situation), and ameliorative (supporting the common interest). Such standards are meant to encourage those who are formally responsible for the process and who should be held accountable to provide information to anyone who is interested and to be rational, integrative, and comprehensive, among other standards. Identifiable standards point the way to best practices for a quality decision process and help prevent the weaknesses that can compound and result in conflicted, ineffective programs (Hohl and Clark 2010).

A third aspect of weak constitutive process, again seen in the Alberta case, is that it seems to allow for special interests to prevail over common interests, or at least significantly so. The overall process seems to favor exclusive interests over inclusive interests of the wider population, who favor more bear conservation. Also, because conflicting special interests are at play, the process does not involve the larger public very directly. These features can damage public order and foster distrust in government, loss of faith in leadership, alienation from government, and defiant acts such as poaching wildlife or illegally using bear habitat.

Finally, the constitutive process should allocate adequate resources, including money, knowledge, skill, and other basic values, appropriate to the scope of the problem. In carnivore conservation this rarely occurs. The provincial government did not do this in Alberta. But money alone is not enough. Leaders must also lend their "authoritative signature" and "control intent" to conserve bears, which they have not done (S. Clark 2002, 64). In addition, the official process must develop a reputation of honesty, integrity, and service to common interests. The constitutive decision-making process has generally been problematic for large carnivore conservation and for other sustainability issues as well (e.g., climate change; Homer-Dixon 2009; Brunner and Lynch 2009).

ORDINARY PROBLEMS

The second major problem concerns the ordinary decision-making process, which is also suboptimal in case after case. Primarily, this is because ordinary decision processes flow from the constitutive process, which has built-in tendencies, interests, and structures that it serves and perpetuates. Here we touch on three aspects of this problem.

One issue is that conservation efforts are frequently reduced to programs that are managed according to fixed bureaucratic rules, roles, and regulations (Primm and Clark 1996). Government agencies typically translate new conservation goals through their existing programs, modes of operation, and technical capabilities, resulting in a variety of professional, organizational, and political interpretations. Consequently, a program's structure, operations, and outcomes may diverge significantly from the original policy aims, which leaves programs highly vulnerable to both external and internal interests that are more concerned with issues other than species conservation (see T. Clark 1997 for a detailed case). Because of this kind of analytic error, a common problem in government, the conservation task is reduced to a program that is narrowly bounded in terms of decision process and that does not meet many of the standards of good decision making. As a result, available science may not be used in a timely way, causing delays and intelligence failures. Programs may focus on addressing immediate concerns and overlook longer and system-wide concerns in implementation (see Parsons 1995). Once institutionalized, these programs preclude consideration of alternatives. Many ensuing errors, which might have been preventable, occur.

A second major weakness of ordinary decision processes is that most organizations, both governmental and nongovernmental, are not well prepared to learn from their own performance. Although it is important to have knowledge that can be used *in* decision making (i.e., content knowledge), it is also essential to have knowledge *of* the adequacy of decision making itself (i.e., process knowledge). Active learning, which rests on accurate feedback or appraisal, is key. Yet in most conservation programs learning seldom goes beyond "single-loop," or shallow learning, to "double-loop," or deep learning, about underlying operating assumptions (see Argryis and Schön 1978; Clark and Westrum 1989; Senge 2006; Clark and Cragun 1994; Harrison and Stokes 1992). Few large carnivore conservation efforts show systematic, explicit appraisal and learning, as illustrated by the cases in chapters 2 through 7, although there are some clear exceptions where active learning does take place (see chapter 6). For strong appraisal and active learning to occur, each decision function needs skilled leaders and staff who are open and capable of quick, effective learning.

Another flaw is suboptimal performance, which stems from the fact that it is difficult to harmonize decision-making functions into an effective overall whole. Among the many reasons for this is the fact that the different functions are typically carried out by different individuals, interest groups, agencies, and organizations, each of which confronts different analytic and political challenges. People in different roles often conflict with one another over values, perspectives, goals, legitimacy, resources, and other important matters. In decision making, individuals and groups are limited in their capacity to grapple with the complexity involved, including their own, often narrow, partisan interests and positions, knowledge holdings, and analytic skills. Leaders seldom step forward to harmonize diverse people and functions into an effective whole, too often assuming that there are bureaucratic structures or organizational arrangements that will facilitate an effective program.

INDIVIDUAL PROBLEMS

The third set of concerns relates to the participating individuals and groups, including government scientists and administrators, environmentalists, citizens, and others. People find it seriously challenging to orient themselves realistically in the flow of events in ways that give them traction on the process, especially in the constitutive process. Carnivore conservation is typically complex, lengthy, technical, and political. Cases may run on for years, even decades, as issues fluctuate, people come and go, and the interest levels of organizations and the public ebb. Leaders also vary in the skills and knowledge they possess to aid the process. Available resources are seldom adequate. All these factors conspire to make it difficult for individuals to participate constructively to find common ground. Here we touch on four aspects of this problem.

One issue is that that the decision-making process is often conceived solely as a technical, scientific problem rather than a problem of communities making decisions based on values. Underappreciation of these dimensions makes it more difficult for individuals to find ways to be valued, helpful, and constructive. Existing processes seldom produce the one best or optimal outcome; often they produce only temporary, partial, and emotionally unsatisfying outcomes for many participants. People are sometimes denied meaningful roles in the process and are left feeling devalued, disrespected, and unsatisfied, feelings that can become hardened into inflexible positions. This makes it harder to find common ground with others who hold different perspectives.

A second aspect is that individuals sometimes fail—or even refuse—to see the context within which the task of carnivore conservation takes place. With an acontextual focus, the animals are viewed as conservation "objects," and the full societal, cultural, mythic, and symbolic context is discounted. For example, now that the official estimate of the number of grizzly bears in Greater Yellowstone is approximately 600, three times bigger than the minimum estimate a few decades ago, some people assert that grizzly conservation has been successful and that the survival of bears is assured (Haroldson 2012). But this focus on a single number at a certain point in time diverts people's attention away from other key elements. An overly narrow biophysical focus precludes a sophisticated examination of other dimensions of conservation, such as the adequacy of decision making, leadership behavior, problem-solving effectiveness, the appropriateness of organizational arrangements, and numerous other concerns—all key determinants in wildlife conservation, especially in the longer term.

A third aspect is that many individuals rely on positivistic science, scientific management, and agency bureaucracies to conserve carnivores (see chapter 9). Others distrust science, feeling that it is biased and used against them. As we have seen, however, large carnivore conservation is a pragmatic problem of governance, rather than a technical scientific problem. Good scientific knowledge is essential, but over-relying on science to achieve carnivore conservation is a misplacement of people's expectations and confidence. "Scientizing" politics devalues the use of science and information in decision making by using science as a surrogate to argue value disputes about carnivores, rather than addressing those value disputes directly (Pielke 2007). Scientific facts thus become associated with the political perspective of a particular interest group, which makes it easy for opposing groups to dismiss that information as political posturing. This only worsens the mismatch between what scientists know and what the public believes. All sides selectively pick facts that support their political perspectives (Sarewitz 2004). But finding common ground is less about deciding who is right or wrong than it is about melding and harmonizing differing perspectives and finding shared interests and understanding. This is a political problem, not a scientific problem. Scientizing politics turns the promotional function of the decision process, which should focus on harmonizing different perspectives, into a debate over whose science is right (S. Clark 2002). The biophysical dimension in large carnivore conservation is clearly challenging: typically, there is less information available on the biology and habitat of the species than is expected, wanted, or ideal. And the traditional approach is to call for more science, biological expertise, and data, assuming that more data will resolve uncertainty, conflict, and upgrade decision making. This was certainly the operative assumption in the Alberta case (D. Clark and Slocombe 2010).

But no amount of scientific data can resolve the value conflict between conserving grizzly bears and developing oil and gas in Alberta. Nor can science resolve the conflict over wolves in Wyoming, where some people want to expand the protected population and others want to restrict or remove the population. Science cannot resolve such disputes over people's differing goals and values (Pielke 2007).

A final aspect that makes it difficult for individuals to find common ground is that "we have lost the language, vocabulary, and ability to talk about the common interest" (Steelman and DuMond 2009, 396). Knowledge about social and decision processes, how they work, how to participate in them responsibly, and how to ensure that they are carried out according to high standards, is seldom central to professional education today, not to mention training for leaders or for citizens (see Sihls 1958; Dryzek 1990; Brunner et al. 2005; Kelman 2006). Individuals who are knowledgeable, skilled in democratic process, and located influentially are key to successful outcomes. Also, very importantly, institutions must encourage and permit this kind of individual action.

In sum, there are many challenges to successful large carnivore conservation that fall into the categories of constitutive, ordinary, and individual problems. As Brunner et al. (2002, 16) noted, "people are often trapped in complex structures of management that cause conflict more than facilitating integration or balancing of interests." Some of the traps, according to Hammond, Keeney, and Raiffa (1999, 189), include working on the wrong problem, failing to identify key objectives, and failing to develop a range of good, creative alternatives. People also find it difficult to orient constructively within the existing process and context; too often they are forced into picking a side that feels right. The usual focus on the biophysical dimensions of large carnivores is absolutely necessary for successful conservation, but it is not sufficient for sound decision-making processes that serve common interests. Sophisticated, knowledgeable attention also needs to be directed toward both the constitutive and ordinary decision processes.

Moving Forward: Lessons and Recommendations

The problem definition we have posed suggests that large carnivore conservation shows a host of weaknesses, many related to the decision-making process and participation in it (see appendix). Four sets of strategies—conceptual, practical, individual, and general—can lead to rapid learning and change in order to overcome these weaknesses. The best way forward is for people jointly to build more constructive processes for knowledge, politics, and interests to interact. Foresight and wisdom are needed. As Etheredge (2005, 297) said, wisdom in public policy is "good judgment about important matters, especially embodying a genuine commitment to the well being of individuals and society as a whole."

CONCEPTUAL RECOMMENDATIONS

Our first set of strategic recommendations is about perspective, that is, getting clear on the concept and the goals of conservation problem solving. "Concerned people must conceive of conservation as a process of human decision making, upgrade this process to achieve better outcomes, and by this means change the human practices that threaten carnivores," notes T. Clark et al. (2001, 223). From a constitutive perspective, the goal is to engage and integrate diverse interests and build a working conception of constitutive as well as ordinary decision processes and how to improve them. Commonly, problems arise in meeting this goal. For example, Hammond, Keeney, and Raiffa (1999) lay out common pitfalls or traps in decision processes, including working on the wrong problem, failing to identify key objectives, failing to develop a range of good, creative alternatives, overlooking crucial consequences of the alternatives, giving inadequate thought to trade-offs, disregarding uncertainty, failing to account for risk tolerance, and failing to plan ahead when decisions are linked over time. Underlying emotional and cognitive dynamics are at the heart of many of these problems (see Kahneman 2011). Lasswell (1971, 96) also identifies a number of functional or structural weaknesses that afflict all the decision functions: a reputation for dishonesty, technical inefficiency, weak loyalty and skill of official personnel, limited effectiveness of impact, rigidity and lack of realism in adjustment to change, and failure to be deliberate and responsible. These are based on widely supported standards for sound decision making.

Improving constitutive dynamics is not, of course, an easy or a straightforward task. Among other things, it requires upgrading ordinary decision process, as ordinary cases can sometimes have major direct and indirect impacts on constitutive process. We recommend learning from the work on grizzly bear management in the Banff National Park region of Canada (Rutherford et al. 2009; Richie, Oppenheimer, and Clark 2012), from the case described by Wilson and colleagues in chapter 6, and from the experiences of many others worldwide.

It is clear that large carnivore conservation is interest-driven, that is, "political." Interests figure prominently in the gathering, processing, dissemination, and use of all knowledge, scientific or otherwise, and hiding special interests under the guise of "science," "objectivity," "peer review," "expertise," "government," or "authority" does not play well in practice (Hawley 2007).

We recommend finding practical ways to expand collaboration among diverse people and interests in the production and use of knowledge in both constitutive and ordinary decision-making processes. We need constitutive changes in the ways that formal science and bureaucracies go about their business, in order to give standing (i.e., allocate competence) to a more complete set of actors and a wider range of approaches. We need to move away from thinking that focuses on narrow rules or investigations that at best might provide answers to parts of problems and move toward practice-based approaches that better solve whole problems in context.

Accommodating people's interests through adequate process is central to effective conservation. The task is to accommodate all valid and appropriate interests consistent with the overriding goal of establishing and maintaining a resource regime or system of public order in which environmental sustainability and human rights can flourish. The practical task for participants, given this goal, is to distinguish common interests from special interests. We recommend using the common interest tests provided in Steelman and DuMond (2009). These include procedural, substantive, and practical issues. It also requires being familiar with the standards for sound constitutive and ordinary decision making.

The fundamental task for people and society is to achieve and maintain a continuous balance of inclusive, partially inclusive, and exclusive interests that serve the long-term aggregate common interest. All participants should reject special interest claims, that is, those that benefit specific individuals or groups at the expense of the common good. Different common interests require protection in different degrees depending on the context of the case at hand. It is vital in the decision process to accommodate all these interests. The content of these terms is not fixed and must be negotiated in practice in real time in each case. These processes should include diverse scientific, local, practical, and value-based knowledge.

PRACTICAL RECOMMENDATIONS

Our next set of strategic recommendations is about making the decision-making process function soundly in day-to-day operations. Lasswell (1971, 96) listed these and discussed weaknesses in decision process with the hope that, once people were aware of them, they could work to avoid them. They include a reputation for dishonesty, technical inefficiency, weak loyalty and skill of official personnel, limited effectiveness of impact, rigidity, a lack of realism in adjustment to change, and not being deliberate and responsible. These cut across the entire spectrum of decision-making processes. Avoiding all of these requires strong skills of observation and management. We urge participants, especially leaders, to learn and practice these skills as they pertain to decision making. The reality of everyday work is such that finding and securing an effective decision-making process is very difficult. The mix of concerns that comes together in decision making—animal and symbol, science and politics, management and policy, and content and process— makes it difficult to find common ground. Integrating these elements into an enduring solution is perhaps the major challenge in large carnivore conservation today. This argues for paying special attention to the decision-making process as a subject of explicit study, observation, and management. Practically, the decision-making process is a means for reconciling or at least managing conflicts––rational, political, and moral—not always an easy task. We recommend learning from experience, harvesting hard-won lessons about what works and what does not, distilling these into usable guides, and disseminating them as "best practices" as widely as possible.

It's important to know about common and foreseeable process problems, such as lack of goal clarity, intelligence failures and delays, or over-control of decision making, so that means can be set up and managed to avoid them before they arise. Ideally, decision-making processes should be inclusive of valid and appropriate interests (not exclusionary), principled (high-minded and practical, and not just expedient), farsighted, and focused on common interests. They should also be open to rapid learning and self-correcting so that they can solve the real problems at hand and build a reputation for honesty.

We recommend paying close attention to matters of both content and procedure in conservation. Principles of content relate to every detail of the process of resource management, claims, and constitutive decision. Principles of procedure tell how relevant content is to be examined. Thus, we suggest using procedural rationality in all aspects of conservation, including the five intellectual tasks of problem orientation: clarification of goals, examination of trends in past decision making, analysis of conditions under which the trends occurred, projection of probabilities into the future, and invention, evaluation, and selection of options to address problems (S. Clark 2002). Experience shows that decision making that incorporates these matters tends to be successful in finding and securing enduring solutions (Brunner et al. 2002, 2005).

RECOMMENDATIONS FOR INDIVIDUALS

Our last set of strategic recommendations is for individuals to become better informed about both constitutive and ordinary processes of which they are a part. It is all too easy for people to get caught up in their own interests, in self-reinforcing symbolic politics, and in the stereotypical roles society offers. Any one of these can be counterproductive to achieving the common interest goals of conservation, sustainability, and democracy.

We would like to see individual attention refocus on the decision-making process itself instead of emphasizing just technical, biological matters. We need to move beyond today's dominant approach whereby, as described by Li (2007, 7), "questions that are rendered technical are simultaneously rendered nonpolitical. For the most part, experts tasked with improvement exclude the structure of political-economic relations from their diagnosis and prescriptions. They focus more on the capabilities of the [technical aspect] than on the practices through which one social group impoverishes another. . . . Experts are trained to frame problems in technical terms. This is their job." It is also a major part of the problem. Sound decision making attends to a number of contextual variables: (1) representative and responsible participation; (2) the perspectives of participants, including their common and special interests, their identities, expectations and demands, and establishment of and access to decision-making arenas; (3) the situations in which they interact, including spatial/temporal dimensions and other characteristics; (4) their values, from which they derive their bases of power, and their authority and control over these; (5) the kinds of strategies and range of activities they pursue as they seek to achieve their goals through decision processes; (6) the outcomes or aggregate consequences of decision making; and (7) the effects of their decision making in terms of change in "value-institutions" and the diffusion or restriction of innovations.

Individuals would benefit from developing a skill set that includes critical thinking, observation, management, and relevant technical matters as a basis for more effective participation in the decision process. *Skill* refers to the ability to use knowledge, respect, influence, and other values strategically and practically to address problems of significance. *Critical thinking* involves carrying out the five intellectual tasks of problem orientation described earlier. *Observation* focuses on what to observe and research. *Management* focuses on points of leverage (i.e., people, their perspectives, the situation, values at stake and in play, and the strategies people are using). Observation and management skills target social and decision processes as well as management of self. Mapping social and decision processes in any policy problem will typically suggest productive lines of investigation and alert participants to aspects of the political landscape that can help in solving and managing problems. Professionals who are the most successful demonstrate these practical skills.

We recommend that participants clarify their standpoints (see S. Clark 2002, 111–26), that is, their understanding of their location and role in both ordinary and constitutive process. Being realistically and contextually grounded is key to choosing how to position oneself to make the maximum contribution. Standpoint clarification requires self-assessment, experience, and insight. Some people assume a standpoint based on the notion that carnivore conservation is solely a scientific, technical, and management effort. It is easy to get lost in convention and uncritically accept prevailing doctrines and formulas, for example, but to do so is to misread the context, the process, and the problem at hand.

GENERAL RECOMMENDATIONS

We end with some additional general recommendations to improve large carnivore conservation through more effective decision making and more secure and responsible participation.

First, knowledgeable and skilled leaders are needed at all levels within a project or conservation arena. As we have discussed, societies have built-in mythic and institutional structures and dynamics that set up how people make meaning for themselves, regulate the use of resources and the allocation of competency, and tell people about their "proper" relationship to nature (which may or may not be adaptive or sustainable). The diversity of these myths and sub-myths, combined with individual subjective differences in people, fuel conflict in resource conservation and management in ordinary and constitutive processes. Traditional administrative leadership supplemented by facilitators or conflict managers will simply not be adequate to manage these complexities. Leaders who are well versed in observing and managing social and decision process and skilled in critical thinking are needed.

Finally, we recommend using adaptive governance to address large carnivore conservation (see Brunner et al. 2005 and Steelman 2010 for description). Some participants in large carnivore conservation are already moving in this direction. See chapter 6, for example, or the recent proliferation of community-scale "Bear Smart" programs in western Canada, in which local people are self-organizing to address bear-human conflicts (Ciarniello 1996). Such programs are often socially and technologically innovative, responsive to unfolding situations, and, interestingly, organized by local people, often women.

Conclusions

Despite the many difficulties of decision making in this pressing environmental and sustainability arena, there are many positive trends and conditions that promise more successful large carnivore conservation in the future. The public worldwide is becoming increasingly aware of the imperative to move toward sustainability, however understood, and making demands for better conservation, better decision making, and better governance. Sadly, this is because environmental problems, including biodiversity loss and extinction rates, are growing in impact and visibility and motivating people everywhere to press for upgraded decision making and better policy and programs. In addition, advocates and diverse groups are organizing for more influence in decision making in many places and sectors of society, and many professional communities are seeking greater contextual understanding of the problems they face and exploring interdisciplinary methods to address problems. As a consequence, decision-making processes are becoming more open, inclusive, factual, and effective. Finally, leaders are moving to the fore in many cases to guide resolution of problems in ways that serve the common good. In all this, large carnivores serve as powerful and popular symbols of the shift toward sustainability.

We need to continue, then, to focus on the adequacy of decision-making processes and seek ways to upgrade them for better outcomes. Existing decision process in carnivore conservation does not yet provide a sound prescription that facilitates active learning and improvement at either constitutive or ordinary levels. Not only is there limited learning within individual cases, there seems to be very little learning across cases. Fortunately, there are many opportunities to upgrade decision making, and these are being field tested in wide-ranging cases, such as those described in this volume. These cases need to be linked together to maximize learning and diffusion of lessons. We now have a rich database from which to harvest lessons for more adaptive governance and better conservation outcomes.

Our recommendations can help harmonize competing approaches to carnivore conservation. Our recommendations seek to minimize socioeconomic disruptions that result from conflicting interests, policies, and institutions; they do not support the creation of one policy or institutional practice for all processes of resource production, distribution, and use in social processes and decision making. Inclusive prescriptions and competent, skilled leaders and decision makers are essential. In the end, we believe that we must fully adopt conservation and sustainability goals and, in doing so, accommodate diverse citizens and interests in large carnivore conservation through more democratic involvement.

Appendix
Common Recurring Weaknesses in Constitutive Decision Making

Common recurring weaknesses in constitutive decision making in large carnivore conservation along with recommended improvements. Many of these problems also apply to the ordinary decision-making process.

INTELLIGENCE (*PLANNING*)
Avoid

Simplistic problem definitions with single dominant objective.

Domination by a single organization or interest.

Assuming that the program will benefit everyone.

Discounting long-term consequences.

Ignoring the need to balance local and central (usually government) participation.

Do

Explore multiple definitions and objectives.

Include several diverse organizations.

Assume instead that the program will not benefit everyone.

Consider long-term consequences.

Be inclusive and involve multiple participants.

PROMOTION (*DEBATE*)
Avoid

Using in-house experts only.

Exclusivity in debating problems and alternatives.

Letting experts override the views of those affected by decisions.

Misappreciation of the benefits and discounting possible costs.

Focusing only on easily measured or easily understood data.

Failure to ask hard questions.

Do

Bring in diverse experts to debate the problem definition and alternatives.

Bring in outsiders right away and minimize their biases.

Balance local and expert views with views of outsiders.

Explore benefits and costs carefully and realistically.

Account for "soft" (qualitative) variables with hard consequences.

Ask hard questions, especially about operating assumptions.

PRESCRIPTION (*RULES*)

Avoid

Failure to coordinate with other participants.

Choosing the alternative most congenial to established views.

Rivals blocking the common interest decision.

Over-control of all aspects of choosing.

Do

Coordinate using "mutual adjustment" to the extent possible.

Explore all options, especially from the viewpoint of others involved.

Ensure that common interests are served.

Be flexible, make adjustments to changing context.

INVOCATION (*ENFORCEMENT*)

Avoid

Inadequate coordination and fair play between strong and weak participants.

Letting others block achievement of common interest goals.

Over-control.

Misspecification or misallocation of the initial application of the goals, rules, contingencies, sanctions, and resources.

Do

Be fair, use genuine coordinators to equalize information and power relations.

Protect against those who seek other goals, perhaps special, exclusive, or expedient interest goals.

Be flexible, learn, and adapt.

Be consistent but attentive to contingencies, expect to modify initial invocation efforts.

APPLICATION (*IMPLEMENTATION*)

Avoid

Delaying or ignoring resolution of conflict and claims.

Biased adjudication.

"Benefit leakage."

Ignoring organizational arrangements.

Succumbing to intelligence failures and delays.

Do

Adjudicate disputes quickly.

Ensure the benefits promised by prescription and recognize the good
   faith that others have invested in the process.

Attend carefully to appropriate organizational arrangements and use
   nonbureaucratic structures.

Seek intelligence from diverse sources, including from those who favor
   other policies.

APPRAISAL (*EVALUATION*)

Avoid

Ignoring criticism, even constructive criticism.

Inadequate monitoring and evaluation.

Inflexibility in sticking with original goals, problem definitions, and
   alternatives.

Defensiveness and ignoring the views of others.

Looking only at performance indicators that are favorable.

Do

Seek outside appraisals and consider these seriously.

Attend carefully to appraisal activities, appraise all seven decision
   functions, including appraisal itself.

Stay flexible, adapt, and learn at individual, organizational, and policy
   levels.

Guard against hubris and arrogance.

Attend to diverse performance indicators, including those offered by
   detractors; indicators should include biophysical, social process, and
   decision-making measures.

TERMINATION (*ENDING*)

Avoid

Dealing unfairly with people.

Assuming that termination means that everyone will see it as a success.

Ignoring the effects of termination on people.

Do

Deal fairly, honestly, and with integrity with all people.

Attend to details, be empirical, talk with those affected, and help them transition.

Be sensitive to the effects on people of ending programs and policies, especially on those who participated in good faith.

NOTE

We want to acknowledge the great many people who have helped us over the last forty years, too many to name individually. We also want to thank our employing organizations for their support of our work, especially the Northern Rockies Conservation Cooperative, executive director Jason Wilmot, and the board of directors. Finally, we thank Denise Casey and two anonymous readers for their critical reviews.

REFERENCES

Alberta Energy. 2010. *About Us*. Accessed December 2, 2010. Available from http://www.energy.gov.ab.ca/About_Us/984.asp.

Allen, M. D. 2010. "Open Letter to Mike Leahy and Kirk Robinson." April 8, 1–4. Available from http://chronicleoutdoors.com/wp-content/uploads/2010/04/RMEF-Letter-To-DOW.pdf.

Argryis, C., and D. A. Schön. 1978. *Organizational Learning*. Cambridge: Blackwell.

Arsanjani, M. H. 1981. *International Regulation of Internal Resources: A Study in Law and Policy*. Charlottesville: University Press of Virginia.

Ascher, W. 2009. *Bringing in the Future: Strategies for Farsightedness and Sustainability in Developing Countries*. Chicago: University of Chicago Press.

Ascher, W., T. A. Steelman, and R. Healy. 2010. *Knowledge and Environmental Politics: Re-Imagining the Boundaries of Science and Politics*. Cambridge, MA: MIT Press.

Bath, A. J. 2000. *Human Dimensions in Wolf Management in Savoie and Des Alpes Maritimes, France. Large Carnivore Initiative for Europe*. Gland: Working Group of IUCN's Specialist Survival Commission.

Bell, M. M. 2011. *An Invitation to Environmental Sociology*. 4th ed. Los Angeles: Pine Forge Press.

Berger, P. L. 1963. *Invitation to Sociology*. Garden City, NJ: Doubleday.

Berger, P. L., and T. Luckmann. 1987. *Social Construction of Reality: A Treatise in the Sociology of Knowledge*. New York: Penguin.

Boitani, L., C. S. Asa, and A. Moehrenschlager. 2004. "Tools for Canid Conservation." In *The Biology and Conservation of Wild Canids*, edited by D. W. MacDonald and C. Sillero-Zubiri, 143–59. Oxford: Oxford University Press.

Bonar, S. A. 2007. *The Conservation Professional's Guide to Working with People*. Washington, DC: Island Press.

Boulanger, J., and G. Stenhouse. 2009. *Demography of Alberta Grizzly Bears: 1999–2009*. Nelson, BC: Integrated Ecological Research.

Brunner, R. D., C. H. Colburn, C. M. Cromley, R. A. Klein, and E. A. Olson. 2002.

Finding Common Ground: Governance and Natural Resources in the American West. New Haven, CT: Yale University Press.

Brunner, R. D., and A. Lynch. 2009. *Adaptive Governance and Climate Change*. Boston: American Meteorological Society.

Brunner, R. D., T. A. Steelman, L. Coe-Juell, C. M. Cromley, C. M. Edwards, and D. W. Tucker. 2005. *Adaptive Governance: Integrating Science, Policy, and Decision Making*. New York: Columbia University Press.

Casey, D., and T. W. Clark. 1996. *Tales of the Wolf: Fifty-One Stories of Wolf Encounters in the Wild*. Moose, WY: Homestead Press.

Charon, J. M. 2007. *Symbolic Interactionism: An Introduction, an Interpretation, an Integration*. Upper Saddle River, NJ: Prentice Hall.

Ciarniello, L. 1996. *Human-Bear Conflict in British Columbia: Draft Discussion Paper*. Victoria, BC: Ministry of Environment, Lands, and Parks.

Clark, D. A., J. Gailus, and M. Gibeau. 2010. "Grizzly Bear Conservation in Alberta: An Explanatory Hypothesis for a Wicked Problem." Paper presented at 24th International Congress for Conservation Biology, Edmonton, AB, July 6.

Clark, D. A., and D. S. Slocombe. 2010. "Grizzly Bear Conservation in the Foothills Model Forest: Appraisal of a Collaborative Ecosystem Management Effort." *Policy Sciences* 44:1–11.

Clark, S. G. 1997. *Averting Extinction: Reconstructing Endangered Species Recovery*. New Haven, CT: Yale University Press.

———. 2000. "Interdisciplinary Problem Solving in Endangered Species Conservation: The Yellowstone Grizzly Bear Case." In *Endangered Animals: A Reference Guide to Conflicting Issues*, edited by R. P. Reading and B. J. Miller, 285–301. Westport, CT: Greenwood Press.

———. 2002. *The Policy Process: A Practical Guide for Natural Resource Professionals*. New Haven, CT: Yale University Press.

———. 2008. *Ensuring Greater Yellowstone's Future: Choices for Leaders and Citizens*. New Haven, CT: Yale University Press.

———. 2009. "An Informational Approach to Sustainability: 'Intelligence' in Conservation and Natural Resource Management Policy." *Journal of Sustainable Forestry* 28:636–62.

Clark, T. W., and J. Cragun. 1994. "Restoration of the Endangered Black-Footed Ferret: A 20-Year Overview." In *Restoration and Recovery of Endangered Species: Conceptual Issues, Planning and Implementation*, edited by M. L. Bowles and C. J. Whelan, 272–97. London: Cambridge University Press.

Clark, T. W., R. Crete, and J. Cada. 1989. "Designing and Managing Successful Endangered Species Programs." *Environmental Management* 13:159–70.

Clark, T. W., P. Paquet, and A. P. Curlee, eds. 1996. "Conservation of Large Carnivores in the Rocky Mountains of North America." Special Issue, *Conservation Biology* 10:936–1058.

Clark, T. W., A. P. Curlee, S. C. Minta, and P. M. Kareiva, eds. 1999. *Carnivores in Ecosystems: The Yellowstone Experience*. New Haven, CT: Yale University Press.

Clark, T. W., and A. M. Gillesberg. 2001. "Lessons from Wolf Restoration in Greater Yellowstone." In *Wolves and Human Communities: Biology, Politics, and Ethics*, edited by V. A. Sharpe, B. Norton, and S. Donnelley, 135–49. Washington, DC: Island Press.

Clark, T. W., and D. J. Mattson. 2005. "Making Carnivore Management Programs More Effective: A Guide for Decision Making." In *Coexisting with Large Carnivores: Lessons from Greater Yellowstone*, edited by T. W. Clark, M. B. Rutherford, and D. Casey, 271–76. Covelo, CA: Island Press.

Clark, T. W., D. J. Mattson, R. P. Reading, and B. J. Miller. 2001. "Interdisciplinary Problem Solving in Carnivore Conservation: An Introduction." In *Carnivore Conservation*, edited by J. Gittleman, S. M. Funk, D. MacDonald, and R. K. Wayne, 223–40. Cambridge: Cambridge University Press.

Clark, T. W., and M. B. Rutherford. 2005. "The Institutional System of Wildlife Management: Making It More Effective." In *Coexisting with Large Carnivores: Lessons from Greater Yellowstone*, edited by T. W. Clark, M. B. Rutherford, and D. Casey, 211–53. Washington, DC: Island Press.

Clark, T. W., M. B. Rutherford, and D. Casey, eds. 2005. *Coexisting with Large Carnivores: Lessons from Greater Yellowstone*. Covelo, CA: Island Press.

Clark, T. W., and R. Westrum 1989. "High Performance Teams in Wildlife Conservation: A Species Reintroduction and Recovery Example." *Environmental Management* 13:663–70.

Clayton, S., and S. Opotow. 2003. *Identity and Natural Environment: The Psychological Significance of Nature*. Cambridge, MA: MIT Press.

Craighead, F. C., Jr. 1979. *Track of the Grizzly*. Washington, DC: Sierra Club.

Craighead, J. J., J. S. Sumner, and J. A. Mitchell. 1995. *The Grizzly Bears of Yellowstone: Their Ecology in the Yellowstone Ecosystem, 1959–1992*. Washington, DC: Island Press.

Cromley, C. M. 2000. "The Killing of Grizzly Bear 209: Identifying Norms for Grizzly Bear Management." In *Foundations of Natural Resources Policy and Management*, edited by T. W. Clark, A. R. Willard, and C. M. Cromley, 173–220. New Haven, CT: Yale University Press.

Dryzek, J. S. 1990. *Discursive Democracy: Politics, Policy, and Political Science*. Cambridge: Cambridge University Press.

Eagleton, T. 2007. *The Meaning of Life: A Very Short Introduction*. Oxford: Oxford University Press.

Erhlich, P. R., and R. E. Ornstein. 2010. *Humanity on a Tightrope: Thoughts on Empathy, Family, and Big Changes for a Viable Future*. New York: Rowman and Littlefield.

Etheredge, L. S. 2005. "Wisdom in Public Policy." In *Wisdom: Psychological Perspectives*, edited by R. Sternberg and J. Jordan, 297–328. New York: Cambridge University Press.

Festa-Bianchet, M. 2010. *Status of the Grizzly Bear* (Ursus arctos) *in Alberta: Update 2010*. Alberta Wildlife Status Report No. 37. Edmonton: Alberta Sustainable Resource Development/Alberta Conservation Association.

Flores, A., and T. W. Clark. 2001. "Finding Common Ground in Biological

Conservation: Beyond the Anthropocentric vs. Biocentric Controversy." Yale School of Forestry and Environmental Studies *Bulletin Series* 105:241–52.

Freeden, M. 2003. *Ideology: A Very Short Introduction*. Oxford: Oxford University Press.

Gailus, J. 2010. *A Grizzly Challenge: Ensuring a Future for Alberta's Threatened Grizzlies*. Calgary: Alberta Wilderness Association. Accessed October 27, 2010, available from http://www.sierraclub.ca/en/action-grizzly-bear/publications/grizzly-challenge -ensuring-future-albertas-threatened-grizzlies.

Gittleman, J. L., S. M. Funk, D. W. MacDonald, and R. K. Wayne, eds. 2001. *Carnivore Conservation*. Cambridge: Cambridge University Press.

Hammond, J. S., R. L. Keeney, and H. Raiffa 1999. *Smart Choices: A Practical Guide to Making Better Decisions*. Boston: Harvard Business School Press.

Haroldson, M. A. 2012. "Assessing Trend and Estimating Population Size from Counts of Unduplicated Females." In *Yellowstone Grizzly Bear Investigations: Annual Report of the Interagency Grizzly Bear Study Team, 2011*, edited by F. T. van Manen, M. A. Haroldson, and K. West, 10–15. Bozeman, MT: US Geological Survey.

Harrison, R., and H. Stokes. 1992. *Diagnosing Organizational Culture*. San Francisco: Jossey-Bass/Pfeiffer.

Harrison, T., and G. Laxer. 1995. *The Trojan Horse: Alberta and the Future of Canada*. Montreal: Black Rose Books.

Hatch, C. 2010a. "Bear Spurs Closure of Green River Campsites." *Jackson Hole Daily*, July 10–11:3.

———. 2010b. "Grizzly Hit, Killed by Driver in Grand Teton." *Jackson Hole Daily*, June 12–13:2.

Hawley, K. 2007. *Trust: A Very Short Introduction*. Oxford: Oxford University Press.

Hohl, A., and S. G. Clark. 2010. "Best Practices: The Concept, an Assessment, and Recommendations." New Haven: Yale School of Forestry and Environmental Studies Working Paper 24:151–70.

Homer-Dixon, T. 2009. *The Great Transformation: Climate Change as Cultural Change*. Available from www.homerdixon.com.

Honadle, G. 1999. *How Context Matters: Linking Environmental Policy to People and Place*. West Hartford, CT: Kumarin Press.

Hornocker, M., and S. Negri, eds. 2010. *Cougar: Ecology and Conservation*. Chicago: University of Chicago Press.

Johnston, D. 1987. *The International Law of Fisheries: A Framework for Policy-Oriented Inquires*. New Haven, CT: New Haven Press.

Kahenman, D. 2011. *Thinking, Fast and Slow*. New York: Penguin Books.

Karanth, K. U., and R. Chellam. 2009. "Carnivore Conservation at the Crossroads." *Oryx* 43:1–2.

Kellert, S. R. 1997. *Kinship to Mastery: Biophilia in Human Evolution and Development*. Washington, DC: Island Press.

Kellert, S. R., M. Black, C. R. Rush, and A. J. Bath. 1996. "Human Culture and Large Carnivore Conservation in North America." *Conservation Biology* 10:977–90.

Kelman, H. C. 2006. "Interests, Relationships, Identities: Three Central Issues for

Individuals and Groups in Negotiating Their Social Environment." *Annual Review of Psychology* 57:1–26.

Koger, S. M., and D. D. Winter. 2010. *The Psychology of Environmental Problems*. New York: Psychology Press.

Lasswell, H. D. 1971. *A Pre-View of Policy Sciences*. New York: American Elsevier.

Lasswell, H. D., and A. Kaplan. 1950. *Power and Society: A Framework for Political Inquiry*. New Haven, CT: Yale University Press.

Lasswell, H. D., and M. S. McDougal. 1992. *Jurisprudence for a Free Society: Studies in Law, Science, and Policy*. New Haven, CT: New Haven Press.

Leadbetter, D., ed. 1984. *Essays on the Political Economy of Alberta*. Toronto: New Hogtown Press.

Li, T. M. 2007. *The Will to Improve: Governmentality, Development, and the Practice of Politics*. Durham, NC: Duke University Press.

MacDonald, D. W., and A. J. Loveridge, eds. 2010. *Biology and Conservation of Wild Felids*. Oxford: Oxford University Press.

MacDonald, D. W., and C. Sillero-Zubiri, eds. 2004. *Biology and Conservation of Wild Canids*. Oxford: Oxford University Press.

Manfredo, M. J., D. J. Decker, and M. D. Duda. 1998. "What Is the Future for Human Dimensions of Wildlife?" *Transactions of 63rd North American Wildlife and Natural Resource Conference*: 278–92.

Mangelsen, T. 2010. "Grizzly Was a Beloved Star to Many." Guest Shot. *Jackson Hole News and Guide*, May 26:5A.

Mattson, D. J., and N. Chambers. 2009. "Human-Provided Waters for Desert Wildlife: What Is the Problem?" *Policy Sciences* 42:113–36.

McDougal, M. S., H. D. Lasswell, and L. Chen. 1980. *Human Rights and World Public Order: The Basic Policies of an International Law of Human Dignity*. New Haven, CT: Yale University Press.

McDougal, M. S., H. D. Lasswell, and W. M. Reisman. 1981. "The World Constitutive Process of Authoritative Decision." In *International Law Essays: A Supplement to International Law in Contemporary Perspective*, edited by M. S. McDougal and W. M. Reisman, 191–286. New York: Foundation Press.

McDougal, M. S., H. D. Lasswell, and I. Vlasic. 1963. *Law and Public Order in Space*. New Haven, CT: Yale University Press.

Middleton, A. D. 2012. *The Influence of Large Carnivore Recovery and Summer Conditions on the Migratory Elk of Wyoming's Absaroka Mountains*. PhD Diss. University of Wyoming.

Neary, B. 2010. "Wyoming Hunters Blame Wolves for Elk." *Casper Star Tribune*, March 16. Available from http://trib.com/news/state-and-regional/article_cd15e5e3-cb55-5e79-bbcf-8578364d33da.html.

Nie, M. A. 2003. *Beyond Wolves: The Politics of Wolf Recovery and Management*. Minneapolis: University of Minnesota Press.

Nielsen, S. E., G. B. Stenhouse, and M. S. Boyce. 2006. "A Habitat-Based Framework for Grizzly Bear Conservation in Alberta." *Biological Conservation* 130:217–29.

Parsons, W. 1995. *Public Policy: An Introduction to the Theory and Practice of Policy Analysis.* Lyme, CT: Edward Elgar.

Pielke, R. A., Jr. 2007. *The Honest Broker: Making Sense of Science in Policy and Politics.* Cambridge: Cambridge University Press.

Primm, S. A. 1996. "A Pragmatic Approach to Grizzly Bear Conservation." *Conservation Biology* 10:1026–35.

———. 2000. "Real Bears, Symbol Bears, and Problem Solving." *NRCC News* 13:6–8.

Primm, S. A., and T. W. Clark. 1996. "Making Sense of the Policy Process for Carnivore Conservation." *Conservation Biology* 10:1036–45.

Ray, J. C., K. H. Redford, R. S. Steneck, and J. Berger, eds. 2005. *Large Carnivores and the Conservation of Biodiversity.* Washington, DC: Island Press.

Reading, R. P., B. J. Miller, A. L. Masching, R. Edward, and M. K. Phillips, eds. 2010. *Awakening Spirits: Wolves in the Southern Rockies.* Golden, CO: Fulcrum.

Richie, L., J. D. Oppenheimer, and S. C. Clark. 2012. "Social Process in Grizzly Bear Management: Lessons for Collaborative Governance and Natural Resource Policy." *Policy Sciences* 45:265–91.

Robinson, M. J. 2005. *Predatory Bureaucracy: The Extermination of the Wolf and the Transformation of the West.* Boulder: University of Colorado Press.

Rocky Mountain Elk Foundation (RMEF). 2009. "Elk Chart." Accessed April 27, available from http://www.rmef.org/NR/rdonlyres/706B1DE4-E6DF-45F1-A96C -2C8E6974D264/0/ElkChart.pdf.

Rosen, T., and A. Bath. 2009. "Transboundary Management of Large Carnivores in Europe: from Incident to Opportunity." *Conservation Letters* 2 (2):1–6.

Rutherford, M. B., M. L. Gibeau, S. G. Clark, and E. C. Chamberlain. 2009. "Interdisciplinary Problem Solving Workshops for Grizzly Bear Conservation in Banff National Park, Canada." *Policy Sciences* 42:163–88.

Sahurie, E. J. 1992. *The International Law of Antarctica.* New Haven, CT: New Haven Press.

Sarewitz, D. 2004. "How Science Makes Environmental Controversies Worse." *Environmental Science and Policy* 7:385–403.

Senge, P. 2006. *The Fifth Discipline: The Art and Practice of the Learning Organization.* New York: Doubleday.

Shils, E. 1958. "Ideology and Civility." *Sewanee Review* 66:450–80.

Smith, C. A. 2011. "The Role of State Wildlife Professionals under the Public Trust Doctrine." *Journal of Wildlife Management* 75:1539–42.

Staff Reporters 2010a. "Wyo. Gov Wants Wildlife Issues Litigated in State." *Jackson Hole Daily*, June 7:7.

———. 2010b. "Judge to Hear Wolf Case." *Jackson Hole Daily*, June 15:1.

Stanford, J. 2007. "Wolves Not Slaughtering Elk." *Jackson Hole Underground*. Accessed February 27, available from http://www.newwest.net/index.php/city/article /shocker_elk_not_being_slaughtered_by_wolves/C101/L101/ Stanford, J. 2007. Wolves Not Slaughtering Elk.

Steelman, T. A. 2010. *Implementing Innovation: Fostering Enduring Change in Environmental and Natural Resource Governance*. Cambridge, MA: MIT Press.

Steelman, T., and M. E. DuMond. 2009. "Serving the Common Interest in US Forest Policy: A Case Study of the Healthy Forest Restoration Act." *Environmental Management* 43:396–410.

Stenhouse, G. B., M. Boyce, and J. Boulanger. 2003. *Report on Alberta Grizzly Bear Assessment of Allocation*. Hinton, AB: Foothills Model Forest.

Stumpf-Allen, R. C. G., B. L. McFarlane, and D. O. Watson. 2004. *Managing for Grizzly Bears in the Foothills Model Forest: A Survey of Local and Edmonton Residents*. Hinton, AB: Foothills Model Forest.

Torgerson, D. 1985. "Contextual Orientation in Policy Analysis: The Contribution of Harold D. Lasswell." *Policy Sciences* 18:241–61.

Wilkinson, K., S. G. Clark, and W. R. Burch. 2007. "Other Voices, Other Ways, Better Practices: Bridging Local and Professional Environmental Knowledge." *Yale School of Forestry and Environmental Studies Report* 14:1–57.

Wilkinson, L., and the Alberta Grizzly Bear Recovery Team. 2008. *Alberta Grizzly Bear Recovery Plan 2008–2013. Alberta Species at Risk Recovery Plan No. 15*. Edmonton: Alberta Sustainable Resource Development.

Wilkinson, T. 1998. *Science under Siege: The Politician's War on Nature and Truth*. Boulder, CO: Johnson Books.

———. 2001. "Gambling with Grizzlies: Is This the Right Time to Take Yellowstone Grizzlies off the Endangered Species List?" *Wildlife Conservation Magazine* Nov/Dec: 28–56.

Yalom, I. D. 1980. *Existential Psychotherapy*. New York: Basic Books.

9

The North American Model of Wildlife Conservation An Analysis of Challenges and Adaptive Options

SUSAN G. CLARK AND CHRISTINA MILLOY

Introduction

Supporters of the North American Model of Wildlife Conservation claim that its historic principles and established practices offer the best means to conserve wildlife (including large carnivores) in the foreseeable future. Its basic premise is that government wildlife agencies, funded by revenues from public hunting and fishing, use scientific knowledge and expertise to manage wildlife for the public good. Promoted as a century-long success story, the model establishes how wildlife will be used and who gets to decide, founded largely on concepts of public trust, scientific management, and single and multiple use formulas administered by government (e.g., Wildlife Society 2007; Prukop and Regan 2005; Geist 1995; Aldrich n.d.; Crane 2009). As such, it is clearly a constitutive process with institutional and organizational dimensions and many ordinary decision-making outcomes (see chapter 8).

There are concerns, however, about the role and adequacy of the model. Attributing its successes to hunters and hunting, some of its proponents want to enhance the role of hunting and fishing and encourage the urban public to participate in activities that support the model (e.g., Mahoney 2004a, 2004b, 2006; Mahoney et al. 2008; US Fish and Wildlife Service 2008). Other interests want to include the broader public in wildlife management, believing that diverse perspectives and values beyond those of hunters and anglers should be involved (e.g., Jacobson et al. 2010; Steelman 2010). In addition, we suggest that the model's actual conservation record over the past century is not as impressive as is sometimes suggested.

Despite restoration of turkeys, white-tailed deer, and a few other species, populations and species at many locations have lost ground and face increasing losses from: (1) environmental changes (e.g., habitat degradation and loss, endangerment and extinction of nongame species, and global change); (2) broad social changes (e.g., economic upheaval, industrialization and energy development, agricultural expansion, urbanization); and (3) declining hunter and angler populations, along with an expanding urban population little connected to wildlife and nature. Much of wildlife conservation is based on positivism (Romesburg 1981, 1989). There is growing criticism of scientific management and positivism, of experts and their roles in society, and of the state and federal agencies that manage wildlife and land (e.g., Nie 2004; Ascher, Steelman, and Healy 2010). So what is the future of the model in managing fish and wildlife? Stemler (2008) also considered this question in writing about the White House Conference on Wildlife Policy. The Wildlife Society (2007), National Rifle Association, Rocky Mountain Elk Foundation, and Sierra Club are among the groups that have also invited reflection about the model's adequacy. We hope our analysis adds to this growing discourse and improves wildlife conservation.

We describe here the model's doctrine and formula and some problems it faces. We examine key elements of the social context, the status of wildlife, and the decision process in order to assess how the model has performed in practice. Building on this assessment, we suggest a problem definition based on public trust and common interests and offer three strategic options to adapt the model to new circumstances and enhance wildlife conservation.

Our analysis, empirical and data based, meets widely recognized standards of "meta" analysis (see Lasswell 1971). We seek a comprehensive, fundamental understanding of the model and its context using well established and widely used interdisciplinary theory and methods (e.g., Lasswell and McDougal 1992). This approach focuses on the dynamics of individuals, values, beliefs, practices, and institutions (Lasswell and Kaplan 1950). It requires users to establish their observational standpoints, clarify public order and human dignity goals, carry out procedurally rational tasks, map the full context, understand decision making, and attend to matters of content and procedure. Data came from online sources, professional journals, literature from associations, federal and state fish and wildlife agencies, academic institutions, and museums (e.g., peer-reviewed literature, Yale Peabody Museum, Heinz Foundation's Millennium Report 2005). We also reviewed other sources that provide insight into the model and its operation (e.g., Orion: The Hunter's Institute [no date], White House Conference on Wildlife Policy

2008). We consulted knowledgeable practitioners in universities, nongovernmental organizations, and agencies.

Concerning our standpoint, we have worked to improve wildlife management and conservation issues for over forty years in the United States, Canada, and elsewhere. Our experience includes field biology, scientific research, management, development, policy analysis, and education. We have worked for and with state and federal fish and wildlife agencies, NGOs, the private sector, landowners, industry, elected officials, and universities. We acknowledge that the subject of this chapter poses many analytic challenges, including understanding interactive ordinary, governance, and constitutive dimensions, and that there are reasonable contrasting interpretations. There are also other analyses that concur with ours (e.g., Jacobson and Decker 2006; Jacobson, Decker, and Carpenter 2007; Jacobson et al. 2010; Leopold 2010; Organ, Mahoney, and Geist 2010). Our observations constitute a "hypothesis-schema," open for further research and testing, which we encourage (see Lasswell and Kaplan 1950). In the end, we seek to enhance wildlife conservation for the common good via open, inclusive, comprehensive, targeted, fair, cooperative, and effective means.

The North American Model

This section describes the history and principles of the model, its institutional and organizational setting, problems noted by its supporters and others, and its social context.

HISTORY AND DOCTRINE

Antecedents of the model go back to Anglo-Saxon law and the Magna Carta of 1215 (historical discussion in Lund 1980; Roth and Boynton 1993; Tilleman 1995; Bean and Rowland 1997). The following description of the model's seven-part doctrine is based on Geist, Mahoney, and Organ (2001).

The first element of the doctrine is that "wildlife is a public trust resource." The idea that wildlife could not be privately owned is evident in Biblical writing and in the inclusion of wildlife among things that could not be owned as codified in law by Roman Emperor Justinian in A.D. 529 (Geist et al. 2001, 176). In the early part of the eleventh century, Danish King Canute altered this code in England and made wildlife the property of the king and elite landowners (Adams 1993). Two hundred years later the Magna Carta reaffirmed the rights of the public in fish and wildlife. The first major case in the United States addressing what is now "The Public Trust Doctrine" was a Supreme Court decision in 1842, *Martin v. Waddell*, in which a landowner claimed that King Charles II had made a land grant in 1664 that gave him exclusive rights to harvest oysters (Geist, Mahoney, and Organ 2001, 177). Chief Justice Roger Taney ruled, based on the Magna Carta, that the king had held fishing rights as a public trust. In colonial times, since New Jersey at that time held fishing rights for the oysters, these rights were considered to be in the public trust (Bean and Rowland 1997). This set the stage for later government in the United States, at federal and state levels, to maintain oversight of wildlife in the public trust. The extinction or near loss of passenger pigeons, bison, elk, Carolina parakeets, and other species caused sportsmen and conservationists to lay the foundation for the model.

Second is the "elimination of markets for wildlife." History demonstrates that wildlife can be rapidly depleted when dead animals are considered valuable (Hewitt 1921; Matthiessen 1959). The passenger pigeon, once the most populous bird in the United States, was hunted to extinction as a cheap food source. The last remaining passenger pigeon died in the Cincinnati Zoo in 1914. Some trade in wildlife remains today (Organ, Gotie et al. 1998; Prescott-Allen and Prescott-Allen 1996).

Third is the "allocation of wildlife by law." From a consumption perspective, "surplus wildlife" is allocated to the public via permits and licenses for fishing, hunting, and trapping. Such allocations are not dictated by the market, landowners, or special privilege. Additionally, the public should have input into wildlife conservation (Geist 1995; Geist, Mahoney, and Organ 2001, 177–78).

The fourth principle is that "wildlife can only be killed for a legitimate purpose." As a public trust resource, wildlife can only be killed for legitimate reasons, including food, fur, self-defense, or protection of property (Geist 1988, 1995). Conservationist George Grinnell developed the "Code of the Sportsman," which included the idea that wildlife killed by sport hunters should not be wasted (Organ, Muth et al. 1998).

Fifth is that "wildlife are considered an international resource." Wild animals do not recognize territorial or political boundaries in their movements. The Migratory Bird Treaty of 1916 was the first significant treaty between the United States and Canada to protect wildlife (Hewitt 1921; Jahn and Kabat 1984). Other international agreements followed.

The sixth element is that "science is the proper tool for discharge of wildlife policy." Aldo Leopold was a proponent of the use of positivistic science for allocating natural resources. This is a component of the "Roosevelt Doctrine," which depends on scientific management (Leopold 1930, 1933; see also Romesberg 1981, 1989; Morcol 2001; S. Clark et al. 2010). These concepts are central to the wildlife profession today (Geist, Mahoney, and Organ 2001, 178; Gill 1996).

Seventh is the "democracy of hunting." This means that hunting should not be restricted to those of means, upper class, or any other special interest. In Aldo Leopold's words, a key element of hunting is the "democracy of sport" (Meine 1988, 169). Hunting is considered a democratic activity because all citizens have the right to engage in it. Sport hunting is distinguished from market hunting, which does not ascribe to the code of "fair chase" (Posewitz 1994; Kerasote 1994).

FORMULA AND INSTITUTIONS

In practice, this doctrine is carried out through a formula that is structured by the use of science, state and federal government, and interested associations. These practices and organizational networks constitute the "institution" of wildlife management. Institutions are the "well established and structured pattern of behavior or of relationships that is accepted as a fundamental part of a culture" (*Webster's* 1994), or "the shared concepts used by humans in repetitive situations organized by rule, norms, and strategies" (Ostrom 1996, 86; see also Young, King, and Schroeder 2009).

Along with the doctrine, the formula establishes an authoritative and controlling structure for the model (see Lasswell and McDougal 1992). The formula is complex and includes the 1930 "American Game Policy" developed by Aldo Leopold and others, who called for stable funding and for training of wildlife professionals. Following the Dust Bowl of the 1930s, Congress passed the Federal Aid in Wildlife Restoration Act in 1937 (16 USC §§ 669-669k), which is also known as the Pittman-Robertson Wildlife Restoration Act, or PRWR. Wildlife programs at the federal and state levels ensued. Since Franklin D. Roosevelt's signing of the PRWR in 1937, some species have been restored, including the wild turkey, white-tailed deer, pronghorn antelope, wood duck, beaver, black bear, giant Canada goose, American elk, desert bighorn sheep, bobcat, mountain lion, and bald eagle (USFWS Southeast Region: Federal Aid Division website).

The formula is further specified in other federal prescriptions. The Sport Fish Restoration Program (SFR), authorized by the Dingell-Johnson Act (16 U.S.C. §§ 777-777l), was passed in 1950. Both PRWR and SFR generate revenue by taxing consumptive users, that is, hunters, anglers, boaters, and shooters. Funds are held in public trust controlled by the US Fish and Wildlife Service (2008, 2009a, 2009b, 2009c). These acts authorize the Secretary of the Interior to cooperate with the states, through their fish and wildlife departments, in implementing sport fish and wildlife restoration projects. The funds are apportioned annually based on a formula that includes land area and the number of licensed hunters and anglers in a state (http://wsfrprograms.fws.gov/Subpages/GrantPrograms/WR/WR_AppnFormula .pdf). In addition to the federal government, state wildlife agencies are also prominent in the wildlife institution, as are user associations, such as Ducks Unlimited.

The total amount of funds generated by SFR from 1952 through 2010 was $6,583,260,439 and for PRWR 1939-2010 was $6,411,069,221. The total amount of SFR funds apportioned in 2010 was $389,552,973; for PRWR it was $472,719,710. In comparison, the State Wildlife Grants Program, which does not derive funds from excise taxes, uses federal grant funds for programs that benefit wildlife and their habitats, including species not hunted or fished. The amount apportioned in 2010 under the State Wildlife Grants Program was $73,767,660 (USFWS Wildlife and Sport Fish Restoration Program Website 2010). Clearly, funds for fish and wildlife conservation have been significant.

Resources, monetary and otherwise, are always a key concern in policy. The model's formula is set up so that consumptive users of a resource pay for access, the federal government administers funds, and state agencies manage the wildlife "resource." This arrangement creates a convergence of interests among sportsmen (clients), government fish and wildlife agencies (suppliers), and other interests, including the industries that manufacture the products that are taxed. However, the model is viewed by some as only marginally responsive to their interests, including many people who are concerned about animal rights and welfare, nonconsumptive recreationists such as wildlife watchers, and the general public, who are not significant contributors to or participants in the model as institutionalized (see Mattson and Clark 2010). Also, chapters 2 through 7 document how the formula is implemented and what the outcomes are in practice.

PROBLEMS IN PRACTICE

The model is a shared worldview or paradigm about wildlife and our relation to it (and to nature) that has served as a "working specification of the common interest" for the last century. This model coalesced shared beliefs and institutional arrangements that have conserved, restored, and maintained many wildlife species and populations, but not without problems and growing challenges. For example, at various times and places, the model has been used to justify extermination of large carnivores for purposes such as increasing populations of ungulate game species (e.g., Robinson 2005).

Diverse interrelated problems have been articulated pertaining to wildlife conservation, the context, the agencies, and the model itself. For example, the Association of Fish and Wildlife Agencies and the US Fish and Wildlife Service assert that wildlife and the agencies are facing new challenges, including shrinking funding, global climate change, urban sprawl and encroachment, and an increasingly urbanized society disconnected from the natural environment. According to Aldrich (n.d.), hunting is the "glue" for success of the model, and Stemler (2008, 1) noted that the decline in hunting and hunters is problematic, as is the loss of "access to quality hunting opportunities." Further, Semcer (2008, 4) wrote that "communication and coordination" are problematic for the model. Importantly, the model has not been extended beyond North America, and damaging markets for wildlife have not been eliminated globally.

Consider the problem of shrinking funding. US society and outdoor recreation are changing in ways that have reduced the user groups that supply the revenue for wildlife conservation under the model. According to the 2006 National Survey of Fishing, Hunting, and Wildlife-Associated Recreation (USFWS 2006, 4), there has been a downward trend over the last few decades in the United States in the number of hunters (12.5 million) and anglers (30 million), with a combined total estimated at 33.9 million today. Simultaneously, there has been an increase in the number of wildlife watchers, which includes nonconsumptive activities such as bird watching or photography, to approximately 71.1 million (USFWS 2006, 4). This strongly suggests a need to include other types of participants in the model beyond hunters and anglers.

An additional problem is the conservation of species and populations other than game. "Nongame" makes up the vast majority of wildlife species, and its conservation has been largely ignored or minimized by state agencies (Nie 2004). Although many nongame species, such as some types of bats, are eligible for funding and restoration through the model, these species are not receiving adequate assistance because of traditional thinking in the agencies that funds should first benefit species directly related to hunting and fishing. Additionally, state agencies' thinking, in too many cases, is itself problematic when it comes to addressing large carnivore conservation, ecosystem management, and climate change (Nie 2004). In fact, the increasing number of threatened and endangered species could be viewed as a growing failure of the state game and fish organizations, as well as the overall institution of wildlife management and the model itself (Clark and Rutherford 2005; Jacobson and Decker 2006; Jacobson et al. 2010). Wildlife conservation is further hindered by overlapping, competing, and contradictory policy sectors (e.g., housing, jobs, healthcare). Priority given to job creation, for example, may override considerations for wildlife habitat conservation. There are numerous larger national issues, such as national security, economy, and energy concerns that intersect with conservation and compete for time, attention, and resources. Such conflicts with wildlife conservation are expected to intensify.

"State wildlife governance" is also problematic (Jacobson and Decker 2006; Jacobson, Decker, and Carpenter 2007; Jacobson et al. 2010). Nie (2004, 199) is critical of present arrangements, which center on states' rights, scientific management and positivism, and bureaucracy, including state wildlife commissions, game-dependent agency budgets, and the role of both in ongoing political conflict. He says that the "state wildlife management paradigm is characterized by scientism and agency capture," issues that have been recognized since the 1970s (e.g., Tober 1989; Mangun, O'Leary, and Mangun 1992; Kellert 1996). Consequently, nonconsumptive users have challenged the client or user-based paradigm of state programs (e.g., Hagood 1997). These people have criticized the commission framework and have argued that the current formula runs counter to the idea of wildlife as a public trust because most Americans do not hunt, shoot, fish, or trap (e.g., Pacel 1998). These interests have also charged that wildlife commission members often have conflicts of interest because they have financial stakes in consumptive uses of wildlife, such as ranching and hunting (e.g., see Alaska Wildlife Alliance 2000–2001).

The model is also sometimes weak in practice. For example, Mattson and Chambers (2009, 113) found that management failed to provide water for Arizona's desert wildlife as planned. They also found that the decision process used by the agencies was "shaped by the precepts of scientific management" (i.e., positivism, objectivism, and instrumentalism, experts always know best) and "thus largely failed to foster civility, common ground, and a focus on common interests, and instead tended to exacerbate deprivations of dignity and respect" for many people who care deeply about wildlife conservation. The use of ballot initiatives is one way around the problem of state wildlife commissions and agency bureaucracies not addressing issues of importance to the public (Nie 2004; Mattson and Clark 2010); the public simply puts an issue on the ballet, votes on it, and bypasses the commissions and agencies.

Some of the problems identified here focus on *content*, that is, biophysical issues such as habitat and ecosystem degradation and threatened and endangered species problems. Other problems focus on *procedure*, or social and decision processes, such as policy influences of traditional hunting and wildlife interests or the privatization of wildlife through game ranching and commercial hunting. The cases in chapters 2 through 7 illustrate these problems.

The Decision-Making Process

In order to appraise the utility of the model and offer a comprehensive problem definition, we examine the status of wildlife conservation briefly with regard to context, wildlife, and decision making that must be addressed if wildlife is to be effectively conserved in the future.

CONTEXT

To understand the model's context, we mapped its social process. In other words, we identified participants who are involved, their perspectives, their value demands and claims, and their strategies (see Muth and Bolland 1983; Anderson 1999). There are many participants involved with how wildlife is managed in North America and who gets to decide, including international, national, and state governmental agencies, scientific experts, nongovernmental organizations, special interests, transnational public and private groups, and individuals. Each consists of people with particular perspectives (voiced in their claims), values, and preferred strategies—all of which play out in the debate and determine the outcomes and effects in institutional terms.

The social process of the wildlife management arena is complex, dynamic, and full of conflict. There are supporters and opponents of the model. Supporters include, for example, hunters and anglers, the Association of Fish and Wildlife Agencies, Ducks Unlimited, National Rifle Association, Rocky Mountain Elk Foundation, part of the general public, many wildlife professionals, many elements within state and federal agencies, National Audubon Society, Defenders of Wildlife, National Wildlife Federation, businesses and industry groups related to sporting supplies and equipment, and outdoor and recreational interests. Opponents are also a diverse group that includes some landowners (particularly ranchers and farmers), many nonconsumptive users, some development and exploitation businesses (e.g., oil and gas industry), and organizations devoted to animal rights, such as People for the Ethical Treatment of Animals and the Humane Society of the United States. Some of these groups are utilitarian, others conservation- or preservation-oriented. Some promote animal rights and oppose all consumptive use of wildlife. Others are simply indifferent and passive. Much of the general public—unconnected to wildlife except for the squirrel or rabbit in the yard, the occasional viewing of the "Animal Planet" channel on TV, or a family trip to the zoo—is unaware of the model's existence and its influence on the biological and political landscape.

Each of these people or groups has certain expectations about how the world works and makes demands or claims for certain value outcomes based on those expectations. Claims give us insight into the subjective perspectives of people, including how they identify themselves, for example, as a "woman," "hunter," "wildlife agency employee," or "scientist." For example, some supporters consider the model one of the most significant environmental success stories (e.g., Ball 1985; Kallman 1987; Mansell 2000; Leopold 2010; Miller 2010; Organ, Mahoney, and Geist 2010), claiming that it reversed the exploitation of wildlife and turned wildlife management into a "triumph of the commons" (see Hardin 1968 for discussion on commons; Geist 1988, 1995), that it is the envy of the world, and that it has restored white-tailed deer, wild turkeys, and other game species through the payment of excise taxes by anglers, boaters, hunters, and shooters (Geist, Mahoney, and Organ 2001).

Knowing about claims allows an analyst to understand why people behave as they do. Claims appeal to several generalized value categories— power, wealth, respect, knowledge, skill, affection, rectitude, and well-being (with no rank or order intended; Lasswell 1971; Clark and Rutherford 2005, 220–23). The appendix shows a sample of supporters' claims concerning the model taken from the literature cited in this chapter. The dominant claim, in terms of values, is for rectitude: the claimants see a moral imperative or responsibility to manage wildlife for future generations to enjoy.

Also key to understanding the social process is to look at the symbols people appeal to and their politics (see chapter 8). Both supporters and opponents invoke symbols in support of their claims. Supporters of the model invoke heroes such as Theodore Roosevelt, Gifford Pinchot, and Aldo Leopold, all figures who were instrumental in developing the model, essentially part of the original "wise use doctrine" (Trefethen 1975; Reiger 2001). They also single out hunters and hunting (and less so anglers and fishing) as key to the success of the model, connecting outdoorsmen with the North American pioneer spirit, which they claim is best evoked and nurtured through the hunting experience once frontiers disappeared (see, e.g., Slotkin 1973, 1985,

1992 on the American frontier myth; Brown 2007a, 2007b, 2008 on the history of wildlife conservation). Supporters' lore is also full of tales of successfully plucking species back from the brink of extinction. Furthermore, they claim that the model is a system of "sustainable development," wherein wildlife is a renewable, harvestable natural resource (Geist, Mahoney, and Organ 2001). From an analytic standpoint, these symbols are the relatively popular expressions of people's basic beliefs or worldview. Lore, stories, legends, heroes, and other symbols function to reinforce views of the success of the model, provide insight into supporters' dominant narratives, and divert attention away from content and practical matters.

"Scientific management" is one of the most important symbols in wildlife management (see Romesberg 1981, 1989 for descriptions). It is valuable, without a doubt, but it is proving in many cases to have harmful consequences for management, policy, and democracy (review in Brunner et al. 2002, 2005). As Merkle (1980, 244, 1998) noted, "scientific management, translated into politics, advocated the development of the state as an organ of national planning and allocation according to a rationally derived system of priorities; it glorified a monolithic rational-technical order in place of the weak democratic forum that compromised among the interests of power groups." The harms of scientific management, especially as used by state game and fish agencies and some units of the federal government, have been well documented (e.g., Odum 1989; Daily 1997). In general, people tend to pay little attention to social and decision-making processes and rely too heavily on standardized, bureaucratic, one-size-fits-all, technical rationality, and the dominance of experts (see Saul 1993, Scott 1998 for criticisms of this approach).

It is clear that wildlife conservation is a complex social and decision-making process involving diverse groups with opposing perspectives, value claims, and centers of attention. However, a major problem exists in that there is no adequate arena or process to meld conflicting interests into a shared view of the challenges to wildlife conservation and the practical actions needed to ensure a healthy future for wildlife. Another significant contextual problem is a lack of self scrutiny within the agencies and nongovernmental groups, both of which lack attention to their own operating assumptions, practices, and consequences (see Cherney 2011). Future research can help map the context more fully, aid understanding of participant interactions, and thus yield greater insight into the context of the institution of wildlife management and the model.

We also looked at the changing status of wildlife and its environments to see if the model, especially its organizational and institutional manifestations, is as effective as some supporters suggest. Regarding the status of species, for example, in 2000 a joint effort by the Nature Conservancy (2001) and the Association for Biodiversity Information evaluated the status of species in the United States. The total number of vertebrate species was estimated to be 2,497. Of these, 1,883 species (75 percent) were considered to be secure or apparently secure. The status of five species was unknown. Of species not considered secure or apparently secure, 135 were reptiles or amphibians, which, along with invertebrates, are not included for conservation funding under the model as currently instituted. This leaves 474 mammal, bird, or freshwater fish species that are presumed extinct, possibly extinct, critically imperiled, imperiled, or vulnerable (Stein, Kutner, and Adams 2000, 104). Since the Endangered Species Act was enacted in 1973, there has been an increase in the rate of listings, most attributed to the addition of numerous plants (Stein et al. 2000, 107). The status of many species is unknown. Furthermore, as of October 2013, there are 1,010 vertebrate animal species and populations worldwide listed as "threatened" or "endangered" by the US Fish and Wildlife Service and protected under the Endangered Species Act. Of these, 361 are mammals, 318 birds, and 167 fishes. The vast majority of the world's reptiles and amphibians that occur in the United States, for example, are not eligible for funding from the Sport Fish or Wildlife Restoration Acts because they are not huntable "game species" (USFWS Endangered Species Program website). As well, large carnivores over the last 100 years have generally not faired well under the model and its structural manifestations. These data suggest that the model should be evaluated, revised, and made more effective to reverse the growing trend in species and populations at risk.

An increasing number of university-based ecologists and conservation biologists have investigated species status, threats, and likely future trends. For example, Pimm et al. (1995) found that worldwide extinction rates are 100 to 1,000 times prehistoric levels and increasing, and they estimate that future extinction rates will be ten times greater than recent rates with many species not now threatened becoming extinct. Concerning climate change, Walther et al. (2002) noted that there is ample evidence of ecological impacts of recent climate change from polar terrestrial to tropical marine environments. These changes will have huge consequences on the status and viability of species. Already, the decline of US species and populations in recent decades is evidence that the model is less than fully successful in those cases.

Wildlife depends on suitable habitats and ecosystems. Habitat quality and ecosystem integrity affect and are affected by biodiversity (Chapin, Kofinas, and Folke 2009, 36–38). We now know that more than half of the ecosystem services that humans need for survival and quality of life have been degraded as people struggle to meet their material desires and needs (Heinz Foundation 2005). Many ecosystems are on the decline, and climate change is expected to affect them all dramatically (Steffen et al. 2004; Foley et al. 2005; Heinz Foundation 2005; also Turner et al. 1995; Redman 1999; Jackson 2001; Diamond 2005). There is a coherent pattern of ecological change in flora and fauna in response to these community and ecosystem trajectories. From this and other research, it is clear that we are in the early stage of massive change (Scheffer and Carpenter 2003; Myers 1996; Constanza et al. 1997). Global biodiversity scenarios are being generated now (see Sala et al. 2000; Tylianakis et al. 2008) along with many suggestions about what to do (e.g., Soulé et al. 2005; Sinclair and Byrom 2006; Hobbs and Harris 2001). These data suggest that the model, and its structural and cultural dimensions, has not done enough to slow or reverse the growing trend in ecosystem degradation and loss.

Human impacts on wildlife, habitats, and ecosystems have changed over time, and affected ecosystems are changing more rapidly now than at any other time in human history. The present downward projections in the status of wildlife and ecosystems bring into sharp focus questions about whether the model is capable of conserving wildlife and ecosystems into the future without major adaptations.

DECISION PROCESS

Often people do not conceive of wildlife conservation as a decision process that requires balanced attention to matters of content and procedure. Nevertheless, the outcomes of the decision-making process determine what happens to wildlife. The decision process is our means of reconciling, or at least managing, conflict through politics in order to find a working specification of the community's common interests. Although most wildlife literature does not address this subject systematically or explicitly, we drew on the few available examples (e.g., Mattson and Chambers 2009), the case studies in this volume, and our own experience in the following appraisals.

Decision making is generally considered to have seven interrelated functions: (1) intelligence gathering and planning; (2) debate and promotion about the nature and status of the problems; (3) deciding on a plan to solve the problems (in other words, setting new rules); (4) invoking the new rules in specific cases; (5) applying the rules through administrative and judicial activities; (6) appraising progress or lack of it; and finally (7) terminating the rules when they no longer apply (Lasswell 1971; S. Clark 2002). These functions do not necessarily occur in order; several functions may be ongoing at the same time in any single program. Claimants in the literature that we reviewed targeted one or more parts of the decision process for comment. For example, the Wildlife Society (2007) made five claims (see appendix), some of which called attention to the promotional activity in the decision-making process (i.e., advancing a perspective in the debate about wildlife). Other claims have to do with appraisal (evaluation), implementation, and intelligence (data gathering and planning). Most of the supporters profiled in the appendix focused on appraisal and promotional activities, but none called attention to the entire process or the need to upgrade it. Clearly, the focus of attention of participants is selective and incomplete. Having a more comprehensive understanding of the decision-making process can ensure a more complete picture of wildlife conservation. Useful models of decision making exist that are already helping to reconfigure wildlife conservation.

Even though few people think in terms of decision process, some do in fact conceive of wildlife conservation as a process, at least implicitly. For example, Prukop and Regan (2005, 375–76), representing the Association of Fish and Wildlife Agencies, called for upgrading the intelligence function in their appeal for "more effective and widespread use of human dimensions information." They sought an upgrade in the promotion function when they cited the need for "new or improved conservation alternatives, with a better understanding of them, their costs, and benefits" (see T. Wilkinson 1998). They were actually calling for upgrading the prescription function when they called for "broadening the conservation agenda to reflect a diversity of values, users, and their desires." They were also looking for better application of conflict management when they sought "more direct approaches for dealing with moral and ethical issues." Finally, they sought better appraisal when they said that "more effective evaluation of our efforts" was needed. Overall, their recommendations—which call for upgrading six of what we identify as the seven functions of decision making—suggest that the decision process needs to be made more open to everyone, more factual about the entire context, more creative in finding facts about the social, economic, and political environment, more comprehensive, integrative, and effective, more timely and prompt, more focused on achieving common interests, more constructive, and more contextual or practical. Recurring weaknesses that characterize each of the decision process functions, such as expert biases, poor government coordination, agency rivalries, overcontrol, "benefit leakage," intelligence failures and delays, inadequate coordination and appraisal, inappropriate organizational arrangements, insensitivity of decision makers to valid and appropriate criticism, and failure to end programs and move to new ones, all must be preempted (e.g., S. Clark 2008, 139–71).

To complicate matters, there are two types of decision processes that are rarely distinguished, as discussed in chapter 8. *Ordinary* processes are those in which decisions are made regarding everyday problems; they focus on content choices about population size, for instance, or habitat quality. *Constitutive* decision processes set the rules, procedures, and norms that govern ordinary processes; these are decisions about how to make decisions (McDougal, Lasswell, and Reisman 1981; McDougal 1992–1993). They focus on procedures about how policy should be made and, by implication, who ought to be involved in the decision process (S. Clark 2002, 71). Constitutive processes transcend everyday operations to determine how organizations, institutions, analytic methods, and participants ought to be structured or selected (see Lasswell 1971, 98–111). When seeking to improve wildlife conservation, both ordinary and constitutive processes should be targeted for improvement. As Geist (2004, 1) observed, "Thus [the model is] the product of innumerable political discussions—acrimonious or otherwise. It is not the product of a single mind but expresses the collective wisdom of nearly a century of continent-wide debate and hard bargaining. It has retained what has worked. It therefore has a deep wisdom and could not have been invented by any single mind." He goes on to state that the model has "been examined by a number of symposia and has been discussed in the popular press and on the [I]nternet" (1). Many people involved in wildlife conservation seem to pay most attention to ordinary process and focus on technical content details while overlooking or underattending to constitutive process considerations. Since most wildlife professionals are involved at the ordinary level, most published cases are told from that perspective, typically uncritically promoting and reinforcing the symbols, doctrine, and formula of the model. However, a growing number of people appear interested in the constitutive process, even though they do not use these concepts, distinctions, or language.

Our findings about the context, the status of wildlife, and decision making have practical consequences for the future of the model and for wildlife conservation. The practical task now is to use these findings to construct a realistic problem definition of the conservation challenge at hand. A good problem definition will provide feasible opportunities to improve the decision-making process, activity by activity, and adapt the model's doctrine and formula (see, e.g., Brunner 2010). Traditional science alone offers no such method, at either ordinary or constitutive levels, for creating robust problem definitions that will be useful to society in solving the problems of conserving wildlife and ecosystems.

A PROBLEM DEFINITION AND THE COMMON INTEREST

Although the model has served the American public fairly well over the last century, it shows growing problems of content and procedure today as evidenced in the literature and in practice. A growing number of wildlife species and populations and their habitats and ecosystems are under stress created directly or indirectly by human action. There is also growing demand for more effective social and decision-making processes. These problematic trends and conditions are accelerating. These content and procedural problems—part of a single complex system—as well as the system-wide and institutional nature of the challenges have been recognized, and efforts of varying quality and success have been undertaken to address them.

Yet additional problems also exist. The North American Model is a response to historic cases and incidents (such as the decline of bison, turkeys, deer) and a product of the overall dynamics of our society and culture and its use of natural resources (such as the rise and establishment of bureaucracy and scientific management) since the late 1800s through the mid-1900s. It seems, however, that ongoing constitutive dynamics—efforts to adapt in response to changing circumstances—have been limited to ordinary, conventional means, such as "educating" the public, enlarging bureaucracy, undertaking "campaigns," litigation, expanding positivistic science, or seeking influence and money. But employing ordinary, conventional perspectives and means to address constitutive challenges will not advance our understanding of the complex social and decision-making processes involved in wildlife conservation, nor will it result in success, that is, adaptive responses to the growing content and process challenges over the mid to long term.

Another problem is self-limited learning by supporters of the model. If supporters' loyalty to the model and its institutional forms, however admirable, leads to promoting it uncritically and recycling its symbols as substitutes for addressing actual content and process problems, then potential improvements at both ordinary and constitutive levels will be blocked. Successful use and adaptation of the model requires learning and flexibility at individual, organizational, and policy levels, far beyond the conventional methods in use today (see S. Clark 2002, 166–72; Jacobson, Decker, and Carpenter 2007, 107; Ascher, Steelman, and Healy 2010, 196–197).

Finally, it is debatable whether the model and its operational dimensions, as presently constituted, administered, and implemented, adequately serve common interests. Finding and securing common ground has always been a struggle in wildlife conservation. The overriding goal, as we understand it, is to conserve wildlife: this is a common interest that is shared by members of the community and benefits the community as a whole. In order to understand the diverse participants' connections to the model and the validity and appropriateness of their claims in terms of the common interest, it will be necessary to examine competing claims about the model's success and the need to adapt it.

A major step toward formulating a realistic problem definition would be a full appraisal of the model in terms of common interest concerns, using procedural, substantive, and pragmatic criteria (see Steelman and DuMond 2009). There is clearly no single and objective standard for such judgments, but currently, much of the decision making in wildlife conservation focuses on procedural and legal requirements or on satisfying special interest demands (see Lawrence and Daniels 1997), but these do not adequately address actual problems or fully meet the growing demands for more broad-based, common interest outcomes. Although we cannot offer here the comprehensive appraisal that is needed, the model at present appears to be a complex mix of common and special interests, principled and expedient interests, and inclusive and exclusive interests (see Lasswell and McDougal 1992). This makes it difficult for people to orient themselves to the arena, and it inhibits appraisal, learning, and adaptation activities (see chapter 8). What is needed is wildlife conservation that favors common, principled, and inclusive interests.

The overall problem is an inability to adapt the model's doctrine, formula, and symbolic representations to the changing context and to integrate all valid and appropriate interests into wildlife policies that advance common interests. From a content point of view, the problem is clear—highly visible threats loom over wildlife and ecosystems. From a procedural point of view, the problem is an inability to satisfy demands for more inclusive participation in decision making by a non-consumptive and non-utilitarian public. These dual problems are remediable, at least to an important degree, with proper attention, organization, and leadership. Overall, we perceive that there is abundant opportunity to adapt the model, in both ordinary and constitutive ways, to advance wildlife conservation (see chapter 8). There are some adaptations underway currently around securing funding shortfalls at the state level, but these efforts fall far short of what is needed in terms of adaptations in organizational culture and operations of state game and fish departments to broaden the funding base, meaningfully include non-traditional interests in decision making, and meet the demands of society for expansion of services (Jacobson and Decker 2006).

Moving Forward: Lessons and Recommendations

We offer three strategies to adapt the model to address these ongoing challenges, respecting the admonition of Boitani, Asa, and Moehrenschlager (2004, 159) to view large carnivore conservation "as a complex system of decision making that requires an interdisciplinary approach." Modifications must take into account the need for not only scientific and technical input into decision making, but also social, political, and institutional considerations, all in an integrated fashion. Some supporters of the model are also calling for such changes, albeit in different language (e.g., see the six recommendations of Prukop and Regan 2005, sixteen strategies and six recommendations of the Sporting Conservation Council 2008, and others). Overall, a genuinely interdisciplinary, fully contextual approach is needed (see Clark and Wallace 2012).

ADAPT THE MODEL'S PRINCIPLES

We recommend modifying two of the seven doctrinal principles of the model to better reflect demands of the current conservation context, without jeopardizing the model's established value and potential for future success. The first modification is to the principle that "science is the proper tool for discharge of wildlife policy." The principle of scientific management came into prominence decades ago at a time when positivistic science, "a science of the parts," was the dominant paradigm (see Romesberg 1981, 1989 for a description of positivism; Merkle 1998). Many people still accept positivism as the only means of discovering truth and want to maintain its historic role in wildlife management. They presuppose that science should precede management and policy (see Pielke 2007 on the "linear science-to-policy model"). As valuable as positivism has been and remains, it is a paradigm with many drawbacks critiqued by Brunner et al. (2002, 2005), Brunner and Lynch (2009), Chapin, Kofinas, and Folke (2009), Pielke (2007), and others. It limits our understanding of what constitutes reliable knowledge, it overly objectifies our epistemology and privileges technical experts, and it excludes citizens directly affected by decision making from the process, among other drawbacks (see K. Wilkinson, Clark, and Burch 2007). Its assumptions cause problems in wildlife conservation and in democracy (see Mattson and

Chambers 2009). Despite growing evidence of these limitations, positivistic conceptions of science, management, and policy still dominate the wildlife profession and the institution of wildlife management—its culture, the university programs that prepare professionals, and the management agencies and conservation organizations that employ them. Yet there are those who insist that the "science" is OK, that it's just "politics" that is the problem (see Pielke 2007).

In contrast, a growing number of people are calling for alternatives to positivism. "Ecosystem management" and "resilience-based ecosystem stewardship" are among the several options put forward (e.g., Chapin, Kofinas, and Folke 2009, 5; a survey is in Clark et al. 2010, chapters 1–3). But on closer examination, many articulations of these concepts are just new ways of repackaging, elaborating, and renaming scientific management while holding on to a positivistic epistemology. Nevertheless, the movement away from positivism is accelerating as evidenced, for example, in calls for a new interdisciplinarity for professional certification of wildlife biologists (e.g., the Wildlife Society 2007; see Bammer 2005, and cases in Brunner et al. 2002, 2005). Adaptive governance, an alternative that we support, is an integrative approach based on "a science of the whole." It does not discard positivism, but puts it in its contextual place, such that science is necessary but not sufficient for good decision making. It calls for an explicit, systematic, empirical, and fully problem-oriented, contextual, and multi-method approach—a genuine interdisciplinarity (see Clark and Wallace 2012). This approach is quite different than the multidisciplinary approach called for by some authors (e.g., Jacobson et al. 2010).

Adaptive governance enables the kind of high-order integration that supporters of the model expect now, an approach not presently possible through scientific management. The role of adaptive governance in decision making has been shown to be practical in wildlife conservation. As a result of its successes, this alternative is backed by a growing number of supporters of the model. For example, Prukop and Regan (2005, 376) call for a broader science that will allow for a "stronger integration of human dimension and communication science in our decision-making processes" (see also Sporting Conservation Council 2008). Adaptive governance offers means for better problem solving, integration of knowledge and action, theory and practice, and on-the-ground gains. It rests on an analytic framework and methods, far beyond positivism, that can be taught, learned, and applied (Clark and Wallace 2012), although it has yet to be fully embraced by the wildlife conservation community. Adaptive governance requires meaningful public involvement as well as expert input. It is an open, fully transparent decision-making process. It involves a process of problem solving that is much more open, grounded, and practical than traditional approaches. When used, it has proven move effective than traditional agency approaches.

The second principle of the model that we recommend be changed is the idea of the "democracy of hunting." The contribution of consumptive users such as hunters and anglers is necessary for the model to function as well as it does and should be encouraged (see Mattson et al. 2006). However, these are special interests. Many more people who do not hunt or fish could be directly included in decision making about wildlife (e.g., nonconsumptive users such as wildlife viewers, outdoor recreationalists, and proponents of animal welfare). Inclusion of these people would enfranchise and create a bigger and more diverse public base of support, politically and financially, for wildlife conservation (e.g., if excise taxes were implemented on goods used by these other interest groups). In fact, the "democracy of hunting" appears to conflict with two other principles of the model, that wildlife is a "public trust resource" and that it is an "international resource." These larger interests—the general public and the international public—should be construed to represent the larger common interest.

This recommendation is also supported by others. For example, Crane (2009, 1) says that the model needs to appeal to the "direct interest" of a broader audience, including nonhunters and nonanglers. Supporters need to "aggressively engage hunters and anglers, wildlife enthusiasts, other conservationists, and the general public about the need to maintain wildlife" (Prukop and Regan 2005, 376). Together, the Association of Fish and Wildlife Agencies and the US Fish and Wildlife Service (2006, 6) noted that supporters "must take positive steps to encourage and nurture [public] interest in the natural world." Modifications could be made to include more diverse interests (see Pimbert and Pretty 1995). McLaughlin, Primm, and Rutherford (2005) offer guidelines for successful inclusive, participatory projects. K. Wilkinson, Clark, and Burch (2007) provide an overview of the kind of joint problem solving that bridges across diverse people, perspectives, and interests. This change would make the model and its tangible elements significantly more democratic than hunting alone and more reflective of the current social context.

In sum, adapting these two elements of the doctrine would not compromise the goal of the model—enduring wildlife conservation—but would indeed enhance its ability to meet this goal.

ADAPT THE MODEL'S FORMULA

We also recommend modifying the model's formula, which rests on scientific management, bureaucracy, and a specific relationship with the public. First, scientific management provides a formulaic direction for implementing wildlife conservation as well as a doctrinal foundation. The formula, as noted by K. Wilkinson, Clark, and Burch (2007), relies on scientific theories, experts, and "progress" to address environmental problems. There are many well-documented cases of scientific management leading to disastrous consequences for nature and people (e.g., Scott 1998; Botkin 1990). Assumptions and reductionist scientific theories about how to define and solve problems often neglect to include "the indispensable role of practical knowledge, informal processes, and improvisation in the face of unpredictability" that have proven vital to a functional and healthy society (Scott 1998, 6). The model's formula, including the practices of state fish and wildlife departments, parts of the federal government, and some nongovernmental entities, could be improved by changing its reliance on scientific management.

Second, at present, the model is typically implemented through top-down federal and state bureaucracies that tend to centralize functions and enforce uniform rules and regulations (Galbraith 1977; Wilson 1989; Daft 1995). Bureaucracy is a specific kind of organizational form known for standardization, rigidity, and limited learning (see Etheredge 1985; Argyris and Schön 1978). Bracken (1984, 221) noted that bureaucratic organizations "have certain built-in tendencies, directions they naturally move toward when subjected to different constraints and levels of excitement," which are hard to change. Bureaucratic programs are typically technical and acontextual, which causes inflexibility, limits participation, and elevates the technical expert to the role of decision maker. These bureaucratic elements play out in conservation efforts and have significant consequences for effectiveness and democratic process (see Mattson and Chambers 2009).

To compound matters, in some regions, there are persistent conflicts between federal and state bureaucracies (e.g., when states claim absolute authority and control over all resident wildlife, even federally listed endangered species such as the black-footed ferret; T. Clark 1997). As Perrow (1970, 128) observed, "it is not difficult for an administrator to conclude that cooperation with another agency, unless it is on his own terms, will threaten his autonomy and threaten the wisdom of his approach or program" (see Miller, Reading, and Forrest 1996; Thomas 2003). Much has been written about the limitations of bureaucracy, its formulaic responses in general (e.g.,

Wilson 1989), and its limitations in wildlife conservation in particular (e.g., Nie 2008). Many of the challenges the model now faces are the consequence of these formulaic (bureaucratic) conditions.

We recommend debureaucratizing programs and better managing inter-bureaucratic conflict. One way to overcome bureaucratic problems is to employ more practice-based, innovative, prototyping efforts. A proven strategy to enhance performance, a prototype is a trial or model, official or unofficial, from which something can be learned or copied. Prototypes are not fixed in structure or procedure in advance of beginning a project, but instead are designed and adapted to encourage learning and creativity as the project unfolds. Successful prototypical examples are evident in the field as people work in unique local contexts to address the limitations of bureaucracies (e.g., Wilson and Clark 2007; D. Clark, Lee et al. 2008; D. Clark, Tyrrell et al. 2008; Rutherford et al. 2009). What these practice-based efforts have in common is that they all work outside of agency bureaucratic operations to some degree, but at the same time they work with the agencies. They are much more contextual and participatory than traditional bureaucratic, scientific management (Brunner 2010). Other options to improve agency performance were surveyed by T. Clark (1997, 188–207), including reconstructing organizations, both in structural design and cultural dimensions, improving information processing and organizational learning opportunities, using high-performance teams (staffing, cognitive and emotional characteristics, and team leadership), and using "parallel organizations."

We also recommend more effective, genuine public involvement. K. Wilkinson, Clark, and Burch (2007) looked at the challenges of inclusive civic discourse and democratic process, including building social capital, achieving civility in dialogue, and ensuring democratic process (see Ascher 2009; Ascher, Steelman, and Healy 2010). They concluded that successful processes require participants to be "aware of, and alive to, all interests involved, all relevant cultural perspectives, and how individuals relate to families, neighbors, social groupings generally and patterns of governance" (O'Riordan and Stoll-Kleemann 2002, 89). Pimbert and Pretty (1995) and Arnstein (1969) offer schemes and levels of public participation and recommend a high level of public involvement in all cases. Finally, improvements in leadership are key to overcoming the limitations of bureaucracy.

In sum, adapting these formulaic elements would further the goal of the model, not compromise it.

UPGRADE CAPACITY—LEADERSHIP AND SKILLS

Last, we recommend upgrading the capacity of leadership to accelerate adaptation of the model. Leaders, in partnership with followers, should bring effective responses to problems of content and procedure in interactive ordinary and constitutive processes (see Ascher and Hirschfelder-Ascher 2005). Adapting the doctrine and formula and innovating through practice-based prototypes both require effective leadership. The challenge for leaders is to be problem-oriented and as realistically contextual as is practical, in both strategic and tactical senses. Leaders should attend to both high-order thinking (i.e., actively shaping the future) and fundamental philosophic matters (i.e., doctrine and formula) in addition to administering routine operations. These skills can be learned and used to good effect (Clark and Wallace 2012).

Some options to help leaders were offered by S. Clark (2008, 172–88) in the context of managing the Greater Yellowstone ecosystem. First, leaders should target themselves for self-improvement, that is, learn to think beyond bureaucracy, positivism, and standard operating procedures, and better clarify organizational goals, both official and unofficial. Clarifying goals is important because goal inversion or displacement is common (see S. Clark 2008). Leaders can also upgrade their problem-solving skills by formalizing and using interdisciplinary problem-solving methods; this requires them to be explicitly and fully problem-oriented (Clark and Wallace 2012). Leaders also need to select staff appropriately, work though partnerships, use high-performance teams, and employ tailored problem-solving exercises. These options promise leaders greater effectiveness, when employed with skill and vision.

Second, leaders can upgrade management policy processes by explicitly and systematically focusing on how the decision-making process works (S. Clark 2008, 189–208). Understanding the interactive activities of decision making, standards for good decision making, and common pitfalls can be tremendously helpful. Leaders should also be able to determine whether decisions are in the common interest from procedural, substantive, and practical standpoints (see chapter 1, this volume, and Steelman and DuMond 2009). They also need to be familiar with widely accepted standards for adequate intelligence gathering, processing, and dissemination and for organizing effective arenas for problem solving (see S. Clark 2009; Cherney, Bond, and Clark 2009). Finally, high-caliber leadership requires the capacity to learn actively through prototyping, drawing lessons from experience, and "double-loop" learning (i.e., examining operating assumptions and reassessing underlying goals; Schön 1983).

Third, leaders should move people and practices toward sustainability

(S. Clark 2008, 209–21) by accentuating positive trends, shifting the focus of attention toward more contextual approaches, and by becoming skilled change agents. Ascher (2009) argues that leaders are generally not farsighted enough in all these areas. There are many causes of shortsightedness, including, he argues, "pure impatience," "selfishness," "analytic limitations and uncertainty," and "vulnerability" (29–43). Leaders need to find ways to gain traction to overcome these obstacles. Ascher offers tools for improving leadership. First, he suggests restructuring rewards and risks in management and policy (65–150). Included in these are ways to create and reschedule tangible benefits and costs and social and psychological rewards, ways to improve performance evaluation, and ways to encourage self-restraint on the part of selfish, powerful special interests. Second, he suggests ways to improve analytic frameworks, such as upgrading the rigor and comprehensiveness of analytic exercises and deepening problem definitions (151–88). Third, he offers better ways to frame appeals for support and enhance communication effectiveness of farsighted appeals (189–230). Finally, he offers ways of changing the decision-making or policy process by empowering and insulating farsighted leaders and sustaining decision-making processes in the face of resistance (231–58). These options hold great promise. There is no lack of proven methods readily available to aid leaders' vital work (e.g., Steelman 2010).

These three changes--adapting the doctrine, adapting the formula, and upgrading leadership capacity and skills--require changes in people and their practices. A good place to begin is with current supporters of the model, who are most committed to the most effective wildlife conservation. These three recommendations, if successfully carried out, promise to aid wildlife conservation, serve common interests, and prepare the model for a more successful future.

Conclusions

For more than a century, North American society has pursued a goal of conserving wildlife as a public trust, a goal that has been structured and implemented through the North American Model of Wildlife Conservation, which today encompasses many diverse organizations, associations, and interests. At continental as well as local scales, wildlife conservation is a dynamic process of decision making involving people, values, demands, and the reconciling of conflict. Functionally, the model's doctrine (principles) and formula (rules to implement the doctrine) guide current decision making about wildlife; they dictate how decisions are made, by whom, and for what purposes. It is this decision-making process that must be used to adjust the doctrine and formula of the model and its institutionalized structure to improve performance and adapt to changing circumstances.

Achieving the goal of conservation effectively in practice has been a struggle historically. Enduring wildlife conservation, especially for large carnivores, is becoming more difficult as more people compete for limited natural resources and other societal goals in today's rapidly changing world. At present, some supporters of the model and some organizations and institutions of wildlife management want to maintain it largely as it is, educate the public to its benefits, and enlarge traditional funding means. Others want to adapt the model in various ways and degrees. The basic question is whether the model and the institutional system for conserving wildlife can be adapted to keep pace with contextual changes.

We offer three options to make adaptations. A critical first step is to improve understanding of the context—the people involved, their perspectives, the values they seek and the claims they make, their strategies, and the outcomes they seek. This information can be used to upgrade the wildlife institution, make its decision-making processes more inclusive, and ensure a fair, representative, and balanced approach. Decision making can be upgraded to be more factual, timely, comprehensive, and effective. The common interest, as we see it, is to conserve wildlife for future generations in equitable, efficient, and inclusive ways. These are basic criteria for sound governance in a democracy and the most likely means for securing a biologically diverse world for all people to enjoy. Clarifying and securing this common interest is the key to a healthy future for wildlife.

Appendix
Claims, Values, and Decision Activities in the Policy Process of the North American Model of Wildlife Conservation

Claimants and claims, values demands, and decision process activities in the policy process of the North American Model of Wildlife Conservation. See text for explanation of value demands and decision process activities.

THE WILDLIFE SOCIETY (2007, 1–2)
Promote and support adherence to the seven core concepts of the model (see text).
Value demand—Rectitude (ethics, moral responsibility)
Decision process activity—Promotion (debate)

Foster educational opportunities to increase societal awareness of the model.

> *Value demand*—Enlightenment
> *Decision process activity*—Promotion

Support critical review of the model.

> *Value demand*—Enlightenment
> *Decision process activity*—Appraisal

Support further refinement of the model.

> *Value demand*—Skill
> *Decision process activity*—Invocation/application (= implementation)

Support the identification of threats and challenges to the model.

> *Value demand*—Enlightenment
> *Decision process activity*—Intelligence

ALBERTA WILDERNESS ASSOCIATION (GEIST 2004, 1–2)

The model is credited with halting Hardin's "tragedy of the commons."

> *Value demand*—Rectitude
> *Decision process activity*—Appraisal/promotion

The model is one of the "great achievements of North American culture."

> *Value demand*—Rectitude
> *Decision process activity*—Appraisal/promotion

The model is based on "raw grassroots democracy."

> *Value demand*—Rectitude
> *Decision process activity*—Appraisal/promotion

The model promotes "public involvement with wildlife."

> *Value demand*—Rectitude
> *Decision process activity*—Appraisal/promotion

CONGRESSIONAL SPORTSMEN'S FOUNDATION (CRANE 2009, 6)

"The model is the system of conserving wildlife that 'is the most successful in history.'"

> *Value demand*—Rectitude
> *Decision process activity*—Appraisal/promotion

GEIST, MAHONEY, AND ORGAN (2001, 175)

The model has become "a system of sustainable development of a renewable natural resource that is without parallel in the world."

> *Value demand*—Rectitude
> *Decision process activity*—Appraisal/promotion

The model is "held together by a brotherhood of blue-collar hunters and anglers."
Value demand—Rectitude/affection
Decision process activity—Appraisal/promotion

MAHONEY ET AL. (2008, 8–17, 21)

Hunting has been critical to the success of the model (8).
Value demand—Skill/knowledge
Decision process activity—Appraisal/promotion
State and provincial governments.
Value demand—Power
Decision process activity—Appraisal/promotion
Hold wildlife in trust for the public (10).
Value demand—Power
Decision process activity—Appraisal/promotion
Democratic rule of law (12).
Value demand—Power
Decision process activity—Appraisal/promotion
Democracy of hunting (13).
Value demand—Power
Decision process activity—Appraisal/promotion
Science is the primary basis for wildlife policy (17).
Value demand—Rectitude/knowledge
Decision process activity—Appraisal/promotion
The North American Model of Wildlife Conservation has been responsible for a remarkable resurgence in wildlife as well as a staggering and diffuse economy that has enabled wildlife to "pay its way" across a vast and diverse continent (21).
Value demand—Rectitude/wealth
Decision process activity—Appraisal/promotion

SEMCER (2008), SIERRA CLUB

The model is "one of the most extensive infrastructures for the security of wildlife populations and the habitat on which they depend."
Value demand—Rectitude
Decision process activity—Appraisal/promotion

NOTE

We benefited from hundreds of conversations, site visits, and project participation over the years. We received critical advice from Denise Casey (Northern Rockies Conservation Cooperative), Glen Salmon and John Organ (US Fish and Wildlife Service), Ron Regan (Association of Fish and Wildlife Agencies), Martin Nie (University of Montana), and two anonymous reviewers. We thank our home organizations for their support.

REFERENCES

Adams, D. A. 1993. *Renewable Resource Policy*. Washington, DC: Island Press.

Alaska Wildlife Alliance. 2000–2001. "Hunter/Trapper Monopoly." *The Spirit* 19:1–4.

Aldrich, E. n.d. "North America's Wildlife Conservation Model." *Orion: The Hunters Institute*. Available from http://www.huntright.org/heritage/AldrichConservation model.aspx.

Anderson, R. N. L. 1999. *Managing the Environment, Managing Ourselves: A History of American Environmental Policy*. New Haven, CT: Yale University Press.

Argyris, C., and D. A. Schön. 1978. *Organizational Learning: A Theory of Action Perspective*. Reading, MA: Addison-Wesley.

Arnstein, S. R. 1969. "Ladder of Citizen Participation." *Journal of American Planning* 4:216–24.

Ascher, W. 2009. *Bringing in the Future: Strategies for Farsightedness and Sustainability in Developing Countries*. Chicago: University of Chicago Press.

Ascher, W., and B. Hirschfelder-Ascher. 2005. *Revitalizing Political Psychology: The Legacy of Harold D. Lasswell*. Mahwah, NJ: Lawrence Erlbaum Associates.

Ascher, W., T. Steelman, and R. Healy. 2010. *Knowledge and Environmental Policy: Re-Imagining the Boundaries of Science and Politics*. Cambridge, MA: MIT Press.

Ball, G. 1985. *The Monopoly System of Wildlife Management of the Indians and the Hudson's Bay Company in the Early History of British Columbia*. BC Studies, No 66.

Bammer, G. 2005. "Integration and Implementation Sciences: Building a Specialization." *Ecology and Science* 19 (2):6. Available from www.ecologyandsociety .org/vol10/iss2/art6/.

Bean, M. J., and M. J. Rowland. 1997. *The Evolution of National Wildlife Law*, 3rd ed. Westport, CT: Praeger.

Boitani, L., C. S. Asa, and A. Moehrenschlager. 2004. "Tools for Canid Conservation." In *The Biology and Conservation of Wild Canids*, edited by D. W. MacDonald and C. Sillero-Zubiri, 143–59. Oxford: Oxford University Press.

Botkin, D. 1990. *Discordant Harmonies: A New Ecology for the 21st Century*. New York: Oxford University Press.

Bracken, P. 1984. *The Command and Control of Nuclear Forces*. New Haven, CT: Yale University Press.

Brown, R. D. 2007a. "A Brief History of Wildlife Conservation in the United States, Part I." *Fair Chase* 22 (4): 28–33.

———. 2007b. "A History of Wildlife Conservation and Research in the United

States—with Implications for the Future." In *Proceedings of the Taiwan Wildlife Association Congress*, edited by H. Li , 1–30. Taipei: Taiwan National University.

———. 2008. "A Brief History of Wildlife Conservation in the United States, Part II. *Fair Chase* 23 (1): 30–35.

Brunner, R. D. 2010. "Adaptive Governance as a Reform Strategy." *Policy Sciences* 43:301–41.

Brunner, R. D., C. H. Colburn, C. M. Cromley, R. A. Klein, and E. A. Olson. 2002. *Finding Common Ground: Governance and Natural Resources in the American West.* New Haven, CT: Yale University Press.

Brunner, R. D., and A. Lynch. 2009. *Adaptive Governance and Climate Change.* Boston: American Meteorological Society.

Brunner, R. D., T. A. Steelman, L. Coe-Juell, C. M. Cromley, C. M. Edwards, and D. W. Tucker. 2005. *Adaptive Governance: Integrating Policy, Science, and Decision Making.* New York: Columbia University Press.

Chapin, F. S., III, G. P. Kofinas, and C. Folke. 2009. *Principles of Ecosystem Stewardship: Resilience-Based Natural Resource Management in a Changing World.* New York: Springer.

Cherney, D. N. 2011. "'Environmental Saviors'—The Effectiveness of Nonprofit Organizations in Greater Yellowstone." PhD Dissertation. Boulder: University of Colorado.

Cherney, D. N., A. C. Bond, and S. G. Clark. 2009. "Understanding Patterns of Human Interaction and Decision Making: An Initial Map of Podocarpus National Park, Ecuador." *Journal of Sustainable Forestry* 28:694–711.

Cherney, D. N., and S. G. Clark. 2009. "The American West's Longest Large Mammal Migration: Clarifying and Securing the Common Interest." *Policy Sciences* 42: 95–112.

Clark, D. A., D. S. Lee, M. M. R. Freeman, and S. G. Clark. 2008. "Polar Bear Conservation in Canada: Defining the Policy Problem." *Arctic* 61:347–60.

Clark, D. A., M. Tyrrell, M. Dowsley, M. Foote, A. L. Freeman, and S. G. Clark. 2008. "Polar Bears, Climate Change, and Human Dignity: Disentangling Symbolic Politics and Seeking Conservation Policies. *Meridian* Fall:1–3.

Clark, S. G. 2002. *The Policy Process: A Practical Guide for Natural Resource Professionals.* New Haven, CT: Yale University Press.

———. 2008. *Ensuring Greater Yellowstone's Future: Choices for Leaders and Citizens.* New Haven, CT: Yale University Press.

———. 2009. "An Informational Approach to Sustainability: 'Intelligence' in Conservation and Natural Resource Management Policy." *Journal of Sustainable Forestry* 28:636–62.

Clark, S. G., A. Hohl, C. Picard, and D. Newsome, eds. 2010. "Large Scale Conservation: Integrating Science, Management, and Policy in the Common Interest." Yale School of Forestry and Environmental Studies, Working Paper 24.

Clark, S. G., and R. Wallace. 2012. "Interdisciplinary Environmental Leadership: Learning and Teaching Integrated Problem Solving." In *Environmental Leadership:*

A Reference Handbook, edited by D. Gallagher, N. Christensen, and R. N. L Andrews, 420–29. Thousand Oaks, CA: Sage Publications.

Clark, T. W. 1997. *Averting Extinction: Reconstructing Endangered Species Recovery*. New Haven, CT: Yale University Press.

Clark, T. W., and M. B. Rutherford. 2005. "The Institutional System of Wildlife Management: Making It More Effective." In *Coexisting with Large Carnivores: Lessons from Greater Yellowstone*, edited by T. W. Clark, M. B. Rutherford, and D. Casey, 211–53. Covelo, CA: Island Press.

Constanza, R., R. D'Arge, R. De Groot, S. Farber, M. Grasso, B. Hannon, K. Limburg, et al. 1997. "The Value of the World's Ecosystem Services and Natural Capital." *Nature* 387:253–60.

Crane, J. 2009. "Guest Editorial: Promoting the North American Model." *The Wildlife Professional* 3(2):6.

Daft, R. L. 1995. *Organization Theory and Design*. New York: West Publishing.

Daily, G. C. 1997. *Nature's Services: Societal Dependence on Natural Ecosystems*. Washington, DC: Island Press.

Diamond, J. 2005. *Collapse: How Societies Choose to Fail or Succeed*. New York: Viking.

Etheredge, L. S. 1985. *Can Governments Learn?* New York: Pergamon Press.

Foley, J. A., R. DeFries, G. P. Asner, C. Barford, and G. Bonanm. 2005. "Global Consequences of Land Use." *Science* 309:570–74.

Galbraith, J. R. 1977. *Organization Design*. New York: Addison-Wesley.

Geist, V. 1988. "How Markets in Wildlife Meat and Parts, and the Sale of Hunting Privileges, Jeopardize Wildlife Conservation." *Conservation Biology* 2:15–26.

———. 1995. "North American Policies of Wildlife Conservation." In *Wildlife Conservation Policy*, edited by V. Geist and I. McTaggert-Cowan, 75–129. Calgary: Detselig Enterprises.

———. 2004. "The North American Model of Wildlife Conservation as a Means of Creating Wealth and Protecting Public Health While Generating Biodiversity." *Wild Lands Advocate* 12 (6): 1–5.

Geist, V., S. P. Mahoney, and J. F. Organ. 2001. "Why Hunting Has Defined the North American Model of Wildlife Conservation." *Transactions of the North American Wildlife and Natural Resources Conference* 66:175–85.

Gill, R. B. 1996. "The Wildlife Professional Subculture: The Case of the Crazy Aunt." *Human Dimensions of Wildlife* 1:60–69.

Hagood, S. 1997. *State Wildlife Management: The Perverse Influence of Hunters, Hunting, Culture and Money*. Washington, DC: Humane Society of the United States.

Hardin, G. 1968. "The Tragedy of the Commons." *Science* 162:1243-1248.

Heinz Foundation. 2005. *The Millennium Ecosystem Assessment Series and Synthesis Reports*. Washington, DC: Island Press.

Hewitt, C. G. 1921. *The Conservation of Wildlife in Canada*. New York: C. Scribner's Sons.

Hobbs, R. J., and J. A. Harris. 2001. "Restoration Ecology: Repairing the Earth's Ecosystems in the New Millennium. *Restoration Ecology* 9:239–46.

Jacobson, C. A., and D. J. Decker 2006. "Ensuring the Future of State Wildlife Management: Understanding Challenges for Institutional Change." *Wildlife Society Bulletin* 34:531–36.

Jacobson, C. A., D. J. Decker, and L. Carpenter. 2007. "Securing Alternative Funding for Wildlife Management: Insights from Agency Leaders." *Journal of Wildlife Management* 71:2106–13.

Jacobson, C. A., J. F. Organ, D. J. Decker, G. R. Batcheller, and L. Carpenter. 2010. "A Conservation Institution for the 21st Century: Implications for State Wildlife Agencies." *Journal of Wildlife Management* 74:203–09.

Jackson, J. B. C. 2001. "What Was Natural in Coastal Oceans?" *Proceedings of the National Academy of Sciences* 98:5411–18.

Jahn, L. R., and C. Kabat. 1984. "Origin and Role." In *Flyways: Pioneering Waterfowl Management in North America*, edited by A. S. Hawkins, R. C. Hanson, H. K. Nelson, and H. M. Reeves, 374–86. Washington, DC: US Fish and Wildlife Service.

Kallman, H., ed. 1987. *Restoring America's Wildlife*. Washington, DC: US Fish and Wildlife Service.

Kellert, S. R. 1996. *The Value of Life: Biological Diversity and Human Society*. Washington, DC: Island Press.

Kerasote, T. 1994. *Bloodties: Nature, Culture, and the Hunt*. New York: Kodansha International.

Lasswell, H. D. 1971. *A Pre-View of Policy Sciences*. New York: American Elsevier.

Lasswell, H. D., and A. Kaplan. 1950. *Power and Society: A Framework for Political Inquiry*. New Haven, CT: Yale University Press.

Lasswell, H. D., and M. S. McDougal. 1992. *Jurisprudence for a Free Society: Studies in Law, Science, and Policy*. New Haven, CT: New Haven Press.

Lawrence, R., and S. E. Daniels. 1997. "Procedural Justice and Public Involvement in Natural Resource Decision-Making." *Society and Natural Resources* 10:577–90.

Leopold, A. 1930. "Report to the American Game Conference on an American Game Policy." *Transactions of the American Game Conference* 17:281–83.

———. 1933. *Game Management*. New York: Charles Scribner's Sons.

Leopold, B. 2010. "Understanding Our Roots: The North American Model of Wildlife Conservation." *Wildlife Society Bulletin* Fall:12.

Lund, T. A. 1980. *American Wildlife Law*. Berkeley: University of California Press.

Mahoney, S. 2004a. "North American Wildlife Conservation Model." *Bugle* 21 (3): 1–3.

———. 2004b. "The Seven Sisters: Pillars of the North American Conservation Model." *Bugle* 21 (4): 1–3.

———. 2006. "The North American Wildlife Conservation Model: Triumph for Man and Nature." *Fair Chase* 21 (Fall): 20–25.

Mahoney, S. P., V. Geist, J. Organ, R. Regan, G. R. Batcheller, R. D. Sparrowe, J. E. McDonald, et al. 2008. "The North American Model of Wildlife Conservation: Enduring Achievement and Legacy." *Sportsmen's Conservation Council* 2008:7–24.

Mangun, J. C., T. O'Leary, and W. R. Mangun. 1992. "Non-Consumptive Wildlife-Associated Recreation in the United States: Identity Land Dimensions." In

American Fish and Wildlife Policy: The Human Dimension, edited by W. R. Mangun, 175–200. Carbondale: Southern Illinois University Press.

Mansell, W. D., ed. 2000. *Proceedings of the 2000 Premier's Symposium on North America's Hunting Heritage*. Eden Prairie, MN: Wildlife Forever.

Matthiessen, P. 1959. *Wildlife in America*. New York: Viking Press.

Mattson, D. J., K. L. Byrd, M. B. Rutherford, S. R. Brown, and T. W. Clark. 2006. "Finding Common Ground in Large Carnivore Conservation: Mapping Contending Perspectives." *Environmental Sciences and Policy* 9:392–405.

Mattson, D. J., and N. Chambers. 2009. "Human-Provided Waters for Desert Wildlife: What Is the Problem?" *Policy Sciences* 42:113–36.

Mattson, D. J., and S. G. Clark. 2010. "People, Politics, and Cougar Management." In *Cougar: Ecology and Conservation*, edited by M. Hornocker and S. Negri, 206–20. Chicago: University of Chicago.

McDougal, M. S. 1992–1993. "Legal Basis for Securing the Integrity of the Earth-Space Environment." *Journal of Natural Resources & Environmental Law* 8:177–207.

McDougal, M. S., H. D. Lasswell, and M. Reisman. 1981. "The World Constitutive Process of Authoritative Decision." In *International Law Essays: A Supplement to International Law in Contemporary Perspective*, edited by M. S. McDougal and M. Reisman, 191–286. New York: Foundation Press.

McLaughlin, G., S. Primm, and M. B. Rutherford. 2005. "Participatory Projects of Coexistence: Rebuilding Civil Society." In *Coexisting with Large Carnivores: Lessons from Greater Yellowstone*, edited by T. W. Clark, M. B. Rutherford, and D. Casey, 177–210. Washington, DC: Island Press.

Meine, C. 1988. *Aldo Leopold: His Life and Work*. Madison: University of Wisconsin Press.

Merkle, J. 1980. *Management and Ideology: The Legacy of the International Scientific Management Movement*. Berkeley: University of California Press.

———. 1998. "Scientific Management." In *International Encyclopedia of Public Policy and Administration*, edited by J. M. Shafritz, 2036–40. Boulder, CO: Westview Press.

Miller, B. J., R. P. Reading, and S. Forrest. 1996. *Prairie Nights: Black-Footed Ferrets and the Recovery of Endangered Species*. Washington, DC: Smithsonian Institution Press.

Miller, D. A. 2010. "Hunting as a Wildlife Management Tool." *Wildlife Society Bulletin* Fall (2010):13.

Morcol, G. 2001. "Positivist Beliefs among Policy Professionals: An Empirical Investigation." *Policy Sciences* 34 (3-4): 381–401.

Muth, R., and J. M. Bolland. 1983. "Social Context: A Key to Effective Problem Solving." *Planning and Change* 14:214–25.

Myers, N. 1996. "Environmental Services of Biodiversity." *Proceedings of US National Academy of Sciences* 93:2764–69.

Nature Conservancy. 2001. *Status of Biodiversity*. In cooperation with the Association for Biodiversity Information. Washington, DC: Nature Conservancy.

Nie, M. 2004. "State Wildlife Governance and Carnivore Conservation." In *People and Predators: From Conflict to Coexistence*, edited by N. Fascione, A. Delach, and M. E. Smith, 197–218. Washington, DC: Island Press.

———. 2008. "The Underappreciated Role of Regulatory Enforcement in Natural Resource Conservation." *Policy Sciences* 41:139–64.

Odum, E. P. 1989. *Ecology and Our Endangered Life-Support Systems*. Sunderland, MA: Sinauer Associates.

Organ, J. F., R. Gotie, T. A. Decker, and G. R. Batcheller. 1998. "A Case Study in the Sustained Management of Beaver in the Northeastern United States." In *Enhancing Sustainability—Resources for Our Future*, edited by H. A. ver der Linde and M. H. Danskin, 125–29. Gland: IUCN.

Organ, J. F., S. P. Mahoney, and V. Geist. 2010. "The North American Model of Wildlife Conservation." *Wildlife Society Bulletin* Fall (2010):12.

Organ, J. F., R. M. Muth, J. E. Dizard, S. J. Williamson, and T. A. Decker. 1998. "Fair Chase and Humane Treatment: Balancing the Ethics of Hunting and Trapping." *Transactions of the North American Wildlife and Natural Resources Conference* 63: 528–43.

O'Riordan, T., and S. Stoll-Kleemann. 2002. "Deliberative Democracy and Participatory Biodiversity." In *Biodiversity, Sustainability, and Human Communities*, edited by T. O'Riordan and S. Stoll-Kleemann, 87–113. London: Cambridge University Press.

Ostrom, E. 1996. "Institutional Rational Choice: An Assessment of the Institutional Analysis and Development Framework." In *Theories of the Policy Process*, edited by P. A. Sabatier, 126–41. Boulder, CO: Westview Press.

Pacel, W. 1998. "Forging a New Wildlife Management Paradigm: Integrating Animal Protection Value." *Human Dimensions of Wildlife* 3:42–50.

Perrow, C. A. 1970. *Organizational Analysis: A Sociological View*. Belmont, CA: Brooks/ Cole.

Pielke, R. A., Jr. 2007. *The Honest Broker: Making Sense of Science in Policy and Politics*. Cambridge: Cambridge University Press.

Pimbert, M., and J. N. Pretty. 1995. "Parks, People, and Professionals: Putting 'Participation' into Protected Area Management." Discussion Paper No. 57. Geneva: United Nations Research Institute for Social Development.

Pimm, S. L., G. J. Russell, J. L. Gittleman, and T. M. Brooks. 1995. "The Future of Biodiversity." *Science* 269:347–50.

Posewitz, J. 1994. *Beyond Fair Chase: The Ethics and Tradition of Hunting*. Helena, MT: Falcon Press.

Prescott-Allen, R., and C. Prescott-Allen. 1996. "Assessing the Impacts of Uses of Mammals: The Good, the Bad, and the Neutral." In *The Exploitation of Mammal Populations*, edited by V. T. Taylor and N. Dunstone, 45–61. London: Chapman and Hall.

Prukop, J., and R. J. Regan. 2005. "In My Opinion: The Value of the North American Model of Wildlife Conservation—An International Association of Fish and Wildlife Agencies Position." *Wildlife Society Bulletin* 33:374–377.

Redman, C. L. 1999. *Human Impact on Ancient Environments*. Tuscon: University of Arizona Press.

Reiger, J. F. 2001. *American Sportsmen and the Origins of Conservation*. Corvalis: Oregon State University Press.

Robinson, M. 2005. *Predatory Bureaucracy: The Extermination of Wolves and the Transformation of the West*. Boulder: University of Colorado Press.

Romesburg, H. C. 1981. "Wildlife Science: Gaining Reliable Knowledge." *Journal of Wildlife Management* 45:293–313.

———. 1989. "More on Gaining Reliable Knowledge: A Reply." *Journal of Wildlife Management* 53:1177–80.

Roth, T., and S. S. Boynton. 1993. "Some Reflections on the Development of National Wildlife Law and Policy and the Consumptive Use of Renewable Wildlife Resources." *Marquette Law Review* 77:71–83.

Rutherford, M. B., M. L. Gibeau, S. G. Clark, and E. C. Chamberlain. 2009. "Interdisciplinary Problem-Solving Workshops for Grizzly Bear Conservation in Banff National Park, Canada." *Policy Sciences* 42:163–88.

Sala, O. E., F. S. Chapin III, J. J. Armesto, E. Berlow, J. Bloomfield, R. Dirzo, E. Huber-Sanwald, et al. 2000. "Global Biodiversity Scenarios for the Year 2100." *Science* 274:1770–74.

Saul, J. R. 1993. *Voltaire's Bastards: The Dictatorship of Reason in the West*. New York: Vintage.

Scheffer, M., and S. R. Carpenter. 2003. "Catastrophic Regime Shifts in Ecosystems: Linking Theory to Observation." *Trends in Ecology and Evolution* 18:648–56.

Schön, D. A. 1983. *The Reflective Practitioner: How Professionals Think in Action*. New York: Basic Books.

Scott, J. C. 1998. *Seeing Like a State: How Certain Schemes to Improve the Human Condition Have Failed*. New Haven, CT: Yale University Press.

Semcer, B. C. 2008. "The North American Model of Wildlife Conservation: Affirming the Role, Strength and Relevance of Hunting in the 21st Century." Paper presented at the 73rd North American Wildlife and Resources Conference, Washington.

Sinclair, A. R. E., and A. E. Byrom. 2006. "Understanding Ecosystem Dynamics for the Conservation of Biota." *Journal of Animal Ecology* 75:64–79.

Slotkin, R. 1973. *Regeneration through Violence: The Mythology of the American Frontier, 1600–1860*. Hanover, NH: Wesleyan University Press.

———. 1985. *Fatal Environment: The Myth of the Frontier in the Age of Industralization, 1800–1890*. New York: Harper Perennial.

———. 1992. *Gunfighter Nation: The Myth of the Frontier in Twentieth-Century America*. New York: Harper Perennial.

Soulé, M. E., J. A. Estes, B. Miller, and D. L. Honnold. 2005. "Strongly Interacting Species: Conservation Policy, Management, and Ethics." *BioScience* 55:168–76.

Sporting Conservation Council. 2008. *Strengthening America's Hunting Heritage and Wildlife Conservation in the 21st Century: Challenges and Opportunities*. Washington, DC: Sporting Conservation Council.

Steelman, T. A. 2010. *Implementing Innovation: Fostering Enduring Change in Environmental and Natural Resource Governance*. Cambridge, MA: MIT Press.

Steelman, T., and M. E. DuMond. 2009. "Serving the Common Interest in US Forest Policy: A Case Study of the Healthy Forest Restoration Act." *Environmental Management* 43:396–410.

Steffen, W. L., A. Sanderson, P. D. Tyson, J. Jager, and P. A. Matson. 2004. *Global Change and the Earth System: A Planet under Pressure*. New York: Springer-Verlag.

Stein, B. A., L. S. Kutner, and J. S. Adams, eds. 2000. *Precious Heritage: The Status of Biodiversity in the United States*. New York: Oxford University Press.

Stemler, J. 2008. *White House Conference on Wildlife Policy Concludes Tasks, Hunting and Conservation Community to Carry Policies Forward*. Washington, DC: Sporting Conservation Council, October 3.

Thomas, C. W. 2003. *Bureaucratic Landscapes: Interagency Cooperation and the Preservation of Biodiversity*. Cambridge, MA: MIT Press.

Tilleman, W. A. 1995. "The Law Relating to Ownership of Wild Animals Is Rather Complicated." In *Wildlife Conservation Policy—A Reader*, edited by V. Geist and I. McTaggert-Cowan, 133–45. Calgary: Detselig Enterprises.

Tober, J. A. 1989. *Wildlife in the Public Interest: Nonprofit Organizations and Federal Wildlife Policy*. New York: Praeger.

Turner, B. L., W. C. Clark, R. W. Kates, J. F. Richards, J. T. Mathews, and W. B. Meyer, eds. 1995. *The Earth as Transformed by Human Action: Global and Regional Changes in the Biosphere over the Past 300 Years*. New York: University of Cambridge Press.

Trefethen, J. B. 1975. *An American Crusade for Wildlife*. New York: Winchester Press.

Tylianakis, J. M., R. K. Didham, J. Bascompte, and D. A. Wardle. 2008. "Global Change and Species Interactions in Terrestrial Ecosystems." *Ecology Letters* 11:1351–63.

US Fish and Wildlife Service, and US Department of Commerce, US Census Bureau. 2006. *2006 National Survey of Fishing, Hunting, and Wildlife-Associated Recreation*. Washington, DC: US Fish and Wildlife Service.

US Fish and Wildlife Service. 2008. *Conservation Heritage Strategic Plan: Wildlife and Sport Fish Restoration Program*. Washington: US Fish and Wildlife Service.

———. 2009a. "Endangered Species Program." Accessed July 14, 2009. Available from http://www.fws.gov/Endangered/wildlife.html#Species.

———. 2009b. "Southeast Region: Federal Aid Division—The Pittman-Robertson Federal Aid in Wildlife Restoration Act." Accessed July 14, 2009. Available from http://www.fws.gov/southeast/federalaid/pittmanrobertson.html.

———. 2009c. *Wildlife and Sportfish Restoration Program*. Accessed July 7, 2009. Available from http://wsfrprograms.fws.gov/.

Walther, G. R., E. Post, P. Convey, A. Menzel, C. Parmesan, T. J. C. Beebee, J. Fromenti, O. Hoegh-Guldberg, and F. Bairlein. 2002. "Ecological Responses to Recent Climate Change." *Nature* 416:389–95.

Webster's New Universal Unabridged Dictionary. 1994. Avenel, NJ: Barnes and Noble.

Wildlife Society. 2007. "Final TWS Position Statement on the North American Model of Wildlife Conservation." Available from http://joomla.wildlife.org/documents /positionstatements/41-NAmodel%20Position%20Statementfinal.pdf.

Wilkinson, K., S. G. Clark, and W. Burch. 2007. "Other Voices, Other Ways, Better

Practices: Bridging Local and Professional Knowledge." Yale School of Forestry and Environmental Studies, Report No. 14.

Wilkinson, T. 1998. *Science under Siege: The Politicians' War on Nature and Truth.* Boulder, CO: Johnson Books.

Wilson, J. Q. 1989. *Bureaucracy: What Government Agencies Do and Why They Do It.* New York: Basic Books.

Wilson, S. M., and S. G. Clark. 2007. "Resolving Human-Grizzly Bear Conflict: An Integrated Approach in the Common Interest." In *Integrated Resource and Environmental Management*, edited by K. S. Hanna and D. S. Slocombe, 137–63. New York: Oxford University Press.

Young, O. R., L. A. King, and H. Schroeder. 2009. "Summary for Policy Makers." In *Institutions and Environmental Change: Principal Findings, Applications, and Research Frontiers*, edited by O. R. Young, L. A. King, and H. Schroeder, xiii–xix. Cambridge, MA: MIT.

10 Complexity, Rationality, and the Conservation of Large Carnivores

DAVID J. MATTSON AND SUSAN G. CLARK

Introduction

Protecting large carnivores in the western United States is complex. This complexity is revealed by the cases in this book, whether they are large in scale such as regional management of mountain lions or smaller endeavors such as watershed-level management of carnivore-livestock conflicts. Even when the focus is on small areas over limited spans of time, the relevant dynamics are guaranteed to be shaped by broader-scale, longer-term conditions (chapters 1, 8, and 9).

Complexity can be understood in various ways. We define complexity as the number and strength of the contingencies that will affect an outcome of interest, compounded by nonlinearities, synergies, scale-dependent structures, and cross-scale flows. Complex systems are dynamic, emergent, evolving, dissipative, and nonequilibrium (Weaver 1948; Funtowicz and Ravetz 1994). Ecologists suggest that different configurations of complex systems engender varying degrees of resistance to change and resilience to perturbation, which, from an ecosystem perspective, are thought to be desirable (Holling 1994; Gunderson and Holling 2002).

At a minimum, the domains of complex conservation systems include not only the biophysical, social, psychological, and policy elements, but also observer and participant subjectivities (Clark 1993, 2002; Harding, Hendricks, and Farqui 2009). States and flows of organisms, entities, materials, information, values, narratives, and identities are all relevant to making sense of the cases presented in this book. But the filters imposed by our individual subjectivities are equally relevant, especially for communications that are aimed at building a shared and useful understanding of the world among those who strive for constructive change (Richards 2001; Charon 2007).

We contend that greater rationality can contribute to making conservation gains in a complex, contingent, and dynamic world. We define rationality as more about process than outcomes, specifically, the process of bringing more aspects of a situation into consciousness as an integral part of discernment, judgment, and choice (Jones 1999). Rationality is an important means by which people learn lessons from experience along with the contingencies of context that will dictate how best to extrapolate and apply those lessons (Mezirow 1996; Miller 1996). Rationality is also a means by which we can see the essentials that will be relevant to developing and applying effective tactics and strategies. Yet we reject the positivistic notion of perfect rationality, especially as an outcome (Simon 1982). If nothing else, rationality is intrinsically bounded by affect, language, and all the other ways that people represent the world to themselves and others, which is partly why we see subjectivity as central to understanding human behaviors (Mellers, Schwartz, and Ritov 1999; Böhm and Brun 2008).

In this chapter we define the problem of complexity and offer some concepts and frames that we find useful for embracing, distilling, and communicating the complexity of conservation systems. We hope to help people reach a level of rationality sufficient for them to engage effectively with one another and the world to achieve their desired ends. The notion of parsimony is central to this endeavor; useful insight depends on recognizing essential patterns without being overwhelmed by specifics, which embodies the eternal tension between comprehensiveness and selectivity (Newell and Simon 1972). We also invoke the standard of sufficiency rather than optimization. In keeping with Herbert Simon, we view humans primarily as satisficers, although what might be sufficient for one person in a particular context may not be sufficient for another (Simon 1956; Webster and Kruglanski 1994; Kuhn 2001).

Our overarching goal here is to foster collective rationality in the service of a commonwealth of human dignity, which we define as a condition where as many people as possible participate in and benefit from the shaping and sharing of essential values, signified by widespread experiences of well-being (Mattson and Clark 2011). We see this commonwealth as critical to achieving the enduring conservation of large carnivores in the West. It is highly unlikely that a society dominated by indignity can support a healthy and sustainable environment or healthy and sustainable carnivore populations

(Homer-Dixon 1999). This goal provides an essential criterion for analyzing and appraising the many aspects of conservation cases. We all pay attention to the things we care about, for which we have aspirations and emotional attachment (Damasio 1994; Dalgleish 2004). Goals are a key part of the frames we use to judge the sufficiency and parsimony of analysis and appraisal. In what follows, we feature not only concepts and frames, but also standards linked directly to the dignity goal, by which we propose to judge the performance of carnivore conservation systems.

The Problem of Complexity

Most people have difficulty comprehending complex systems (Newell and Simon 1972). Even if they are sufficiently oriented to a situation, they may have trouble communicating their insights or the subjective elements that likely shaped their understandings. The potential difficulties of dealing with complexity in a social context are manifold: achieving sufficient insight, communicating that insight, and recognizing and sharing subjectivities (Newell and Simon 1972; Clark 2002; Harding, Hendricks, and Farqui 2009).

In this section we elaborate on the problem of complexity. We start with synopses of two cases that highlight key facets of complexity in large carnivore conservation, featuring management of mountain lions in the West and grizzly bears in the Yellowstone ecosystem. The first case is described in more detail in a separate chapter. The second is not, but has been well documented. We then provide an overview of cognitive phenomena that intrinsically limit how humans deal with complex systems, followed by a brief critique of current theories and methods for managing and talking about nature-human systems. Finally, we describe a useful framework (*sensu* Ostrom 2011) for analyzing, discussing, and representing complexity, especially in appraisals of policy systems.

MOUNTAIN LIONS IN THE WEST

This case highlights the interaction of trends in populations of lions and their prey, scientific and other representations of these dynamics, volatile micro- and meso-scale dynamics of the social process organized around contestations of power, macro-scale social and institutional trends related to perspectives on wildlife and governance, and related stakeholder demands regarding the outcomes of management.

Mountain lions have recently spread into areas from which they were extirpated during the previous century (Beier 2010). Their principal prey, mule deer, has at the same time declined in many areas. Another prey species, bighorn sheep, is also at risk. However, other prey species such as elk and white-tail deer have increased. The causes of mule deer and bighorn sheep declines and elk and whitetail deer increases are vigorously debated, with mountain lions blamed by some for the declines (Ruth and Murphy 2010; Mattson and Clark 2010). Except in national parks, humans cause the deaths of most adult lions, principally through sport hunting (Quigley and Hornocker 2010).

People's representations of these dynamics—scientific and otherwise—have been politicized in ways that are aligned with conflicting worldviews and interests, with little prospect of proving cause and effect in highly variable and contingent natural systems (Mattson and Clark 2010; Ruth and Murphy 2010). Interacting effects are commonplace, replication of study conditions is difficult and expensive, and other prospects for control are minimal. Sorting out the effects of weather, habitat change, prey trends, sport hunting, human infrastructure, and predation is difficult. At the same time, it is comparatively easy for people to construct narratives that appeal to self-evident "truths" and assumptions and do not require a definitive empirical basis (Stone 1989). These partisan explanatory narratives appear to interact synergistically with the persistent uncertainty of scientific explanations to fuel conflict (chapters 1 and 2; Mattson and Clark 2010).

Volatile social dynamics are also galvanized by power struggles that are linked to the deferential service of hunters' interests by state wildlife managers. Those who place an intrinsic value on lions feel disenfranchised, especially in contrast to those who prize domination and hunting (Gill 1996, 2001; Nie 2004a, 2004b; Clark and Rutherford 2005; Jacobson and Decker 2006; Mattson and Clark 2010; chapter 2). State-level wildlife management in the West is a rigid system that has emerged because power has been held exclusively in the hands of those with predominantly a single worldview and related set of interests. Hunters, the primary clients, directly or indirectly provide most of the funding in a system where those with authority (commissioners and agency personnel) share their preferences and perspectives—a synergistic and self-reinforcing system. These arrangements have led marginalized stakeholders, typically those who advocate animal protection, to resort to agitating strategies to gain access to decision making, including litigation, legislation, and the creation of focusing events (Mattson and Clark 2010). Focusing events, which typically involve the negative reactions of urban stakeholders to routine lethal management tactics of the wildlife agencies, are a prime example of pronounced nonlinear dynamics (chapters 2 and 3; Mattson and Clark 2012).

Many tensions have accumulated in this management system because of broad-scale, long-term societal trends. Novel worldviews, featuring intrinsic or aesthetic valuations of lions, have emerged since the early 1900s, but have not had substantive expression in the structures of state wildlife management institutions (Gill 2010; Mattson and Clark 2010). These new worldviews are correlated with geographically uneven societal changes, such as urbanization and higher education, that have resulted in the coupling of demographic profiles with different demands regarding the outcomes of lion management (Mattson and Clark 2010). This linking has fueled destructive ingroup-outgroup dynamics, with each group negatively constructing those who have different interests (cf. Alon and Omer 2006). Overall, the dynamics of state-level wildlife management are a classic example of macro-scale change interacting with institutional arrangements and stakeholders' narratives to shape conflict at smaller scales. These dynamics seem to be largely opaque to most of the participants in this system, who show little evidence of having examined their own subjectivities (Clark and Rutherford 2005; Mattson and Clark 2010).

GRIZZLY BEARS IN THE YELLOWSTONE ECOSYSTEM

Management of grizzly bears in the Yellowstone ecosystem of Montana, Wyoming, and Idaho reveals dynamics similar to those of mountain lion management, but also different from them. In particular, grizzly bear management highlights the interaction of surprising and emergent biophysical changes with the methods and norms of politicized science, scientized politics, and scale-dependent geopolitical dynamics.

Although grizzly bears in Yellowstone have exhibited well-described behavioral tendencies, such as heavy consumption of whitebark pine seeds at high elevations, their behavior has also exhibited surprising and dramatic changes (Mattson, Blanchard, and Knight 1991). Human garbage disappeared as a major food in the early 1970s, followed by the unexpected emergence of cutthroat trout and army cutworm moths as major foods during the 1970s and 1980s (Craighead, Sumner, and Mitchell 1995; Reinhart and Mattson 1990; Mattson et al. 1991). This was followed during the 2000s by the virtual extirpation of both whitebark pine (by an unprecedented, climate-driven outbreak of mountain pine beetles; see Macfarlane, Logan, and Kern 2012) and cutthroat trout (by a nonnative introduced predator, lake trout; see Haroldson et al. 2005), with an apparent compensatory shift by bears toward ever greater reliance on meat from elk and bison (Haroldson et al. 2004). Meanwhile, the effects of reintroduced wolves on the availability of meat are yet to be fully realized, although episodic and largely unpredictable uses of alternative foods such as mushrooms and various roots and insects might be expected to continue (Vucetich, Smith, and Stahler 2005; White and Garrott 2005; Mattson 2000). Importantly, most of these changes have been caused by human activities, and virtually all were unexpected.

Meanwhile, the intrinsic limits of scientific methods, interacting with preferential scientific norms and related explanatory narratives, have prevented convergence on widely-accepted conclusions regarding the effects of all these habitat changes on bear numbers and behaviors (Mattson and Craighead 1994; Mattson et al. 1996b; MacCracken and O'Laughlin 1998). The burden of proof is allocated differently by different people, analytic frames are contested, and the objectivity of technical specialists, including scientists, is called into question (Mattson and Craighead 1994; Mattson et al. 1996b; MacCracken and O'Laughlin 1998; Wilkinson 1998; Peacock and Peacock 2009; Kevin 2011). These and other phenomena signal politicized science, which is a natural outcome of the extent to which politics in this case have been scientized, that is to say, typified by the invocation of science to justify and advance special interest agendas that are rooted in different interests and worldviews (e.g., Mattson and Chambers 2009; Mattson and Clark 2010). More to the point of this chapter, the grizzly bear case illustrates the extent to which people's conflicting stories can compound the complexity that already exists in human-nature relationships. Self-reinforcing dynamics and synergies are amply evident (Mattson et al. 1996b).

Finally, the grizzly bear case illustrates the extent to which symbolic stakes get entangled with contests of power, coupled to scale-dependent geopolitical dynamics. Since the 1970s grizzly bears have been managed under the auspices of the federal Endangered Species Act, which unambiguously empowers not only the federal government, but also people who value grizzly bears for intrinsic rather than instrumental reasons (Servheen 1998; Kuehl 1993; Primm and Murray 2005). Federal management gives some degree of standing to people throughout the United States who prize grizzly bears for their beauty and mere existence. By contrast, state-level management gives primacy to hunters and others regionally who hold utilitarian perspectives, as illustrated by the mountain lion case (Mattson and Clark 2010; chapter 9). All of this fuels conflict between those who are aligned with state management versus those who are aligned with federal management, amplified by negative ingroup-outgroup dynamics organized around worldviews, identity, and place of residence (Primm and Clark 1996; Primm and Murray 2005). As in the mountain lion case, most participants appear to be confined by largely unexamined mental models and behaviors that perpetuate dynamics that are potentially destructive to civil society and human dignity (Servheen 1998; Mattson et al. 2006; Mattson and Clark 2011; e.g., Wilkinson 1998).

The Limits of Human Cognition

Given this complexity, it is worth examining the intrinsic capacity of people to deal with information about contingent, nonlinear, and nonequilibrium systems. Our cognitive capacities are central to the question of whether we will remain trapped in the emergent systems that have self-organized around negative human behaviors or whether, through analysis, we will gain insights that allow us to prescribe and implement policies and practices leading to more positive societal outcomes.

Humans have limited abilities to process information in real time (Cowan 1995). Studies have shown that we can hold only 7 ± 2 chunks or bits of information in short-term memory and related buffers (Miller 1956; Kareev 2000), that our so-called span of judgment is limited to between 2 and 3 binary structures (Halford, Wilson, and Phillips 1998; Birney and Halford 2002), that we can efficiently process no more than 3 to 5 chunks of information in parallel (Schneider and Detweiler 1987); and that most of us can accurately interpret no more than 3-way relational interactions (Halford et al. 2005). Attempts by people to interpret 5-way interactions typically yield results comparable to what one would expect by chance. Related problem-solving time and use of cognitive resources increase geometrically with increasing problem complexity, especially beyond 2 by 2 problem structures (Ashcroft 1992; Hirschi and Frey 2002; Kello et al. 2010). Visual working memory exhibits similar limitations (Ware 2004). Compounding all of this, people differentially retain and attend to information and experience based on its recency, primacy, salience, and consistency with prior beliefs (Simon 1982; Cowan 1995), to the extent, for example, that people who are uncertain about climate change are more likely to believe in global warming during unseasonably hot days than during unseasonably cold days (Li, Johnson, and Zaval 2011).

Given these constraints, people exhibit a number of adaptations to deal with the complexities of our environment. First and foremost, we combine bits of information into larger and larger conceptual packages, which is the fundamental imperative behind our pervasive categorization of the world (Halford, Wilson, and Phillips 1998; Simon 1974). Much of our success in adequately orienting to the world depends on whether the categories we devise efficiently capture fundamental structures (Newell and Simon 1972; Simon 1982). We also sequence tasks so that we can engage in serial as well as parallel processing (Zhang and Norman 1994; Halford et al. 1998). The efficacies of our sequencing depend on how well our goal orientation and declarative memory modules (e.g., mental models) maintain self-coherence and orientation in time and space (Anderson et al. 2004). Increasingly, we resort to external aids such as physical and digital scratch pads and other storage devices, which are more or less effective depending on how we've indexed information (Newell and Simon 1972; Zhang 2000). But, in the end, most of us seem to deal with complexity by ignoring information and defaulting to simpler representations of reality, which may allow us to manage cognitive dissonance, but at the potential cost of effective orientation (Elman 1993; Halford et al. 1998; Birney and Halford 2002).

Most cognitive scientists seem to accept a dual processing model of human cognition (Evans 2003, 2008). The two types of processing are variously called reflexive versus reflective, heuristic versus analytic, automatic versus deliberative, or type I versus type II (Evans 2008; Glöckner 2008; Keysers et al. 2008; Stanvoich 2011). Although some argue that these are merely two ends of a continuum rather than differentiated processes, the key distinction is between responses that are, on the one hand, instantaneous, parallel, and largely subconscious versus, on the other hand, more prolonged, serial, and overtly conscious (Newell 2005; Evans 2008; Glöckner and Witteman 2010). The reflexive mode is thought to have arisen from the evolutionary imperative for fast and frugal, or effort-reducing, decision-making processes (Gigerenzer 2004; Shah and Oppenheimer 2008). This mode is thought to employ what are commonly called heuristics, or prepackaged responses, accessed by environmental cues interacting with patterns stored in long-term memory (Gigerenzer, Todd, and ABC Research Group 1999). According to this model, heuristics are triggered based on probabilistic matches of perceived current patterns to past experiences, and they are fine-tuned over time by the perceived success of each in addressing the demands that are presented (Gigerenzer et al. 1999; Rieskamp and Otto 2006). Some research has suggested that "fast and frugal" is more successful than "conscious and analytic" in addressing commonplace or rapidly evolving situations (e.g., Goldstein and Gigerenzer 2002). But appraising or planning strategies for intervening in complex systems such as large carnivore management is a fundamentally different kind of context (Harding, Hendricks, and Farqui 2009).

These insights into human cognition clarify elements that are germane to upgrading our abilities to engage rationally with complex conservation systems. Most participants in the cases featured in this book appear to be caught in fast and frugal decision making based on simple heuristics or a need for rapid cognitive closure. One could argue that this is fundamentally problematic because these heuristics and related mental models have led people to behave generally in ways that are problematic for the commonwealth of human dignity. Rationality is a primary means by which people can reflect on their mental models and related habitual behaviors (e.g., heuristics) in order to foster intentional change (Kegan 1994). Key elements are the ways in which we have conceptualized, or chunked, the world, our models of relations among chunks, and our individual and collective abilities to analyze and interrelate analytic tasks serially (Newell and Simon 1972; Halford, Wilson, and Phillips 1998). But whatever our aspirations, our limited ability to process information instantaneously will always be a constraint (Baddeley 1986).

The Limits of Existing Complexity Frameworks

Different people and different intellectual disciplines have come up with different conceptual systems for talking about and gaining insight into complexity. Some key and often repeated notions are self-organizing, emergent, nonlinear, and self-reinforcing (Cilliers 1998). Chaos theory features the mathematics of nonlinearities giving rise to bifurcations and attractors (Stewart 2002). Resilience theory features resistance, resilience, thresholds, and associated conservation and release of energy and materials (Holling 1994). Agency theory, network theory, operations research, and artificial intelligence and computational design employ similar but also different concepts.

Without presenting an exhaustive critique, all of these models have common limitations. None are comprehensive in the sense of *explicitly* covering the domains of biophysical, social, psychological, policy, and subjective phenomena germane to understanding the complex systems that policy analysts confront. None provide a pragmatic language for framing insights that directly and unambiguously lead to prospective action. All posit the existence of control parameters deriving from underlying rules, expressed as generalizable theoretical (versus meta-theoretical) propositions.

This cursory critique invokes meta-theory as a basis for judging theory in the service of policy analysis and intervention. Our meta-theoretic premises are postmodern. We adhere to the notion that useful explanation, as a guide to action in complex systems, is highly contextual. In the language of Cilliers (1998), representation is always distributed specific to individual systems and in defiance of generalizable algorithms or simple theory. This perspective is fully consistent with the centrality of relational contingency in nonequilibrium complex systems, which is the essence of the principle of con-figurative contextuality (Lasswell 1971a; Clark 2002). We also adhere to the notion that language and concept systems are an inescapable filter that we humans impose on the world (McAdams 1996; Koltko-Rivera 2004). System boundaries and features emerge as much from the application of symbolic constructions as from any underlying objective reality (Cilliers 1998). And, finally, in the spirit of Dewey, we adhere to pragmatism, which aligns with postmodernism and consequentialism (Dewey and Bentley 1949). We judge theory by whether it helps people define problems in ways that provide in-sights leading to actions likely to increase their well-being. By this standard, chaos theory, network theory, agency theory, resilience theory, operations research, and artificial intelligence and computation design all have severe limits when applied to the analysis of conservation systems.

The Problem

The problem of complexity in conserving large carnivores of the West can be succinctly stated. As a class, these cases are complex in all the ways we understand this notion and beset by high levels of conflict organized around elevated symbolic stakes. Most participants appear to be locked into unre-flective behaviors that are problematic for civil society, and they are often inattentive to the highly contingent and evolving nature of the biophysical, social, and policy environments. The comparatively inflexible institutions that have emerged to govern carnivore conservation also tend to exacer-bate rather than ameliorate conflict. Faced with these problems for the com-monwealth of human dignity, most people lack access to conceptual frame-works and social processes that would otherwise allow them to assess and constructively engage with the complex situations of modern-day carnivore management. Given the analysis presented here, we speculate that improve-ments could come from upgrading the rationality of as many participants as possible, organized around conceptual frames that foster highly contextual problem solving in the service of human dignity.

A Framework for Contextuality

In this section we offer a logically comprehensive framework that we have found to be helpful when appraising or otherwise consciously orienting to complex conservation cases. This framework is attentive to the consid-erations and meta-theoretic criteria introduced in preceding sections. A rationality-enhancing framework should foster attention to all potentially important features of a system, yet encourage parsimony. It should also feature concepts that link in obvious ways to prospective concrete actions. As a bottom line, any derivative explanations should refer to specific situations (according to the principle of situational reference; Lasswell and Kaplan 1950), be locally valid (Cilliers 1998), and offer practical insights.

We find it useful to differentiate the biophysical, social, and policy domains, along with the process of orientation itself. An ideal framework would "chunk" aspects of these domains into categories that efficiently capture structures and functions that are likely to be relevant to understanding most conservation systems. Individual schematics should have around 7 ± 2 categories that can be logically linked for serial analysis to match human cognitive limits. The framework should also correspond to documented structures of human cognition, including distinctions between perceptual, goal, and declarative modules of the in-vogue ACT-R model of cognition, as well as distinctions made by scholars such as Herbert Simon between problem solving and decision making (Anderson et al. 2004; Newell and Simon 1972). In what follows we describe schematics that are part of a framework pertaining to observer orientation, social process, policy process, and the biophysical arena.

THE INTELLECTUAL TASKS OF ORIENTATION
Background
Orientation is foundational to how well people deal rationally with complexity. It affects how people define problems and undertake tasks, which are explicitly or implicitly about solving problems (D'Zurilla and Goldfried 1971). Viewed this way, tasks can be understood as intentional efforts by people to change the world from how they perceive it to be to something more closely approximating what they want (Rochefort and Cobb 1994). This view, which links tasks to problems, makes human aspirations or goals central. However, given a goal, reaching the desired end depends on developing a causal model of the relevant system, explicating important static and dynamic features and dimensions, and, based on the new insights, developing and selecting strategies that offer prospects of success (D'Zurilla and Goldfried 1971; Newell and Simon 1972; Beach and Mitchell 1978). These are all seminal elements of rationality, which is to say that the efficacy of people's actions often depends on attending to all of these intellectual tasks consciously, empirically, and efficiently (e.g., Nezu and D'Zurilla 1981). People often deploy ineffective or even counterproductive strategies because they neglect or underattend to some orientation tasks (Clark 2002).

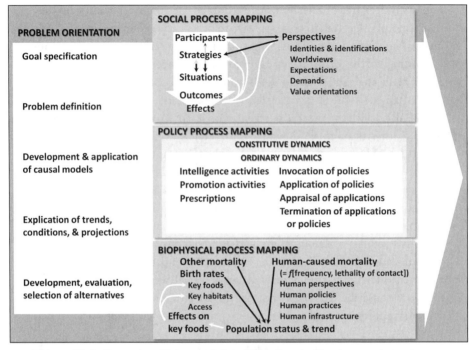

Figure 10.1. Synopsis of framework for efficiently orienting to complex conservation cases. This framework is largely derived from Lasswell (1971b), Clark (2002), and Mattson (1997, 2004).

A Schematic

Given this conception of problem-related orientation, a few universal tasks can be identified (Lasswell 1971b; Clark 2002; fig. 10.1). First is goal specification. Given an identified discrepancy between one's aspirations and the world as it exists, one can then identify "the problem," at least provisionally (Dery 1984). Having identified the problem, at least in terms of outcomes, one can then develop and apply causal models that help clarify important dynamic and static features of the relevant human-biophysical system. Once identified, these features then need to be examined relationally in the context of time as past trends, current conditions, and future projections. Then, with a useful, locally valid account of the system in hand, alternatives for solving the problem can be developed, evaluated, and selected. Thus the elements can be identified as: (1) goal specification; (2) problem definition; (3) development and application of causal models; (4) explication of past trends, current conditions, and future projections; and (5) developing, evaluating, and selecting alternatives (D'Zurilla and Goldfried 1971; Lasswell 1971b; Newell and Simon 1972; Beach and Mitchell 1978; Clark 2002). Having described these elements in sequential order, all can be carried out at any point in time, with the potential of leading to both revised understandings of "the problem" and goals.

This schematic relates logically to common conceptions of cognition. Goal specification and problem definition explicitly link to the goal module of cognition posited by the ACT-R model (Anderson et al. 2004). Developing a conception of the relevant system, located in time, space, and other key dimensions, relates to functioning of the ACT-R's declarative module and Simon's conception of analysis, distinct from decision making (Newell and Simon 1972; Kotovsky, Hayes, and Simon 1985). And the process of creating and selecting problem-solving alternatives entails Simon's "decision making" as well as the deontological, consequentialist, and affective procedural mechanisms of the ACT-R model (Anderson et al. 2004). All of these phenomena can be understood as key elements of how people maintain a sense of intentional coherence in the world.

The intellectual tasks of orientation provide a way of thinking and talking about the process of rationality. But the framing of these tasks does not offer much insight into what aspects of the world warrant at least provisional attention or, within broad domains, how to construe complexity in terms that are functional, efficient, and practical. For that, we need to consider schematics and concepts covering social, policy, and biophysical processes.

SOCIAL PROCESS
Background
Human interactions can be thought of as social process, which strongly configures how we behave and the decisions we make (Clark 2002). Construed this way, social process is about people in relationships and, as a consequence, about all the factors affecting how people view and relate to themselves and others (Charon 2007). Given the centrality of interaction, exchanges and the currencies of exchanges are key notions. Although people exchange materials and energy, they also create and exchange information and values. Information is inextricably bound to language, narrative, worldviews, and other uniquely human symbolic processes (Charon 2007). Values are linked to peoples' aspirations and beliefs—the things we want from the world (Lasswell and McDougal 1992, 375–587).

Although the notion of values can be understood in all sorts of ways, we find that conceiving of values functionally and in relation to more or less abiding individual orientations is particularly useful (Mattson, Karl, and Clark 2012). This perspective of values is intrinsic to schematics developed by Harold Lasswell (1948) and Shalom Schwartz (1994). Lasswell posited that humans orient to a handful of key values—power, wealth, respect, skill, rectitude, well-being, enlightenment, and affection—created and exchanged largely through the functioning of institutions. Schwartz posited power as well, but also achievement, hedonism, stimulation, self-direction, universalism, benevolence, tradition, conformity, and security. These complementary schematics can be used to describe and gain insight into people's motivations, self-assessments, and perspectives on "problems" (Mattson et al. 2012). Notice, too, that these schematics more or less conform to the 7 ± 2 rule.

With these basic notions about social process in mind, some features stand out as likely elements of a stable framework of inquiry into social process. Participants and participant attributes are a fundamental unit, regardless of whether considerations of scale dictate that participants be dealt with as individuals or aggregated to the level of generic identities (Lasswell and Kaplan 1950). Participant perspectives self-evidently shape all sorts of things that happen in social process, including expectations of how the world works, demands placed on the world (e.g., "interests"), strategies employed, and behaviors exhibited (Lasswell 1971b; Clark 2002).

With perspectives at center stage, several meta-concepts are germane. Perspectives are closely related to the notion of worldviews, which are linguistic and narrative in structure, which is to say that worldviews are given meaning through the symbolic and affective resonance of language (McAdams 1996; Koltko-Rivera 2004). Worldviews encompass mental models and normative premises, the descriptive as well as prescriptive, the hows as well as the oughts (Koltko-Rivera 2004). Identifications—one's constructed and labeled self—are also part of worldviews, with identifications turned to identity through one's own self labels or through the labeling done by others (Lasswell and McDougal 1992, 350–51). Identifications and worldviews are the basis for community, reinforced by shared patterns of comparatively stable expectations and demands, often manifesting as shared "interests."

The dynamics of social process emerge from the strategies employed by participants and participant communities in response to perceptions of environmental threats and opportunities (Lasswell 1971b). Strategies and related behaviors can arise from either conscious reflection or subconscious reflexive responses (Stanovich 2011). The environment consists not only of biophysical elements but also, perhaps more importantly, of institutions. Institutions are human abstractions that have major effects on the creation, allocation, and exchange of materials, information, and values (Lasswell and McDougal 1992, 375–84). Institutions are, in essence, established patterns of interaction configured by norms maintained by a certain degree of societal consensus, organized around the allocation of specific resources. Institutions largely determine how people interact to get what they want and whether those interactions destroy or build the commonwealth of human dignity (Lasswell and McDougal 1992, 725–89; Acemoglu and Robinson 2012). Formal and informal dispositions of power—consisting of authority, control, responsibility, and accountability—are central to the functioning of institutions (Lasswell and Kaplan 1950). In practice, power is often the means by which institutional patterns are changed or maintained, typically through socially sanctioned threat or coercion.

A Schematic

Putting these concepts of social process together yields some factors that warrant universal attention among those who are trying to understand and engage with complex cases such as large carnivore management (Lasswell and Kaplan 1950; Lasswell 1971b; Clark 2002; fig. 10.1). First are (1) *participants* and (2) their *perspectives*. Perspectives can be broken down into (i) *identities* and *identifications* and (ii) *worldviews* that encompass mental models. Perspectives engender (iii) *expectations* about how the world works in general or might unfold in specific situations, and also configure (iv) *value orientations*, which, together with mental models, catalyze (v) specific *demands* placed by participants on the world, often expressed as interests. Participants use (3) different *strategies* born of conscious deliberation or heuristic-driven reactions, employing different value resources, to achieve demands. All of this plays out in (4) specific *situations* or arenas typified by institutional arrangements, of which disposition of power is an important element. Finally, social process yields (5) immediate *outcomes* and (6) longer-term *effects*. All of these factors warrant being specified in enough detail to appraise the significance of each in context and to understand emergent system properties.

POLICY PROCESS

Background

Shared stable expectations about goals and the structure of the world are central to collective human action (Mathieu et al. 2000; Richards 2001). Stable expectations are expressed as habits, heuristics, norms, and policies. At the subconscious level of the individual, comparatively stable expectations can be seen as central to the triggering of heuristics in response to pattern matching (Gigerenzer, Todd, and ABC Research Group 1999). Individual expectations about the world are shaped and shared through processes such as acculturation, public discourses, and collective analysis (Thompson and Fine 1999; Stout et al. 1999; Koltka-Rivera 2004).

Policies are perhaps the most formal codifications of shared expectations, typically stabilized and legitimized through socially sanctioned processes (Lasswell 1971b; Clark 2002). Policies are the culmination of collective problem orientation in that policies are typically a response to, or a solution for, a perceived social problem, which is why problem definition is central to politics (Clark 2002). Policies allocate societal resources to achieve specific aims, which is the essential outcome of politics (Dery 1984). Whose problem definition "wins" largely determines what policies are crafted, empowered, and resourced, which translates into the "winners" more often getting what they want from the world (Weiss 1989; Rochefort and Cobb 1993). This is why the nature of policy processes matters so much, particularly whether they tend to polarize and divide or ameliorate and unify.

All the key cognitive functions—discerning, analyzing, judging, and choosing—are part of policy processes. However, decision making, consisting largely of judging and choosing, is central (Lasswell 1971b; Clark 2002). Decisions are made regarding what information is relevant, how to process and analyze it, what problems warrant being addressed, what policies address putative problems, when and where to apply policies, how to apply them, what to learn from the process of application, and when to end policies. The makeup of specific processes also entails decisions about who should have power, who has competence and standing to participate, and how participants should interact (Lasswell and McDougal 1992, 1131–53; chapter 8).

A Schematic

With these considerations in mind, the policy process can be logically categorized with two different scales and seven different functions (fig. 10.1). The first distinction is between decisions made at a larger scale about constituting or creating a policy process and decisions made at a smaller scale as part of a process already in place, in other words, the difference between constitutive and ordinary dynamics (Clark 2002; chapter 8).

The second series of distinctions pertains primarily to functions that are part of ordinary dynamics, although they can apply to any policy process at any scale. The first three functions of a policy process are integral to the process of orientation, which we described earlier (D'Zurilla and Goldfried

1971; Lasswell 1971b; Beach and Mitchell 1978; Lasswell and McDougal 1992, 1155–201; Clark 2002). Information that might be relevant to a suite of inter-related and potentially competing problem definitions needs to be identified, assembled, and analyzed as part of (1) *intelligence activities*. Optimally, the intelligence gained is used to inform (2) *promotional activities*, by which participants with different interests and related problem definitions advocate and debate their preferred policies. Promotional activities end when (3) a *prescription* has been authoritatively adopted as a basis for policy—the culmination of creating, evaluating, and choosing alternatives.

The remaining four functions pertain to decisions and activities that typically follow adoption of a policy (Lasswell 1971b; Lasswell and McDougal 1992, 1203–59; Clark 2002). These include, first, deciding where and when to (4) *invoke a policy*, which is its initial use in specific situations where there appears to be a match between the observed situation, embedded problems, and the prescribed scope and intent of the policy as a presumed solution. Then (5) *the policy is applied*, which entails resolving how the generalities of a policy translate into actions specific to a given situation, including determining the legitimacy of different interpretations, sometimes through appeal to judicial reviews. At some point, the people involved engage in (6) *appraisal*, a process of judging the efficacies of policies and related invocations and applications, ideally as a basis for learning lessons that can result in fine-tuning existing policies or creating better ones. Sometimes appraisal is the basis for (7) *terminating* either entire policies or specific applications. Regardless, termination at both scales inevitably happens as perceived problems are solved or redefined, or better solutions are imagined and implemented.

Although we presented the functions in sequence, in reality they often occur nonsequentially and embedded as functions within other functions (Clark 2002). Scale is an important consideration when thinking about policy dynamics. An additional note: these conscious functions are similar to the ways that heuristics are thought to be subconsciously developed, triggered, and refined, perhaps as "policies" embedded in the subconscious (Enkawa and Salvendy 1989; Sterman 1994; Gigerenzer, Todd, and ABC Research Group 1999; Rieskamp and Otto 2006).

We use these seven functions or facets of the policy process to frame our appraisal of conservation cases presented in this book. One virtue of these functions is that they can all be linked unambiguously to the goal of attaining and maintaining a commonwealth of human dignity (Lasswell 1971b; Lasswell and McDougal 1992, 1155–259; Clark 2002). When they are carried out poorly, in a way that leads to despotic outcomes, we lose the commonwealth.

Performed well, in a way that leads to inclusively equitable outcomes, we are more likely to foster widespread experiences of human dignity. The end of this chapter focuses on standards of appraisal linked to human dignity.

BIOPHYSICAL PROCESS

Background

The notion of social process organized around and in policy process is a powerful heuristic for understanding and engaging with complex collective undertakings such as managing large carnivores. Attending to the tasks of orientation can also help people improve their rationality, both as participants and observers. But, inasmuch as people involved with large carnivores are attempting to maximize values and maintain desired affective states, they are also trying to achieve certain outcomes in the biophysical environment, which recommends having some way to break down complex biophysical processes into categories of factors that will provide insights needed to craft effective interventions.

Even though carnivore behaviors are increasingly a focus of attention in conservation, outcomes are still largely reckoned in terms of ultimate measures such as population sizes and growth rates (Mattson et al. 1996a, 1996b; Mattson 2004). Population growth is, of course, a function of birth, death, immigration, and emigration rates. Barring phenomena such as demographic variation, which can show up as a lack of female impregnation because of low densities of males, birth rates are governed largely by the condition of reproductive females, which is governed in turn by access to and abundance of food (Mattson 2004). Food abundance is reckoned for carnivores in terms of prey densities and population trends and for omnivores somewhat more complexly in terms of the full suite of vegetal and animal foods and the habitats that support them.

Death rates are almost always usefully broken down into deaths attributable to human causes versus deaths attributable to nonhuman causes (Mattson et al. 1996a; Mattson 2004). Human causes are often the focus of management attention, including sport hunting, poaching, responses to conflicts and depredations, and collisions with vehicles. Human-caused deaths can be understood as the outcome of the frequency of human/carnivore encounters and the odds that any encounter will turn lethal for the animal (Mattson et al. 1996a, 1996b; Mattson 1997). Frequency of encounters is determined largely by densities of humans and infrastructures (e.g., roads), carnivore densities, the nearness of infrastructures to intrinsically attractive carnivore habitats, and the extent to which humans and human-related foods (e.g., garbage or livestock) attract carnivores (Mattson 1997, 2004). The lethality of these encounters is determined largely by whether people are armed, their recent history with carnivores, immediate circumstances, their perspectives, and other motivations (Sillero-Zubiri and Laurenson 2001; Mattson 1997, 2004). People's practices, perspectives, and policies naturally have major effects on levels of human-caused carnivore mortality, which links the practical challenges of managing carnivore populations to the domains of social and policy processes (Primm 1996; Sillero-Zubiri and Laurenson 2001).

A Schematic

With these concepts in mind, a suite of factors can be identified that will likely be relevant to understanding almost all cases where conservation of large carnivores is the focus of attention (Mattson et al. 1996a, 1996b; Sillero-Zubiri and Laurenson 2001; Mattson 1997, 2004; fig. 10.1). In order to analyze the outcomes of (1) *human-caused mortality*, human (i) *perspectives*, (ii) *policies*, (iii) *practices*, and (iv) *infrastructure* that have direct or readily identified indirect effects on either the (a) *frequency* or (b) *lethality of contact* between people and carnivores need to be clarified and understood. (2) *Nonhuman causes of mortality* also need to be identified; these are often most relevant to understanding recruitment of juveniles to adults. Of relevance to (3) *birth rates*, status and trends of (i) *key foods* (most often prey species) and (ii) *associated habitats* need to be clarified, as well as (iii) *factors that affect access to these foods*, especially by reproductive females. Also, (4) the *effects of the focal carnivore on its foods* need to be understood because these are relevant both to the carnivores and relations with humans who may be competing for the same prey. Finally, (5) some reckoning of the *status and trend of the focal carnivore population* is needed, referenced not only to births and deaths, but also, if relevant, to emigration and immigration.

A Framework for Appraisal

In this final section we introduce standards for appraising the functioning of the social, policy, and biophysical realms applicable to any conservation endeavor, but with a focus here on large carnivores. Standards, of course, follow from goals, which is a fact often missed by people who adhere to standards as ends in themselves, which can mean that the goals are tacit, unexamined, and otherwise opaque to those who invoke the standards. As stated earlier, we adhere to the overarching goal of a commonwealth of human dignity. Moreover, there is ample evidence that most people share the goal of human dignity, in whatever ways they may understand or express it (Mattson and Clark 2011). Given this goal, standards would logically help people judge whether the commonwealth of human dignity was being sustained or destroyed. It is self-evident that policies and the related process of developing, applying, and ending policies have profound effects not only on the management of large carnivores, but also on civil society. Value dynamics are, in turn, a seminal manifestation and diagnostic of how well social and policy processes are performing.

We also subscribe to the goal of conserving large carnivores, and thus the status and trends of populations, prey, and habitats are relevant to judging the performance of social and policy systems. However, this animal-focused goal is typically not widely shared among participants in large carnivore conservation cases, or, even if shared, not given the same priority. For this reason and because others elsewhere have dealt at length with how to judge biophysical outcomes for large carnivores (e.g., Gittleman et al. 2001; Maehr, Noss, and Larkin 2001; MacDonald and Loveridge 2010), we focus next on standards applicable to social and policy processes.

STANDARDS
Social Process
A commonwealth of human dignity can be said to exist when as many people as possible experience a sense of well-being arising from widespread access to and enjoyment of important values (Mattson and Clark 2011). This implies an equitable distribution of value costs and benefits, and it implies social and policy processes that ultimately create rather than destroy values. Using Lasswell's (1948) schematic of values (power, wealth, respect, skill, rectitude, enlightenment, affection, and well-being) allows for somewhat greater specificity. (1) People should experience empowerment in their lives, in particular control, but also access to authoritative policy processes that might affect them. (2) They should also have access to wealth sufficient to maintain physical well-being, not be subject to inequitable monetary costs attached to public policies, nor benefit unduly at the expense of others. (3) Their sense of morality should not be routinely violated by public acts of others, especially public officials, but in order to be compatible with a commonwealth of dignity, people's sense of right should be inclusive rather than condemn differences among people. (4) Respect should be widely valued and shared, both in public and private. (5) People should have ready access to the affection afforded by interactions with family and community. (6 and 7) And, finally, people should have widespread access to and opportunities for expressing skills and enlightenment (Lasswell and McDougal 1992, 725–59, 1017–27).

Applied to carnivore conservation, the preceding value-based standards would suggest problematic dynamics in both social and policy processes if most people felt unempowered or lacked access to policy processes that meted out consequences of great importance to them; if they felt routinely deprived of respect; if they were subject to blatantly unfair economic costs, when others were given unfair economic advantages; if their families or communities were disrupted by public policies and related sanctioned practices; if their universalist and other inclusive rectitude were being routinely violated, especially by public officials; or if life-enhancing opportunities to exercise their skills or to gain and employ enlightenment were routinely denied to them. First and foremost, widespread experiences of disrespect would be a prime indicator of problematic dynamics, primarily because respect is often an integrative experience by which people assess their experiences of the world (McDougal, Lasswell, and Chen 1980; Mattson and Clark 2011). Most of the cases presented in this book show evidence of widespread exhibitions and experiences of disrespect.

Policy Process

Although destructive value dynamics can arise from the effects of culture, historical conflict, or broad-scale social trends, the effects of such intrinsic drivers can either be considerably amplified or dampened by the design and execution of policy processes (Brunner 2002). How we make collective authoritative decisions either can inflame conflict and widen social-cultural divides or mitigate and ameliorate them, that is, either polarize people around differences or help unite them around common ground. Policy processes are, in fact, a major determinant of whether societies trend toward or away from the commonwealth of human dignity (Lasswell and Kaplan 1950; Lasswell and McDougal 1992). Positive trends predictably arise from policy processes that are inclusive, fair, accessible, respectful, and informed, from processes focused on helping people identify and secure common interests as part of civil and pragmatic efforts to solve problems by developing strategies for mutual gains. Conversely, negative trends predictably arise from processes that are uncivil, closed to all but a few who fit a certain identity, disrespectful to most of those who have a stake, that blatantly serve limited partisan interests, disregard inconvenient information, and concentrate ever more power and wealth in the hands of those who already have it (Acemoglu and Robinson 2012).

The seven functions of the policy process facilitate the specification of dignity-relevant standards, much like values do for social process (Lasswell and McDougal 1992; Clark 2002). The intelligence, promotion, prescription, invocation, application, appraisal, and termination functions each have standards specific to the tasks and related decisions that they encompass (Lasswell 1971b). We do not present those standards here. A full discussion can be found in chapter 8, and illustrative applications can be found in each case history of this book. In fact, this book is organized around the premise that the quality of policy processes is critically important to explaining the outcomes and effects of each case of large carnivore management. As with values dynamics, most cases in this book show evidence of formal policy processes that are destructive to the commonwealth of human dignity.

Conclusion

Conservation of large carnivores in the West is complex in all the ways we understand this notion, including cross-scale effects, synergies, nonlinearities, self-organization, and emergent properties. Large carnivore cases are typified by large-scale biophysical and social trends interacting with smaller-scale social and policy systems to create volatilities and surprises. The comparatively rigid institutions that have emerged over time have often amplified conflict among stakeholders, with predictable destructive consequences for civil society.

Participants in large carnivore cases often appear caught in problematic behaviors perpetuated by institutional incentives, lack of perceived alternatives, and inattention to self. However, our limited intrinsic ability to deal cognitively with complexity is also a predictable driver, which can lead to filtering and ignoring information, resorting to lower-order conceptualizations, and exhibiting behaviors based on the subconscious triggering of habitual behaviors. Although people are capable of engaging with complexity by consciously breaking down serial tasks according to conceptual frames, most bodies of theory about complexity are of limited use to people who are trying to understand the real-life complexity of conservation cases as a guide to constructive action.

We offer a set of concepts and related schematics that meet meta-theoretic criteria for helping people engage pragmatically with complexity, including fostering a balance between comprehensiveness and selectivity, covering all relevant domains, attending to people's limited ability to process information in parallel, making the tasks of orientation explicit, and employing frames that are amenable to serial analysis. The schematics we suggest cover social, policy, and biophysical processes. Equally important, we offer a set of criteria for appraising cases in view of the normative goal of attaining and sustaining the commonwealth of human dignity (see appendix). We adopt this goal because it promises widespread support among stakeholders, regardless of their perspectives on large carnivores, and because we hypothesize that meeting this goal undergirds any realistic prospects for long-term conservation (Mattson and Clark 2011).

We close by observing the obvious—that we emphasize the role of rationality in addressing problematic social and policy dynamics. We see no other realistic way for humanity simultaneously to conserve large carnivores and to foster a commonwealth of human dignity in a highly dynamic and rapidly evolving world. Conversely, we see subconsciously driven behaviors framed by the paradigm of scientized management as a recipe for perpetuating existing behaviors and institutions, with deleterious effects on both people and carnivores (chapters 1, 3, and 9). The open question is whether people can collectively overcome much that is hard-wired by evolution in order to marshal the requisite collective rationality, all the while managing the destructive impulses unleashed by our individual existential fears (*sensu* Yalom 1980; Hanson and Mendius 2009).

Appendix
Queries and Graphics for Synoptically Appraising Conservation Cases

One challenge in appraisal is that of assembling the many dimensions of an evaluation into a format that creates a useful synopsis of a system. Different formats work better for different people, but graphic devices are often highly efficient at coalescing and distilling information (Card, Mackinley, and Schneiderman 1999; Chen 2004). As the expression goes, a picture is worth a thousand words, or a hundred dimensions. But the picture needs to be logically, ideally tightly, coupled with the underlying information, and the design needs to make it easy to understand multiple dimensions of a complex system from shapes and colors on a flat surface (Ware 2004). Another challenge is to preserve as much of the complexity of a system as possible without resorting to lower-order representations or dumping large amounts of information.

The use of graphics to convey complexity efficiently and effectively has received considerable attention among researchers involved with human factors, operations design, situation analysis, and display technology. There are many possible graphic schemes, assessments of which have been based on the time required to register information, the accuracy of that registration process, and the efficiencies of eye movement (Goldberg and Kotval 1999; Chen and Yu 2000; Morse and Lewis 2000). Although certain graphic schemes tend to perform better than others, performance is also affected by the perceptual and cognitive abilities of observers (Chen and Yu 2000; Chen 2004). Of the many schemes, radar charts (also known as star charts or spider graphs) have received widespread use (e.g., Conroy and Soltan 1998; Pereira, Quintana, and Funtowicz 2005; Forlines and Wittenburg 2010). A radar chart consists of multiple axes arrayed uniformly around a common point of origin, like spokes of a wheel (Chambers et al. 1983). Values for different variables are plotted along each axis and connected to create a polygon. There have been critiques of this method, but it conveys information well when there are a moderate number of dimensions ($c. \leq 25$), when the axes are arrayed according to a conceptual gradient, when the axes use the same scale, and when fill colors are used (Chambers et al. 1983; Kosslyn 2006).

We developed a radar chart format for presenting results of a relatively comprehensive standardized approach to appraising conservation cases with the intent also of systematically disclosing aspects of the observer's subjectivity. Our method invokes the standards of appraisal that we present in this book and in chapter 10, and it builds on a series of queries covering trends and conditions in the biophysical, social process, and policy domains (tables 10.1, 10.2, and 10.3). Observers of a system answer each query by scoring the performance of the focal function within the system on a scale of 0 to 10, with 0 being poor and 10 being excellent. The scoring is subjective, and so additional scores are solicited for each query, including how confident the observer is in his or her judgment, the extent to which he or she thinks participants in the case generically pay attention to the particular factor, his or her confidence, again in that judgment, and finally, the extent to which the observer thinks he or she pays attention to the factor. All scales are 0 to 10, expressing either little or no confidence to high confidence in a judgment, or little or no attention to lots of attention given to a particular factor. In this way we attempt to capture not only an observer's judgment regarding overall system performance, but also aspects of subjectivity, including his or her confidence in judgments, as well as effects arising from the extent to which both the observer and system participants are or are not paying attention to specific factors.

The final step of the process entails taking scores and plotting them on six different radar charts. There are three pairs, one pair each for biophysical factors, social process, and policy process. The larger chart of each pair

Table 10A.1.

Queries and ratings (0–10) related to biophysical trends and conditions

Axis	Factor	Rating	Confidence
1	Current size of carnivore population(s) relative to demographic risks	_____	_____
	Extent to which this factor is a focus of attention for participants	_____	_____
	Extent to which this factor is a focus of attention for rater (self)	_____	
2	Current size of population/connectivity relative to genetic risks	_____	_____
	Extent to which this factor is a focus of attention for participants	_____	_____
	Extent to which this factor is a focus of attention for rater (self)	_____	
3	Current trajectory of carnivore population(s)	_____	_____
	Extent to which this factor is a focus of attention for participants	_____	_____
	Extent to which this factor is a focus of attention for rater (self)	_____	
4	Current size of carnivore population(s) relative to carrying capacity	_____	_____
	Extent to which this factor is a focus of attention for participants	_____	_____
	Extent to which this factor is a focus of attention for rater (self)	_____	
5	Current status of primary food(s)	_____	_____
	Extent to which this factor is a focus of attention for participants	_____	_____
	Extent to which this factor is a focus of attention for rater (self)	_____	
6	Current status of secondary food(s)	_____	_____
	Extent to which this factor is a focus of attention for participants	_____	_____
	Extent to which this factor is a focus of attention for rater (self)	_____	
7	Current trajectory of primary food(s)	_____	_____
	Extent to which this factor is a focus of attention for participants	_____	_____
	Extent to which this factor is a focus of attention for rater (self)	_____	
8	Current trajectory of secondary food(s)	_____	_____
	Extent to which this factor is a focus of attention for participants	_____	_____
	Extent to which this factor is a focus of attention for rater (self)	_____	
9	Current impacts of predation by carnivore on its prey population(s)*	_____	_____
	Extent to which this factor is a focus of attention for participants	_____	_____
	Extent to which this factor is a focus of attention for rater (self)	_____	
10	Current trend of impacts of predation on prey population(s)*	_____	_____
	Extent to which this factor is a focus of attention for participants	_____	_____
	Extent to which this factor is a focus of attention for rater (self)	_____	
11	Current level of depredation by carnivore on livestock*	_____	_____
	Extent to which this factor is a focus of attention for participants	_____	_____
	Extent to which this factor is a focus of attention for rater (self)	_____	
12	Current trend of livestock depredation*	_____	_____
	Extent to which this factor is a focus of attention for participants	_____	_____
	Extent to which this factor is a focus of attention for rater (self)	_____	
13	Current status of human encroachment on primary habitat*	_____	_____
	Extent to which this factor is a focus of attention for participants	_____	_____
	Extent to which this factor is a focus of attention for rater (self)	_____	
14	Current trend of human encroachment on primary habitat*	_____	_____
	Extent to which this factor is a focus of attention for participants	_____	_____
	Extent to which this factor is a focus of attention for rater (self)	_____	
15	Current level of threat by carnivore to human safety*	_____	_____
	Extent to which this factor is a focus of attention for participants	_____	_____
	Extent to which this factor is a focus of attention for rater (self)	_____	

continued

Table 10A.1.

continued

Axis	Factor	Rating	Confidence
16	Current trend in threat to human safety*	___	___
	Extent to which this factor is a focus of attention for participants	___	___
	Extent to which this factor is a focus of attention for rater (self)	___	
17	Current effectiveness of depredation management and prevention	___	___
	Extent to which this factor is a focus of attention for participants	___	___
	Extent to which this factor is a focus of attention for rater (self)	___	
18	Current trend of effectiveness in depredation management and prevention	___	___
	Extent to which this factor is a focus of attention for participants	___	___
	Extent to which this factor is a focus of attention for rater (self)	___	
19	Current effectiveness of human safety management	___	___
	Extent to which this factor is a focus of attention for participants	___	___
	Extent to which this factor is a focus of attention for rater (self)	___	
20	Current trend in effectiveness of human safety management	___	___
	Extent to which this factor is a focus of attention for participants	___	___
	Extent to which this factor is a focus of attention for rater (self)	___	
21	Current level of human-caused mortality in carnivore population(s)	___	___
	Extent to which this factor is a focus of attention for participants	___	___
	Extent to which this factor is a focus of attention for rater (self)	___	
22	Current trend in human-caused mortality of carnivore	___	___
	Extent to which this factor is a focus of attention for participants	___	___
	Extent to which this factor is a focus of attention for rater (self)	___	

*Scores are inverted; the estimated score is subtracted from 10.

Table 10A.2.

Queries and ratings (0–10) related to trends and conditions in social process

Axis	Factor	Rating	Confidence
1	Level of symbolic "loading" on the carnivore in the case	___	___
	Extent to which this factor is a focus of attention for participants	___	___
	Extent to which this factor is a focus of attention for rater (self)	___	
2	Trend in symbolic "loading" on the carnivore	___	___
	Extent to which this factor is a focus of attention for participants	___	___
	Extent to which this factor is a focus of attention for rater (self)	___	
3	Adequacy of arena for identifying and securing common interests	___	___
	Extent to which this factor is a focus of attention for participants	___	___
	Extent to which this factor is a focus of attention for rater (self)	___	
4	Trend in adequacy of arena	___	___
	Extent to which this factor is a focus of attention for participants	___	___
	Extent to which this factor is a focus of attention for rater (self)	___	
5	Equitability of access to and distribution of power	___	___
	Extent to which this factor is a focus of attention for participants	___	___
	Extent to which this factor is a focus of attention for rater (self)	___	
6	Trend in equitability of access to and distribution of power	___	___
	Extent to which this factor is a focus of attention for participants	___	___

Axis	Factor	Rating	Confidence
	Extent to which this factor is a focus of attention for rater (self)	_____	
7	Equitability of financial burden among stakeholders	_____	_____
	Extent to which this factor is a focus of attention for participants	_____	_____
	Extent to which this factor is a focus of attention for rater (self)	_____	
8	Trend in equitability of financial burden	_____	_____
	Extent to which this factor is a focus of attention for participants	_____	_____
	Extent to which this factor is a focus of attention for rater (self)	_____	
9	Sufficiency of participant skills for dealing with all aspects of case	_____	_____
	Extent to which this factor is a focus of attention for participants	_____	_____
	Extent to which this factor is a focus of attention for rater (self)	_____	
10	Trend in sufficiency of skills	_____	_____
	Extent to which this factor is a focus of attention for participants	_____	_____
	Extent to which this factor is a focus of attention for rater (self)	_____	
11	Breadth (versus narrowness) of loyalties and affections among participants	_____	_____
	Extent to which this factor is a focus of attention for participants	_____	_____
	Extent to which this factor is a focus of attention for rater (self)	_____	
12	Trend in breadth of loyalties and affections	_____	_____
	Extent to which this factor is a focus of attention for participants	_____	_____
	Extent to which this factor is a focus of attention for rater (self)	_____	
13	Inclusiveness (versus exclusiveness) of ideologies and worldviews	_____	_____
	Extent to which this factor is a focus of attention for participants	_____	_____
	Extent to which this factor is a focus of attention for rater (self)	_____	
14	Trend in inclusiveness of ideologies and worldviews	_____	_____
	Extent to which this factor is a focus of attention for participants	_____	_____
	Extent to which this factor is a focus of attention for rater (self)	_____	
15	Sufficiency of insight and information about all facets of the case	_____	_____
	Extent to which this factor is a focus of attention for participants	_____	_____
	Extent to which this factor is a focus of attention for rater (self)	_____	
16	Trend in sufficiency of insight and information	_____	_____
	Extent to which this factor is a focus of attention for participants	_____	_____
	Extent to which this factor is a focus of attention for rater (self)	_____	
17	Level and scope of respectfulness and civility among participants	_____	_____
	Extent to which this factor is a focus of attention for participants	_____	_____
	Extent to which this factor is a focus of attention for rater (self)	_____	
18	Trend in level and scope of respectfulness and civility	_____	_____
	Extent to which this factor is a focus of attention for participants	_____	_____
	Extent to which this factor is a focus of attention for rater (self)	_____	
19	Level and scope of dignity experienced by participants	_____	_____
	Extent to which this factor is a focus of attention for participants	_____	_____
	Extent to which this factor is a focus of attention for rater (self)	_____	
20	Trend in experiences of dignity	_____	_____
	Extent to which this factor is a focus of attention for participants	_____	_____
	Extent to which this factor is a focus of attention for rater (self)	_____	

Table 10A.3.

Queries and ratings (0–10) related to trends and conditions in the policy process

Axis	Factor	Rating	Confidence
1	Dependability of information used to inform decisions	____	____
	Extent to which this factor is a focus of attention for participants	____	____
	Extent to which this factor is a focus of attention for rater (self)	____	
2	Comprehensiveness of information used relative to important factors	____	____
	Extent to which this factor is a focus of attention for participants	____	____
	Extent to which this factor is a focus of attention for rater (self)	____	
3	Cost-effectiveness of information gathering efforts	____	____
	Extent to which this factor is a focus of attention for participants	____	____
	Extent to which this factor is a focus of attention for rater (self)	____	
4	Timeliness of information availability	____	____
	Extent to which this factor is a focus of attention for participants	____	____
	Extent to which this factor is a focus of attention for rater (self)	____	
5	Availability and accessibility of information to all stakeholders	____	____
	Extent to which this factor is a focus of attention for participants	____	____
	Extent to which this factor is a focus of attention for rater (self)	____	
6	Openness of the arena to all to promote their preferred policies	____	____
	Extent to which this factor is a focus of attention for participants	____	____
	Extent to which this factor is a focus of attention for rater (self)	____	
7	Rationality of debate about preferred policies	____	____
	Extent to which this factor is a focus of attention for participants	____	____
	Extent to which this factor is a focus of attention for rater (self)	____	
8	Opportunities for the creation and promotion of joint-gains strategies	____	____
	Extent to which this factor is a focus of attention for participants	____	____
	Extent to which this factor is a focus of attention for rater (self)	____	
9	Prevalence of joint-gains (versus special interest–focused) policies	____	____
	Extent to which this factor is a focus of attention for participants	____	____
	Extent to which this factor is a focus of attention for rater (self)	____	
10	Extent to which policies are intrinsically ameliorative (versus inflammatory)	____	____
	Extent to which this factor is a focus of attention for participants	____	____
	Extent to which this factor is a focus of attention for rater (self)	____	
11	Degree to which policies effectively match identified issues	____	____
	Extent to which this factor is a focus of attention for participants	____	____
	Extent to which this factor is a focus of attention for rater (self)	____	
12	Degree to which policies are realistic given available value resources	____	____
	Extent to which this factor is a focus of attention for participants	____	____
	Extent to which this factor is a focus of attention for rater (self)	____	
13	Extent to which implementation of policies is consistent with policy intent	____	____
	Extent to which this factor is a focus of attention for participants	____	____
	Extent to which this factor is a focus of attention for rater (self)	____	
14	Extent to which implementation of controversial policies is non-inflammatory	____	____
	Extent to which this factor is a focus of attention for participants	____	____
	Extent to which this factor is a focus of attention for rater (self)	____	

Axis	Factor	Rating	Confidence
15	Extent to which authority and control are sufficient to implement policies	_____	_____
	Extent to which this factor is a focus of attention for participants	_____	_____
	Extent to which this factor is a focus of attention for rater (self)	_____	
16	Timeliness of policy implementations	_____	_____
	Extent to which this factor is a focus of attention for participants	_____	_____
	Extent to which this factor is a focus of attention for rater (self)	_____	
17	Extent to which resolution of conflict is ameliorative	_____	_____
	Extent to which this factor is a focus of attention for participants	_____	_____
	Extent to which this factor is a focus of attention for rater (self)	_____	
18	Extent to which resolution of conflict is consistent with policy intent	_____	_____
	Extent to which this factor is a focus of attention for participants	_____	_____
	Extent to which this factor is a focus of attention for rater (self)	_____	
19	Extent to which policy implementation resolves or prevents problems	_____	_____
	Extent to which this factor is a focus of attention for participants	_____	_____
	Extent to which this factor is a focus of attention for rater (self)	_____	
20	Extent to which appraisal is ongoing	_____	_____
	Extent to which this factor is a focus of attention for participants	_____	_____
	Extent to which this factor is a focus of attention for rater (self)	_____	
21	Independence of appraisal efforts	_____	_____
	Extent to which this factor is a focus of attention for participants	_____	_____
	Extent to which this factor is a focus of attention for rater (self)	_____	
22	Quality of appraisals	_____	_____
	Extent to which this factor is a focus of attention for participants	_____	_____
	Extent to which this factor is a focus of attention for rater (self)	_____	
23	Extent to which termination of policy applications is timely	_____	_____
	Extent to which this factor is a focus of attention for participants	_____	_____
	Extent to which this factor is a focus of attention for rater (self)	_____	
24	Extent to which termination of policy applications is ameliorative	_____	_____
	Extent to which this factor is a focus of attention for participants	_____	_____
	Extent to which this factor is a focus of attention for rater (self)	_____	
25	Extent to which termination of policies appropriately reflects circumstances	_____	_____
	Extent to which this factor is a focus of attention for participants	_____	_____
	Extent to which this factor is a focus of attention for rater (self)	_____	
26	Openness of the termination process to all who are affected	_____	_____
	Extent to which this factor is a focus of attention for participants	_____	_____
	Extent to which this factor is a focus of attention for rater (self)	_____	

depicts judgments regarding trends and conditions within the corresponding domain. The smaller chart depicts the extent to which system participants are judged to be paying attention to each factor. The subdomains addressed by each query are labeled around the perimeter of each larger radar chart, and an overall numeric score is also provided in the center. This number is simply the sum of scores for all judgments divided by the score for a "perfect" system, where all trends and conditions are excellent and where most participants are paying lots of attention to all relevant factors—multiplied by 100.

EXAMPLES

Figures 10A.1 and 10A.2 show examples of radar charts that depict appraisals of the cases of mountain lions in the West and grizzly bears in the Yellowstone ecosystem. One set of charts and related scores is presented for the mountain lion case, and two sets for the grizzly bear case. The two cases illustrate a contrast in systems (fig. 10A.1), and the two scores for the grizzly bear

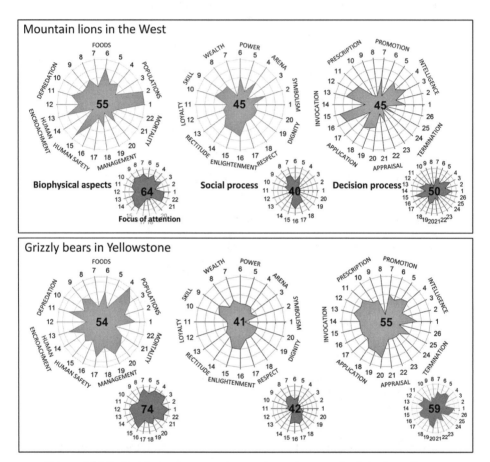

Figure 10A.1. Radar charts synopsizing judgments regarding trends and conditions in biophysical elements, social process, and policy process made by a single observer for the case of mountain lion management in the West and grizzly bear conservation in the Yellowstone ecosystem. Inset radar charts show observer judgments regarding the extent to which each element is a focus of attention for participants in each case.

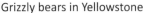

Grizzly bears in Yellowstone

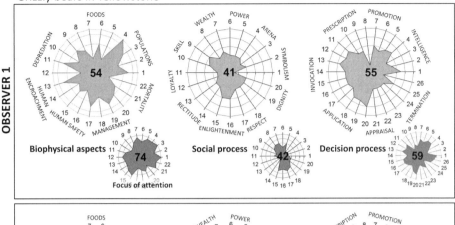

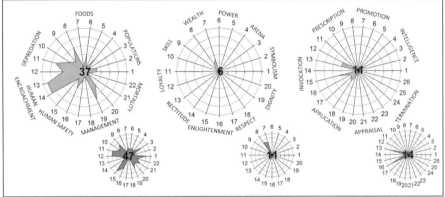

Figure 10A.2. Radar charts synopsizing judgments regarding trends and conditions in biophysical elements, social process, and policy process made by two different observers for the case of grizzly bear conservation in the Yellowstone ecosystem. Inset radar charts show observer judgments regarding the extent to which each element is a focus of attention for participants in this case.

case represent a contrast in judgments by two different observers regarding the same system (fig. 10A.2).

The examples reveal both commonalities and differences in perceptions of these cases by the observers who scored them. In all three appraisals, participants were judged to be paying considerably more attention to biophysical factors than to any others and the least explicit attention to social process. The aggregate state of social process, largely related to value dynamics, was also consistently scored lower in these appraisals compared to the states of biophysical or policy processes. A final commonality was that all domains were rated as par at best, but more often far subpar. Comparing the mountain lion and grizzly bear cases, promotion, appraisal, and termination functions of the policy process were judged to be performing generally poorest in both, and the status of skill and enlightenment values and of carnivore populations, as such, were judged to be best in the realms of biophysical and social processes. Otherwise, power was judged inequitable, the arena inflammatory, and dignity and respect outcomes as deficient in both cases. Overall, the profiles of social process function were quite similar.

Regarding differences, one observer of the grizzly bear case clearly perceived conditions to be much worse than perceived by the other observer, especially in social and policy processes. In fact, the more critical observer judged that conditions could probably not get much worse. The common 0 and 1 scores given by observer 2 for social and policy processes (fig. 10A.2) precluded calculating a meaningful correlation of scores with those of observer 1 in these domains, but the correlation of scores for biophysical elements was modestly positive (r = 0.48), which indicated some measure of agreement here. Contrasting the mountain lion and grizzly bear cases, lion populations were considered to be in better condition than bear populations, whereas bear management and processes of bear policy application were judged to be superior to those affecting lions.

We do not see any one person's judgments regarding performance of a system to be definitive. Rather, these graphics are a way to compare cases from a given subjective standpoint or to compare different subjectivities within a given case. As such, these radar charts provide a potentially valuable snapshot of peoples' perspectives, including perceptions they share and perceptions they don't. This insight can be an opportunity for scorers to reflect on their own perspectives, or the basis for starting a conversation among those with different perspectives, but beginning with better orientation to the views of self and others. The query process itself encourages respondents to reflect more systematically on a broader range of topics than perhaps they had considered before—of a scope that we argue is relevant to fostering human dignity outcomes.

NOTE

The perspectives here do not represent the official views of the US Geological Survey or the US Government, with which D. Mattson was affiliated at the time he contributed to this chapter.

REFERENCES

Acemoglu, D., and J. A. Robinson. 2012. *Why Nations Fail: The Origins of Power, Prosperity, and Poverty*. New York: Crown Business.

Alon, N., and H. Omer. 2006. *The Psychology of Demonization: Promoting Acceptance and Reducing Conflict*. Mahwah, NJ: Lawrence Erlbaum.

Anderson, J. R., D. Bothell, M. D. Byrne, S. Douglass, C. Lebiere, and Y. Qin. 2004. "An Integrated Theory of the Mind." *Psychological Review* 111:1036–60.

Ashcroft, M. H. 1992. "Cognitive Arithmetic: A Review of Data and Theory." *Cognition* 44:75–106.

Baddeley, A. 1986. *Working Memory*. New York: Oxford University Press.

Beach, L. R., and T. R. Mitchell. 1978. "A Contingency Model for the Selection of Decision Strategies." *Academy of Management Review* 3:439–49.

Beier, P. 2010. "A Focal Species for Conservation Planning." In *Cougar: Ecology and Conservation*, edited by M. Hornocker and S. Negri, 177–89. Chicago: University of Chicago Press.

Birney, D. P., and G. S. Halford. 2002. "Cognitive Complexity of Suppositional Reasoning: An Application of the Relational Complexity Metric to the Knight-Knave Task." *Thinking and Reasoning* 8:109–34.

Böhm, G., and W. Brun. 2008. "Intuition and Affect in Risk Perception and Decision Making." *Judgment and Decision Making* 3:1–4.

Brunner, R. D. 2002. "Problems of Governance." In *Finding Common Ground: Governance and Natural Resources in the American West*, edited by R. D. Brunner, C. H. Colburn, C. M. Cromley, R. A. Klein, and E. A. Olson, 1–47. New Haven, CT: Yale University Press.

Card, S., J. Mackinley, and B. Schneiderman. 1999. *Readings in Information Visualization: Using Vision to Think*. San Diego: Academic Press.

Chambers, J. M., W. S. Cleveland, B. Kleiner, and P. A. Tukey. 1983. *Graphical Methods for Data Analysis*. Belmont, CA: Wadsworth.

Charon, J. M. 2007. *Symbolic Interactionism: An Introduction, an Interpretation, an Integration*, 9th ed. Upper Saddle River, NJ: Prentice-Hall.

Chen, C. 2004. *Information Visualization: Beyond the Horizon*, 2nd ed. London: Springer-Verlag.

Chen, C., and Y. Yu. 2000. "Empirical Studies of Information Visualization: A Meta-Analysis." *International Journal of Human-Computer Studies* 53:851–66.

Cilliers, P. 1998. *Complexity and Postmodernism*. London: Routledge.

Clark, S. G. 2002. *The Policy Process: A Practical Guide for Natural Resource Professionals*. New Haven, CT: Yale University Press.

Clark, T. W. 1993. "Creating and Using Knowledge for Species and Ecosystem Conservation: Science, Organization, and Policy." *Perspectives in Biology and Medicine* 36:497–525.

Clark, T. W., and M. B. Rutherford. 2005. "The Institutional System of Wildlife Management: Making It More Effective." In *Coexisting with Large Carnivores:*

Lessons from Greater Yellowstone, edited by T. W. Clark, M. B. Rutherford, and D. Casey, 211–53. Washington, DC: Island Press.

Conroy, G., and H. Soltan. 1998. "ConSERV, a Project Specific Risk Management Concept." *International Journal of Project Management* 16:353–66.

Cowan, N. 1995. *Attention and Memory: An Integrated Framework*. New York: Oxford University Press.

Craighead, J. J., J. S. Sumner, and J. A. Mitchell. 1995. *The Grizzly Bears of Yellowstone: Their Ecology in the Yellowstone Ecosystem, 1959–1992*. Washington, DC: Island Press.

Dalgleish, T. 2004. "The Emotional Brain." *Nature Review of Neuroscience* 5:583–89.

Damasio, A. R. 1994. *Descartes' Error: Emotion, Reason, and the Human Brain*. New York: Avon Books.

Dery, D. 1984. *Problem Definition in Policy Analysis*. Lawrence: University Press of Kansas.

Dewey, J., and A. Bentley. 1949. *Knowing and the Known*. Boston: Beacon Press.

D'Zurilla, T. J., and M. R. Goldfried. 1971. "Problem Solving and Behavior Modification." *Journal of Abnormal Psychology* 78:107–26.

Elman, J. L. 1993. "Learning and Development in Neural Networks: The Importance of Starting Small." *Cognition* 48:71–99.

Enkawa, T., and G. Salvendy. 1989. "Underlying Dimensions of Human Problem Solving and Learning: Implications for Personnel Selection, Task Design, and Expert Systems." *International Journal of Man-Machine Studies* 30:235–54.

Evans, J. St. B. T. 2003. "In Two Minds: Dual-Process Accounts of Reasoning." *Trends in Cognitive Sciences* 7:454–59.

———. 2008. "Dual-Processing Accounts of Reasoning, Judgment, and Social Cognition." *Annual Review of Psychology* 59:255–78.

Forlines, C., and K. Wittenburg. 2010. "Wakame: Sense Making of Multi-Dimensional Spatial-Temporal Data." *AVI 2010*. Available from http://www.merl.com/reports /docs/TR2010-031.pdf.

Funtowicz, S., and J. R. Ravetz. 1994. "Emergent Complex Systems." *Futures* 26:568–82.

Gigerenzer, G. 2004. "Fast and Frugal Heuristics: The Tools of Bounded Rationality." In *Handbook of Judgment and Decision Making*, edited by D. Koehler and N. Harvey, 62–88. Hoboken, NJ: Blackwell.

Gigerenzer, G., P. M. Todd, and ABC Research Group. 1999. *Simple Heuristics That Make Us Smart*. New York: Oxford University Press.

Gill, R. B. 1996. "The Wildlife Professional Subculture: The Case of the Crazy Aunt." *Human Dimensions of Wildlife* 1:60–69.

———. 2001. "Professionalism, Advocacy, and Credibility: A Futile Cycle?" *Human Dimensions of Wildlife* 6:21–32.

———. 2010. "To Save a Mountain Lion: Evolving Philosophy of Nature and Cougars." In *Cougar: Ecology and Conservation*, edited by M. Hornocker and S. Negri, 5–16. Chicago: University of Chicago Press.

Gittleman, G. L., S. M. Funk, D. Macdonald, and R. K. Wayne, eds. 2001. *Carnivore Conservation*. Cambridge: Cambridge University Press.

Glöckner, A. 2008. "How Evolution Outwits Bounded Rationality: The Efficient Interaction of Automatic and Deliberate Processes in Decision Making and Implications for Institutions." In *Better than Conscious? Decision Making, the Human Mind, and Implications for Institutions*, edited by C. Engel and W. Singer, 259–84. Cambridge, MA: MIT Press.

Glöckner, A., and C. Witteman. 2010. "Beyond Dual-Process Models: A Categorization of Processes Underlying Intuitive Judgment and Decision Making." *Thinking and Reasoning* 16:1–25.

Goldberg, J. H., and X. P. Kotval. 1999. "Computer Interface Evaluation Using Eye Movements: Methods and Constructs." *International Journal of Industrial Ergonomics* 24:631–45.

Goldstein, D. G., and G. Gigerenzer. 2002. "Models of Ecological Rationality: The Recognition Heuristic." *Psychological Review* 109:75–90.

Gunderson, L. H., and C. S. Holling. 2002. *Panarchy: Understanding Transformations in Human and Natural Systems*. Washington, DC: Island Press.

Halford, G. S., R. Baker, J. E. McCredden, and J. D. Bain. 2005. "How Many Variables Can Humans Process?" *Psychological Science* 16:70–76.

Halford, G. S., W. H. Wilson, and S. Phillips. 1998. "Processing Capacity Defined by Relational Complexity: Implications for Comparative, Developmental and Cognitive Psychology." *Behavioral and Brain Sciences* 21:803–65.

Hanson, R., and R. Mendius. 2009. *Buddha's Brain: The Practical Neuroscience of Happiness, Love and Wisdom*. Oakland: New Harbinger.

Harding, R., C. M. Hendricks, and M. Farqui. 2009. *Environmental Decision-Making: Exploring Complexity and Context*. Annadale, NSW, Australia: Federation Press.

Haroldson, M. A., K. A. Gunther, D. P. Reinhart, S. R. Podruzny, C. Cegelski, L. Waits, T. Wyman, and J. Smith. 2005. "Changing Numbers of Spawning Cutthroat Trout in Tributary Streams of Yellowstone Lake and Estimates of Grizzly Bears Visiting Streams from DNA." *Ursus* 16:167–80.

Haroldson, M. A., C. C. Schwartz, S. Cherry, and D. S. Moody. 2004. "Possible Effects of Elk Harvest on Fall Distribution of Grizzly Bears in the Greater Yellowstone Ecosystem." *Journal of Wildlife Management* 68:129–37.

Hirschi, N. W., and D. D. Frey. 2002. "Cognition and Complexity: An Experiment on the Effect of Coupling in Parameter Design." *Research in Engineering Design* 13: 123–31.

Holling, C. S. 1994. "Simplifying the Complex: The Paradigms of Ecological Function and Structure." *Futures* 26:598–609.

Homer-Dixon, T. F. 1999. *Environment, Scarcity, and Violence*. Princeton, NJ: Princeton University Press.

Jacobson, C. A., and D. J. Decker. 2006. "Ensuring the Future of State Wildlife Management: Understanding Challenges for Institutional Change." *Wildlife Society Bulletin* 34:531–36.

Jones, B. D. 1999. "Bounded Rationality." *Annual Review of Political Science* 2: 297–321.

Kareev, Y. 2000. "Seven (Indeed, Plus or Minus Two) and the Detection of Correlations." *Psychological Review* 107:397–402.

Kegan, R. 1994. *In Over Our Heads: The Mental Demands of Modern Life*. Cambridge, MA: Harvard University Press.

Kello, C. T., G. D. A. Brown, R. Ferrer-i-Cancho, J. G. Holden, K. Linkenkaer-Hansen, T. Rhodes, and G. C. Van Orden. 2010. "Scaling Laws in Cognitive Sciences." *Trends in Cognitive Sciences* 14:223–32.

Kevin, B. 2011. "Everybody Hates Chuck Schwartz." *Sierra* 96:26–31.

Keysers, C., R. Boyd, J. Cohen, M. Donald, W. Güth, E. Johnson, R. Kurzban, et al. 2008. "Explicit and Implicit Strategies in Decision Making." In *Better Than Conscious? Decision Making, the Human Mind, and Implications for Institutions*, edited by C. Engel and W. Singer, 225–58. Cambridge, MA: MIT Press.

Koltko-Rivera, M. E. 2004. "The Psychology of Worldviews." *Review of General Psychology* 8:3–58.

Kosslyn, S. M. 2006. *Graph Design for the Eye and Mind*. New York: Oxford University Press.

Kotovsky, K., J. R. Hayes, and H. A. Simon. 1985. "Why Are Some Problems Hard? Evidence from the Tower of Hanoi." *Cognitive Psychology* 17:248–94.

Kuehl, B. L. 1993. "Conservation Obligations under the Endangered Species Act: A Case Study of the Yellowstone Grizzly Bear." *University of Colorado Law Review* 64:607.

Kuhn, D. 2001. "How Do People Know?" *Psychological Science* 12:1–8.

Lasswell, H. D. 1948. *Power and Personality*. New York: W.W. Norton.

———. 1971a. "From Fragmentation to Configuration." *Policy Sciences* 2:439–46.

———. 1971b. *A Pre-View of Policy Sciences*. New York: American Elsevier.

Lasswell, H. D., and A. Kaplan. 1950. *Power and Society: A Framework for Political Inquiry*. New Haven, CT: Yale University Press.

Lasswell, H. D., and M. S. McDougal. 1992. *Jurisprudence for a Free Society*. New Haven, CT: New Haven Press.

Li, Y., E. J. Johnson, and L. Zaval. 2011. "Local Warming: Daily Temperature Change Influences Belief in Global Warming." *Psychological Science* 22. Online publication.

MacCracken, J. G., and J. O'Laughlin. 1998. "Recovery Policy on Grizzly Bears: An Analysis of Two Positions." *Wildlife Society Bulletin* 26:899–907.

Macdonald, D. W., and A. J. Loveridge, eds. 2010. *Biology and Conservation of Wild Felids*. Oxford: Oxford University Press.

Macfarlane, W. W., J. A. Logan, and W. Kern. 2012. "An Innovative Aerial Assessment of Greater Yellowstone Ecosystem Mountain Pine Beetle-Caused Whitebark Pine Mortality." *Ecological Applications*. Online first.

Maehr, D. S., R. F. Noss, and J. L. Larkin, eds. 2001. *Large Mammal Restoration: Ecological and Sociological Challenges in the 21st Century*. Washington, DC: Island Press.

Mathieu, J. E., T. S. Heffner, G. F. Goodwin, E. Salas, and J. A. Cannon-Bowers. 2000. "The Influence of Shared Mental Models on Team Process and Performance." *Journal of Applied Psychology* 85:273–83.

Mattson, D. 1997. "Wilderness-Dependent Wildlife: The Large and the Carnivorous." *International Journal of Wilderness* 3:34–38.

———. 2000. *Causes and Consequences of Dietary Differences among Yellowstone Grizzly Bears* (Ursus arctos). PhD dissertation. Moscow: University of Idaho.

———. 2004. "Living with Fierce Creatures? An Overview and Models of Mammalian Carnivore Conservation." In *People and Predators: From Conflict to Coexistence*, edited by N. Fascione, A. Delach, and M. E. Smith, 151–76. Washington, DC: Island Press.

Mattson, D. J., B. N. Blanchard, and R. R. Knight. 1991. "Food Habits of Yellowstone Grizzly Bears, 1977–1987." *Canadian Journal of Zoology* 69:1619–29.

Mattson, D. J., K. L. Byrd, M. B. Rutherford, S. R. Brown, and T. W. Clark. 2006. "Finding Common Ground in Large Carnivore Conservation: Mapping Contending Perspectives." *Environmental Science and Policy* 9:392–405.

Mattson, D. J., and N. Chambers. 2009. "Human-Provided Waters for Desert Wildlife: What Is the Problem?" *Policy Sciences* 42:113–35.

Mattson, D. J., and S. G. Clark. 2010. "People, Politics, and Cougar Management." In *Cougar: Ecology and Conservation*, edited by M. Hornocker and S. Negri, 206–20. Chicago: University of Chicago Press.

———. 2011. "Human Dignity in Concept and Practice." *Policy Sciences*. Online publication.

———. 2012. "The Discourses of Incidents: Mountain Lions on Mt. Elden and in Sabino Canyon." *Policy Sciences* 45:315–43.

Mattson, D. J., and J. J. Craighead. 1994. "The Yellowstone Grizzly Bear Recovery Program: Uncertain Information, Uncertain Policy." In *Endangered Species Recovery: Finding the Lessons, Improving the Process*, edited by T. W. Clark, R. P. Reading, and A. Clarke, 101–29. Washington, DC: Island Press.

Mattson, D. J., C. M. Gillin, S. A. Benson, and R. R. Knight. 1991. "Bear Feeding Activity at Alpine Insect Aggregation Sites in the Yellowstone Ecosystem." *Canadian Journal of Zoology* 69:2430–35.

Mattson, D. J., S. Herrero, R. G. Wright, and C. M. Pease. 1996a. "Designing and Managing Protected Areas for Grizzly Bears: How Much Is Enough?" In *National Parks and Protected Areas: Their Role in Environmental Protection*, edited by R. G. Wright, 133–64. Cambridge: Blackwell Science.

———. 1996b. "Science and Management of Rocky Mountain Grizzly Bears." *Conservation Biology* 10:1013–25.

Mattson, D. J., H. Karl, and S. Clark. 2012. "Values in Natural Resource Management and Policy." In *Restoring Lands: Coordinating Science, Politics, and Action*, edited by H. A. Karl, L. Scarlett, J. C. Vargas-Moreno, and M. Flaxman, 239–59. New York: Springer.

McAdams, D. P. 1996. "Personality, Modernity, and the Storied Self: A Contemporary Framework for Studying Persons." *Psychological Inquiry* 7:295–321.

McDougal, M. S., H. D. Lasswell, and L.-C. Chen. 1980. *Human Rights and World Public Order: The Basic Policies of an International Law of Human Dignity*. New Haven, CT: Yale University Press.

Mellers, B., A. Schwartz, and I. Ritov. 1999. "Emotion-Based Choice." *Journal of Experimental Psychology General* 128:332–45.

Mezirow, J. 1996. "Contemporary Paradigms of Learning." *Adult Education Quarterly* 46:158–72.

Miller, D. 1996. "A Preliminary Typology of Organizational Learning: Synthesizing the Literature." *Journal of Management* 22:485–505.

Miller, G. A. 1956. "The Magical Number Seven, Plus or Minus Two: Some Limits on Our Capacity for Processing Information." *Psychological Review* 63:81–97.

Morse, E., and M. Lewis. 2000. "Evaluating Visualizations: Using a Taxonomic Guide." *International Journal of Human-Computer Studies* 53:637–62.

Newell, A., and H. A. Simon. 1972. *Human Problem Solving.* Englewood Cliffs, NJ: Prentice-Hall.

Newell, B. R. 2005. "Re-visions of Rationality." *Trends in Cognitive Sciences* 9:11–15.

Nezu, A., and T. J. D'Zurilla. 1981. "Effects of Problem Definition and Formulation on Decision Making in the Social Problem-Solving Process." *Behavior Therapy* 12: 100–06.

Nie, M. 2004a. "State Wildlife Governance and Carnivore Conservation." In *People and Predators: From Conflict to Coexistence*, edited by N. Fascione, A. Delach, and M. E. Smith, 197–218. Washington, DC: Island Press.

———. 2004b. "State Wildlife Policy and Management: The Scope and Bias of Political Conflict." *Public Administration Review* 64:221–33.

Ostrom, E. 2011. "Background on the Institutional Analysis and Development Framework." *Policy Studies Journal* 39:7–27.

Peacock, D., and A. Peacock. 2009. *In the Presence of Grizzlies: The Ancient Bond between Men and Bears.* Guilford, CT: Lyons Press.

Pereira, Â. G., S. C. Quintana, and S. Funtowicz. 2005. "GOUVERNe: New Trends in Decision Support for Groundwater Governance Issues." *Environmental Modeling and Software* 20:111–18.

Primm, S. A. 1996. "A Pragmatic Approach to Grizzly Bear Conservation." *Conservation Biology* 10:1026–35.

Primm, S. A., and T. W. Clark. 1996. "Making Sense of the Policy Process for Carnivore Conservation." *Conservation Biology* 10:1036–45.

Primm, S., and K. Murray. 2005. "Grizzly Bear Recovery: Living with Success." In *Coexisting with Large Carnivores: Lesson from Greater Yellowstone*, edited by T. W. Clark, M. B. Rutherford, and D. Casey, 99–137. Washington, DC: Island Press.

Quigley, H., and M. Hornocker. 2010. "Cougar Population Dynamics." In *Cougar: Ecology and Conservation*, edited by M. Hornocker and S. Negri, 59–75. Chicago: University of Chicago Press.

Reinhart, D. P., and D. J. Mattson. 1990. "Bear Use of Cutthroat Trout Spawning Streams in Yellowstone National Park." *International Conference of Bear Research and Management* 8:343–50.

Richards, D. 2001. "Coordination and Shared Mental Models." *American Journal of Political Science* 45:259–76.

Rieskamp, J., and P. E. Otto. 2006. "SSL: A Theory of How People Learn to Select Strategies." *Journal of Experimental Psychology General* 135:207–36.

Rochefort, D. A., and R. W. Cobb. 1993. "Problem Definition, Agenda Access, and Policy Choice." *Policy Studies Journal* 21:56–71.

———. 1994. *Problem Definition: An Emerging Perspective.* Lawrence: University Press of Kansas.

Ruth, T. K., and K. Murphy. 2010. "Cougar-Prey Relationships." In *Cougar: Ecology and Conservation,* edited by M. Hornocker and S. Negri, 138–62. Chicago: University of Chicago Press.

Schneider, W., and M. Detweiler. 1987. "A Connectionist/Control Architecture for Working Memory." *Psychology of Learning and Motivation* 21:53–119.

Schwartz, S. H. 1994. "Are There Universal Aspects in the Structure and Contents of Human Values?" *Journal of Social Issues* 50:19–45.

Servheen, C. 1998. "The Grizzly Bear Recovery Program: Current Status and Future Considerations." *Ursus* 10:591–96.

Shah, A. K., and D. M. Oppenheimer. 2008. "Heuristics Made Easy: An Effort-Reduction Framework." *Psychological Bulletin* 134:207–22.

Sillero-Zubiri, C., and M. K. Laurenson. 2001. "Interactions between Carnivores and Local Communities: Conflict or Coexistence?" In *Carnivore Conservation,* edited by J. L. Gittleman, S. M. Funk, D. W. MacDonald, and R. K. Wayne, 282–312. Cambridge: Cambridge University Press.

Simon, H. A. 1956. "Rational Choice and the Structure of the Environment." *Psychological Review* 63:129–38.

———. 1974. "How Big Is a Chunk?" *Science* 183:482–88.

———. 1982. *Models of Bounded Rationality. Vol. I.* Cambridge, MA: MIT Press.

Stanovich, K. E. 2011. *Rationality and the Reflective Mind.* New York: Oxford University Press.

Sterman, J. D. 1994. "Learning In and About Complex Systems." *System Dynamics Review* 10:291–330.

Stewart, I. 2002. *Does God Play Dice? The New Mathematics of Chaos,* 2nd ed. Malden, MA: Blackwell.

Stone, D. A. 1989. "Causal Stories and the Formation of Policy Agendas." *Political Science Quarterly* 104:281–300.

Stout, R. J., J. A. Cannon-Bowers, E. Salas, and D. M. Milanovich. 1999. "Planning, Shared Mental Models, and Coordinated Performance: An Empirical Link Is Established." *Human Factors* 41:61–71.

Thompson, L., and G. A. Fine. 1999. "Socially Shared Cognition, Affect, and Behavior: A Review and Integration." *Personality and Social Psychology Review* 3:278–302.

Vucetich, J. A., D. W. Smith, and D. R. Stahler. 2005. "Influence of Harvest, Climate and Wolf Predation on Yellowstone Elk, 1961–2004." *Oikos* 111:259–70.

Ware, C. 2004. *Information Visualization: Perception for Design. Second Edition.* San Francisco: Morgan Kaufmann.

Weaver, W. 1948. "Science and Complexity." *American Scientist* 36:536–44.

Webster, D. M., and A. W. Kruglanski. 1994. "Individual Differences in Need for Cognitive Closure." *Journal of Personality and Social Psychology* 67:1049–62.

Weiss, J. A. 1989. "The Powers of Problem Definition: The Case of Government Paperwork." *Policy Sciences* 22:97–121.

White, P. J., and R. A. Garrott. 2005. "Yellowstone's Ungulates after Wolves—Expectations, Realizations, and Predictions." *Biological Conservation* 125:141–52.

Wilkinson, T. 1998. *Science under Siege: The Politician's War on Nature and Truth.* Boulder, CO: Johnson Books.

Yalom, I. D. 1980. *Existential Psychotherapy.* New York: Basic Books.

Zhang, J. 2000. "External Representations in Complex Information Processing Tasks." In *Encyclopedia of Library and Information Science, Volume 68*, edited by A. Kent, 164–80. New York: Marcel Dekker.

Zhang, J., and D. A. Norman. 1994. "Representations in Distributed Cognitive Tasks." *Cognitive Science* 18:87–122.

11 Improving Governance for People and Large Carnivores

MURRAY B. RUTHERFORD AND SUSAN G. CLARK

Introduction

The preceding chapters examine the institutions of governance created by people to produce and allocate values associated with large carnivores and their habitats. The full range of values is at play here, from the wealth generated by resource development, to the enlightenment, respect, and affection that can arise from working together to solve problems, to the rectitude that drives many efforts to conserve carnivores, and so on through all the values that people demand or possess. The authors use a standard framework and criteria to evaluate the decision-making processes of these governance institutions. They assess how effectively decision processes are functioning, whether they are advancing the common interest, and what this means for the people and carnivores involved. We believe that the book accomplishes three main things: First, it harvests hard-won experience and lessons from cases of grizzly bear, wolf, and mountain lion conservation from the Yukon to Arizona. Second, it provides insight on fundamental societal issues concerning our institutions, decision-making practices, and cultural and constitutive makeup. And third, it offers practical recommendations about how to find enduring common ground solutions to the diverse challenges of large carnivore conservation.

CHALLENGES FOR PEOPLE AND GOVERNANCE

Over the last two centuries, the primary objective of large carnivore management in North America has evolved from simply managing abundant animals to satisfy the often narrow value demands of a few people to managing and conserving scarce animals to satisfy the diverse and conflicting value demands of a wide range of people (Clark et al. 2001). As a consequence, many of the problems faced by governance institutions in carnivore management have become less technical and more messy or "wicked" (Rittell and Webber 1973; Xiang 2013). Unlike technical problems, which can be resolved through careful rational analysis, messy problems are compounded by conflicting societal goals and values, uncertainty, nonlinearities, thresholds, interconnections with other problems, and other complications arising from the complexity of the socio-ecological systems in which they are embedded (see chapter 10).

Governance institutions that functioned reasonably well in dealing with the technical problems of carnivore management in the past are struggling terribly to cope with the messy problems of carnivore management in the present. Not only are our institutions poorly designed to cope with such problems, but our cognitive capacity as individuals is bounded and we have few good models for understanding complex socioecological systems (see chapters 1 and 10). Institutional and individual incapacities for modern large carnivore conservation are evident in a range of persistent challenges for governance that run through the case studies and other chapters in this book. In this section we review some of these challenges for governance, and in the next section we turn to a discussion of strategies for improvement.

THE PERILS OF SCIENTIFIC MANAGEMENT

The doctrine of scientific management, which we describe in the first chapter, is identified as an important underlying cause of governance problems in many of the cases. Under this doctrine, messy problems are treated as if they were technical problems and scientific expertise is privileged over other forms of knowledge and other participants. Unfortunately, scientific management is ingrained in wildlife management institutions and in the perspectives of many participants in carnivore conservation. Mattson (chapter 2) describes how the commissions and agencies responsible for mountain lion management in the American West use the model of scientific management, along with their bureaucratic authority and the business model of government, to avoid sharing power with new stakeholders in lion management, with the result that the interests of agency staff and those of their hunting "customers" are advanced. Watters et al. (chapter 3) discuss how a problem definition founded on scientific management has dominated wolf reintroduction and management in Wyoming, contributing to an inability to find common ground in a setting of competing goals and values. Scientific management is, in fact, embedded in the "North American Model of Wildlife Management" and contributes to failure to adapt to the changing context of modern wildlife management, as shown by Clark and Milloy (chapter 9).

Even in local collaborative initiatives, the power of scientific management can dominate thinking and actions. The fladry experiments carried out by researchers in southwestern Alberta (chapter 5) may have been fine examples of scientific experimental design, but ranchers on properties near the ranch on which fladry was tested did not appreciate being used as an "experimental control" and blamed the experiment for depredation by wolves on their livestock. The impact this had on ranchers' trust for the researchers and the project was essentially irreversible, at least in the near term. As Rittell and Webber (1973, 163) observe about wicked problems, "every implemented solution is consequential. It leaves 'traces' that cannot be undone." The full implications of this cogent observation are not always appreciated by those who advocate for policy learning modeled on the scientific method in the field.

MYTHS, PERSPECTIVES, AND SYMBOLIC POLITICS

Another recurring set of challenges to the governance institutions examined in this book arises from the strong myths associated with large carnivores, the symbolic power of these animals, and the related deep divergences in perspectives about carnivores and their management. In Yukon (chapter 4), D. Clark et al. explain how the Aboriginal concept of respect for grizzly bears spurred negative responses to bear research conducted by Kluane National Park, because the research methods were considered disrespectful to the animals. The authors link this negative reaction to the emergence of two competing perspectives about grizzly bears and management in the region. One perspective, rooted in traditional scientific management, sees a precarious population of grizzly bears and a need to implement conservation measures inside and outside the park. The other perspective, rooted in local knowledge, sees an abundant population of bears and an attempt by park bureaucrats to expand their jurisdiction and exert control over locals based on flawed scientific research. Existing institutions have been unable to bridge these perspectives and find common ground.

In another example, Watters et al. (chapter 3) discuss the symbolic politics of wolf management in Wyoming. This debate is imbued with symbols of wildness, wilderness, and functioning ecosystems (for many conservation advocates), control and management of a productive landscape (for many ranchers), and stewardship and responsibility toward a relative and teacher (for many Native Americans). When these participants argue over wolf management, they are arguing from starkly different symbolic conceptions of the subject and its context. Again, common ground across divergent perspectives is hard to find.

POWER AND POLITICIZED DECISION MAKING

Watters et al. define the problem of wolf management in Wyoming in terms of politicization of the values held by different participants, and particularly the values of power, wealth, rectitude, and respect. "Politicization" is a term that describes settings where interactions have become so contested and conflict-ridden that the focus is almost entirely on power and how participants can use various resources to increase their power (Lasswell and Mc-Dougal 1992). Unfortunately, politicization is so prevalent in carnivore conservation that it has almost become the norm. For instance, in the Banff-Bow Valley region of Alberta (chapter 7), social interactions and decision making about grizzly bears became so politicized prior to the first interdisciplinary problem-solving workshop that some stakeholders did not even want to talk to each other. Scientists were accused of conducting biased research under a hidden agenda and individuals had sunk to calling each other names in the local media. Under these circumstances the efforts of park staff to develop an acceptable grizzly bear management plan through the park's traditional institutions and consultation processes had little chance of success.

Even in settings that are not fully politicized, managers in government agencies tend to focus on power. Research shows that agency managers typically seek to maximize their autonomy, including their control over resources and decision making, in order to implement their personal visions and satisfy their value demands (Thomas 2003). In Banff it was the unusual willingness of two superintendents near the end of their careers to share power with stakeholders through collaborative decision making that created the conditions under which the Grizzly Bear Dialogue Group was able to prosper in the short term. When these superintendents were subsequently replaced by individuals with more traditional views of power sharing, the new superintendents reverted to conventional park planning processes and withheld power from the Grizzly Bear Dialogue Group, contributing to its eventual demise (chapter 7).

BOUNDED RATIONALITY AND MUDDLING THROUGH

Chapter 10 discusses the struggles of cognitively limited actors faced with complex socioecological systems. Boundedly rational decision makers muddle through, seeking solutions to symptoms rather than clarifying and pursuing goals, considering only a few of the possible solutions and favoring those they have tried in the past, mixing ends and means in analysis, and adopting other strategies through which they attempt to gain traction in a complex world (Lindblom 1959, 1979). In what appears to be a classic example of muddling through, the Arizona Game and Fish Department (chapter 2) responded to public backlash over its efforts to track and kill "problem" mountain lions by hosting a brief public engagement process to gather feedback on its protocols, conducting a public education campaign, and prescribing minor changes to the regulations for lion hunting. Rather than use the opportunity to revisit goals and perhaps develop innovative new strategies, the agency stuck to its traditional management norms and attempted to solve symptoms of a problem by doing the kinds of things it had done in the past. Incremental muddling may work in some instances in the short term, but it encourages decision makers to overlook evidence of growing problems, risks moving a system across irreversible thresholds, and sets up collapse. "Short-range success is often the parent of long-range failure" (Lasswell 1971, 84).

The problems of bounded rationality and incremental decision making are compounded by decision makers' reluctance to establish appraisal structures and adopt appraisal criteria. None of the decision-making processes examined in this book, including the constitutive level of decision making discussed by Clark et al. (chapter 8), included a strong appraisal function. Yet appraisal is fundamental to learning and adaptation.

MULTIPLE AND HIERARCHICAL DECISION PROCESSES

Decision making takes place in multiple settings and at multiple levels, typically structured in a hierarchy, where the outcomes of higher-level decision processes set the rules for decision making at lower levels (chapter 8). In the restricted decision-making structures for carnivore conservation established by constitutive decisions made in the past, local stakeholders may have difficulty finding an arena and voice to present their interests. With increasingly diverse stakeholders and values affected by decisions about large carnivore conservation, this limited participation challenges efforts to find common ground. Thus, conservation advocates are excluded from Arizona Game and Fish Department decision making about lion management (chapter 2), Native Americans are excluded from federal-state negotiations and agreements about wolf reintroduction in Wyoming (chapter 3), and stakeholders in the Banff-Bow Valley have limited and highly constrained input into traditional park planning processes (chapter 7). Not only must participants at the lower levels of decision making operate within the constraints and opportunities established by higher-level decisions, they must live with the practical outcomes of those decisions. Ranchers of the Upper Green River basin and Native Americans of the Wind River Reservation (chapter 3) have been forced to find ways to deal with wolves that have reappeared in their regions as a result of the federal government's decision to reintroduce wolves to Yellowstone National Park, a decision in which the ranchers and Native Americans had little say.

Furthermore, at all levels of decision making new laws or policies are often layered on top of inconsistent or even directly contradictory laws and policies prescribed in the past (Thielmann and Tollefson 2009). This happens because most decision-making processes pay little attention to the termination function, in which previous prescriptions should be adjusted or ended if they are not working, or have served their purpose, or are being replaced by new initiatives (Geva-May 2001). Wilkinson (1992) calls such outdated laws and policies "Lords of Yesterday," because they still reign over modern management decisions and behavior. Inattention to the termination function is evident throughout the cases in this book.

UNIQUE CONTEXTS

Each decision-making process for large carnivore conservation exists in a unique socioecological context. For the ranchers of the Upper Green River in Wyoming, the various dimensions of the problem of wolf conservation (goals, important trends, conditioning factors causing or shaping trends, and predictions about the future) are substantially different than they are for the Native Americans of the Wind River Reserve a few miles away in the same state, and these contextual dimensions are substantially different again for the ranchers and other participants in wolf conservation in the Oldman River Basin in Alberta, several hundred miles to the north. The context for grizzly bear conservation in the Blackfoot River Valley in Montana is dramatically different from the context for grizzly bear conservation in and around Kluane National Park in Yukon. In order to be effective, strategies to address conservation problems in each setting must be designed to fit the unique context in which they will be implemented. One size does not fit all, yet the bureaucratic structure of agencies responsible for carnivore conservation in North America is designed to rationalize and standardize decisions across contexts.

IMPEDIMENTS TO THE ENDURANCE
AND DIFFUSION OF INNOVATIONS

Proponents of innovations in governance must find opportunity and space for innovations to take hold initially within and around existing institutions and then develop enough financial support, capacity of participants, and other resources for these initiatives to endure. In our cases, the two most striking examples of endurance problems for innovations are the collaborative wolf management initiatives in southwestern Alberta (chapter 5) and the Grizzly Bear Dialogue Group of the Banff-Bow Valley (chapter 7). In each of these cases, what began as a promising innovation in local participatory decision making faded out over time. The loss of key members, capacity, and resources were factors in the inability to carry on. In contrast, the Blackfoot Challenge program in Montana (chapter 6) produced enough early on-the-ground success and built a community constituency that was sufficiently strong for the innovation to endure.

A related challenge is the difficulty of transferring lessons from apparently successful programs such as the Blackfoot Challenge to other settings. In addition to the difficulties of adapting programs and policies across unique contexts, there may be strong resistance from those with power in the new setting to adopt innovations that have worked in other settings (Bardach 2004). The tendency of managers to muddle through, their desire to protect their autonomy, and a host of other factors cause them to resist adopting innovations from elsewhere. When pressured to adopt an innovation, managers may reject or partially incorporate the innovation, claiming that they already have equivalent initiatives in place, or arguing that the innovation is not appropriate because of different local circumstances, or adopting the name or other symbols without taking on the substance (Brunner and Clark 1997; Clark 2002). For example, managers in Kluane National Park initiated a "multi-stakeholder collaborative planning exercise" for grizzly bear management that actually paid little attention to the perspectives of local communities around the park (chapter 4).

Moving Forward: Lessons and Recommendations

Despite this litany of challenges for governance in large carnivore conservation, the cases examined in this book also offer considerable hope that we can do better. We see great opportunity for improvement. Traditional carnivore management is performing poorly, but the cases provide examples of some highly innovative initiatives that have taken advantage of local capacity to make progress for the benefit of carnivores and people. In this section we outline five strategies for improving governance, derived from the evaluations and discussion in the preceding chapters: (1) focus attention on the quality of the decision-making process, (2) build new platforms for dialogue and problem solving (bridge building), (3) develop leadership and capacity (interdisciplinary problem solving), (4) be good company (social process), and (5) be pragmatic and reinforce positive trends and conditions. We do not put these strategies forward as "best practices," to be rigidly applied in any setting (recall the earlier discussion of unique contexts). Instead, we seek to promote what Bardach (2004, 206) calls "smart practices," intended for "adaptation or inspiration" in other settings.

FOCUS ATTENTION ON THE QUALITY OF DECISION-MAKING PROCESSES

Finding, securing, and sustaining the common interest in large carnivore conservation is certainly an important societal goal. But just what should be the focus of our attention in pursuing this goal? In our search for success, should we focus on carnivores, people, both, or something else? Our own answer to this question is embodied in this book: we maintain that the focus should be squarely on decision-making processes.

Historically, attention in large carnivore management has centered almost exclusively on the animals and their biology (see chapters 9 and 10). Accordingly, society enlisted biologists to staff state and federal wildlife management agencies. We relied on these wildlife biologists and, more recently, conservation biologists, to tell us what the problems were and what we should do about them. In recent years, a growing recognition that people are also important in all this has led to the "human dimensions" movement, which adds research on human values, attitudes, and behavior to the mix.

It is now clear, however, that large carnivore conservation requires much more than a good understanding of carnivore biology and human attitudes and behavior; it requires understanding of the broad ecological and sociopolitical systems in which carnivores and people exist and interact. It also requires high-quality decision-making processes to make the right choices about how people should manage themselves in relation to each other and to carnivores. To be successful in carnivore conservation, we need to be much more broadly problem oriented, sensitive to real people and contexts, and focused on understanding and upgrading how we individually and collectively make choices that determine the fate of these animals. We should always include the best biological information about the animals and their habitats, but we should also include information on people and their interactions and examine how well our decision-making processes are helping us or hindering us. In other words, we need to revise and enlarge the subject of attention to include governance as well as carnivores and people.

How do we shift the subject of attention yet continue to include the best biological science? In this book, we shifted the focus to governance by using a framework that identifies each of the functions involved in good decision making and specifies standards for the performance of each function. We asked our authors to discuss decision processes and apply standards for these processes and tests of the common interest. The approach we adopted in this book can be used elsewhere. In our experience, turning participants' attention away from their entrenched disputes about exactly what is happening to carnivores or what is happening to people and why and refocusing them instead on examining the quality of decision-making processes and how they can be improved opens the door for people to interact more civilly and constructively. This increases the chances of redefining problems and finding common ground.

A first step in this direction is to ask participants to consider how the existing decision process fares against tests of the common interest (see appendix). These tests can then be used to revise and improve the process. The rationale for this is simple. Decision making that is inclusive, representative, responsible, reasonable, adaptable, and that accommodates affected interests and meets the other standards is more likely to lead to a good understanding of problems and to identify and implement solutions that advance the common interest (Brunner 2002; Steelman and DuMond 2009). Such decision-making processes enable what Rittel and Webber (1973, 162) call "second generation" planning, where "an image of the problem and of the solution emerges gradually among the participants, as a product of incessant judgment, subjected to critical argument."

BUILD NEW PLATFORMS FOR DIALOGUE AND PROBLEM SOLVING (BRIDGE BUILDING)

Because we are all interested in our environment, both its natural and cultural heritage, and a healthy society for ourselves and our children, we are joined in a social relationship in which it is incumbent on us to strive to take each other seriously and recognize the integrity of different worldviews— even those with which we strongly disagree. At our best, we must understand and appreciate other people's "culture of mind," to use Robert Kegan's term (1994, 311). Quality decision making and civil interactions require us to be willing to suspend our judgments of others who differ from us. We need to go beyond stereotyped constructions of others and beyond our own limited culture of mind to discover the basis for how they create meaning and value for themselves. This can be of great benefit to both parties; under the right circumstances both appreciate the experience and come away feeling respected and more willing to engage constructively in the future.

A forum for such cross-participant understanding of culture of mind was created successfully in the interdisciplinary problem-solving workshops that preceded the Grizzly Bear Dialogue Group in Banff (Rutherford et al. 2009; Chamberlain, Rutherford, and Gibeau 2012; see also chapter 7). This and other experiences show that it is possible to create the appropriate conditions and encourage the necessary introspection and exchange of views. Not everyone has the generosity, personal comfort, or freedom from self-referencing for this kind of exchange. These prerequisites must be cultivated by learning though interaction in settings that foster their development. This is a better way, a qualitatively different way, of seeing conflict and difference and getting past them, at least pragmatically.

The way forward, then, is through processes that help us to appreciate difference but also help us to set aside differences in respectful and cooperative interaction. We need to help people to see themselves not as victims, but as part of a social relationship in a real context. Some degree of conflict is an inevitable byproduct of interactions among people with different beliefs and values. The challenge is to understand this and get beyond it, to escape from being prisoners held hostage by our differences and our artificial constructions of those with whom we differ. We are all engaged in a long conversation with ourselves and with others who differ or agree. We are connected, like the two ends of a rope.

A first step in sound bridge building is to realize that this is about "dialogue across differences" (Burbules and Rice 1991, 393). The approach that we recommend encourages such a dialogue across differences, with the bridge footings anchored on either side of the gorge. On this bridge, we can meet and explore each other's cultures of mind, seeking understanding and common ground. This approach stands in marked contrast to the way many people function now in contentious social interactions and decision making: assuming that a bridge already exists and the problem is that others have not walked over to their side.

What we are looking for is, as Kegan (1994, 318) says, "changes in thinking and feeling . . . improved understanding of the other's and one's own position, altered attitudes about the other's capacity and willingness to understand one's own position, and new thinking about the possibility of developing solutions that preserve the most precious features of each other's positions." In the conflict-ridden world of large carnivore conservation, such changes would be of historic significance with dramatic consequences. Kegan and Lahey (2009) offer a detailed workbook to guide people in overcoming resistance to such personal and organizational change.

In short, we want to construct transformative experiences that bring into light many of the forces and factors behind seemingly intractable conflict, struggling leadership, and problems of knowledge creation and use. We want to encourage a "reconstructive" prescription, and we believe that a focus on quality decision making can provide such a prescription. This involves a different kind of personal orientation. It also involves developing ways of knowing that are more tolerant of difference, appreciative of the sources of conflict, and willing to accommodate others to seek mutual benefit.

DEVELOP LEADERSHIP AND CAPACITY (INTERDISCIPLINARY PROBLEM-SOLVING SKILLS)

To solve conservation problems involving complex systems, we need to develop leaders who have the skills to understand and effectively work in those systems (Clark 2008; Clark et al. 2011a, 2011b). One key to this is to develop skill in using a comprehensive framework to map the context for problems and integrate knowledge across disciplines, theories, and models. Elinor Ostrom (2011, 8) describes the importance of such a conceptual framework for institutional analysis: "Frameworks identify the elements and general relationships among these elements that one needs to consider for institutional analysis and they organize diagnostic and prescriptive inquiry. They provide a general set of variables that can be used to analyze all types of institutional arrangements."

We recommend the integrative interdisciplinary framework of the policy sciences, as it is the most comprehensive and yet sufficiently detailed framework that we have encountered for the purpose (Lasswell 1971; Lasswell and McDougal 1992; Clark 2002). As we discuss in chapter 1, this integrative approach includes tools for clarifying the analyst's observational standpoint, performing the intellectual tasks of problem solving, mapping the social context for the problem, and examining and evaluating the decision-making process. The model of decision functions and the evaluative criteria used by all the authors in this book are components of the policy sciences framework, and all the authors also used the full framework in their analyses.

BE GOOD COMPANY (SOCIAL PROCESS)

Human beings are social creatures, and we interact with each other in a variety of ways and settings. These interactions constitute what Harold

Lasswell called the social process, in which people, seeking to satisfy their values (what they want or aspire to), create and use institutions to produce and allocate values, with effects on resources and the environment (Lasswell 1971; Clark 2002). This simple model of the social process can be expanded to include *participants*, their *perspectives*, the *situations* in which they interact, the *values* and *strategies* they use to pursue their desires, the *outcomes* of their interactions in the short term, and the longer-term *effects* on people, resources, and the environment (see chapter 1). Decision-making processes (governance), which are the focus of this book, are created through social processes. As people interact they develop institutions, which "control human conduct by setting up predefined patterns of conduct" (Berger and Luckmann 1967, 55). Included in these predefined patterns of conduct are the norms and structures for governance, or decision-making processes.

The cases in this book reveal that social interactions concerning large carnivores are suffering from a broad erosion of civility. Yet respectful civic engagement is essential for finding and implementing solutions to public problems. We need to encourage people to reengage with each other in more constructive ways—in essence, to be good company. Watters et al. (chapter 3) say that "creating informal arenas around kitchen tables, over cups of coffee, and in hay meadows is an effective way to begin the process of rebuilding civility, human dignity, and skill among diverse communities." The participatory mapping workshops and "neighbor network" of the Blackfoot Challenge grizzly bear project in Montana (chapter 6), the community-based efforts to manage hotspots for grizzly bear conflicts near Haines Junction, Yukon (chapter 4), and the interdisciplinary problem-solving workshops in Banff (chapter 7) are all good examples of new informal arenas of engagement that helped to build trust and social capacity for better decision making.

BE PRAGMATIC AND REINFORCE
POSITIVE TRENDS AND CONDITIONS

Being pragmatic in carnivore conservation means being practical rather than highly theoretical or ideologically driven. It also means adapting to the local context, taking advantage of existing capacity, and learning from experience. In a previous book we recommended working with local people on local problems to reduce the levels of conflict associated with carnivore conservation (Clark, Rutherford, and Casey 2005). We repeat that pragmatic recommendation here. Working together on small local problems can improve interpersonal relationships and establish a record of successes that can be expanded locally and emulated elsewhere. If a small project fails it can be phased out without major political repercussions or loss of goodwill. Whether the project succeeds or fails, the lessons from experience can inform future efforts. As for taking advantage of existing capacity, the observation of Wilson et al. (chapter 6) is apt: "When this effort started it was clear that it would be foolish to ignore the existing institutional capacity of the [Blackfoot Challenge watershed group]." In addition to an inclusive community group like the Blackfoot Challenge, other examples of existing capacity might include a community leader who is skilled in facilitation, a well-respected local charitable organization, or an agency manager who is strongly oriented toward the common interest.

Conclusion

We undertook this book because of our concern about the current state of governance for people and large carnivores and our belief that insufficient attention has been paid to decision-making processes in wildlife management more generally. Although our case studies are all from North America, we believe on the basis of our own experience and reading and the experiences of the other authors in this book that the problems and lessons that we discuss are relevant for wildlife conservation in many other areas of the world. We hope that our efforts will stimulate more interest and discussion about governance in wildlife conservation, and we look forward to participating in the ongoing dialogue.

Appendix
Tests of the Common Interest

Tests to determine whether a decision process serves the common interest (adapted from Brunner et al. 2002; Brunner and Steelman 2005; and Steelman and DuMond 2009).

PROCEDURAL TEST

Is the decision-making process *inclusive* (i.e., are the participants representative of the community as a whole)?

If some interests are not directly represented in the process, are those interests reflected in the outcomes?

Are the participants *responsible* (i.e., are they willing and able to serve the community as a whole, and can they be held *accountable* for the consequences of their decisions)?

SUBSTANTIVE TEST

Are the expectations of the participants about what will be accomplished reasonable?

Have all valid and appropriate concerns been taken into account?

Have the outcomes been approved by participants representative of the community as a whole, indicating that they believe the outcomes are in the common interest?

Are the outcomes compatible with broad societal goals (e.g., democracy, equity, timeliness)?

Do the outcomes ostensibly solve the problem?

PRAGMATIC (PRACTICAL) TEST

Do the outcomes work in practice, and do they uphold the reasonable expectations of those who participated?

Are decisions adapted over time to deal with changing circumstances?

REFERENCES

Bardach, E. 2004. "Presidential Address—The Extrapolation Problem: How Can We Learn from the Experience of Others." *Journal of Policy Analysis and Management* 23:205–20.

Berger, P. L., and T. Luckmann. 1967. *The Social Construction of Reality: A Treatise in the Sociology of Knowledge.* New York: Anchor Books.

Brunner, R. D. 2002. "Problems of Governance." In *Finding Common Ground: Governance and Natural Resources in the American West*, edited by R. D. Brunner, C. H. Colburn, C. M. Cromley, R. A. Klein, and E. A. Olson, 1–47. New Haven, CT: Yale University Press.

Brunner, R. D., and T. W. Clark. 1997. "A Practice-Based Approach to Ecosystem Management." *Conservation Biology* 11:48–58.

Burbules, N. C., and S. Rice. 1991. "Dialogue across Differences: Continuing the Conversation." *Harvard Educational Review* 61:393–416.

Chamberlain, E. C., M. B. Rutherford, and M. Gibeau. 2012. "Human Perspectives and Conservation of Grizzly Bears in Banff National Park, Canada." *Conservation Biology* 26:420–31.

Clark, S. G. 2002. *The Policy Process: A Practical Guide for Natural Resource Professionals.* New Haven, CT: Yale University Press.

———. 2008. *Greater Yellowstone's Future: Choices for Leaders and Citizens.* New Haven, CT: Yale University Press.

Clark, S. G., M. B. Rutherford, M. R. Auer, D. N. Cherney, R. L. Wallace, D. J. Mattson, D. A. Clark, et al. 2011a. "College and University Environmental Programs as a Policy Problem (Part 1): Integrating Knowledge, Education, and Action for a Better World?" *Environmental Management* 47:701–15.

———. 2011b. "College and University Environmental Programs as a Policy Problem (Part 2): Strategies for Improvement." *Environmental Management* 47:716–26.

Clark, T. W., D. J. Matson, R. P. Reading, and B. J. Miller. 2001. "Interdisciplinary Problem Solving in Carnivore Conservation: An Introduction." In *Carnivore Conservation*, edited by J. Gittleman, S. M. Funk, D. MacDonald, and R. K. Wayne, 223–40. Cambridge: Cambridge University Press.

Clark, T. W., M. B. Rutherford, and D. Casey, eds. 2005. *Coexisting with Large Carnivores: Lessons from Greater Yellowstone*. Washington, DC: Island Press.

Geva-May, I. 2001. "When the Motto Is 'Till Death Do Us Part:' The Conceptualization and the Craft of Termination in the Public Policy Cycle." *International Journal of Public Administration* 24:263–88.

Kegan, R. 1994. *In Over Our Heads: The Mental Demands of Modern Life*. Cambridge, MA: Harvard University Press.

Kegan, R., and L. L. Lahey. 2009. *Immunity to Change: How to Overcome It and Unlock the Potential in Yourself and Your Organization*. Boston: Harvard Business Press.

Lasswell, H. D. 1971. *A Pre-view of Policy Sciences*. New York: American Elsevier.

Lasswell, H. D., and M. S. McDougal. 1992. *Jurisprudence for a Free Society: Studies in Law, Science, and Policy*. New Haven, CT: New Haven Press.

Lindblom, C. E. 1959. "The Science of Muddling Through." *Public Administration Review* 19:79–88.

———. 1979. "Still Muddling, Not Yet Through." *Public Administration Review* 39:517–26.

Ostrom, E. 2011. "Background on the Institutional Analysis and Development Framework." *Policy Studies Journal* 39:7–27.

Rittel, H. W. J., and M. M. Webber. 1973. "Dilemmas in a General Theory of Planning." *Policy Sciences* 4:155–69.

Rutherford, M. B., M. L. Gibeau, S. G. Clark, and E. C. Chamberlain. 2009. "Interdisciplinary Problem-Solving Workshops for Grizzly Bear Conservation in Banff National Park, Canada." *Policy Sciences* 42:163–87.

Steelman, T. A., and M. E. DuMond. 2009. "Serving the Common Interest in US Forest Policy: A Case Study of the Healthy Forests Restoration Act." *Environmental Management* 43:396–410.

Thielmann, T., and C. Tollefson. 2009. "Tears from an Onion: Layering, Exhaustion and Conversion in British Columbia Land Use Planning Policy." *Policy and Society* 28:111–24.

Thomas, C. W. 2003. *Bureaucratic Landscapes: Interagency Cooperation and the Preservation of Biodiversity*. Boston: MIT Press.

Wilkinson, C. F. 1992. *Crossing the Next Meridian: Land, Water, and the Future of the West*. Washington, DC: Island Press.

Xiang, W. 2013. "Working with Wicked Problems in Socio-Ecological Systems: Awareness, Acceptance, and Adaptation." *Landscape and Urban Planning* 110:1–4.

Contributors

AVERY C. ANDERSON
Yale School of Forestry and
    Environmental Studies
Yale University
New Haven, CT 06511

DAVID N. CHERNEY
Northern Rockies Conservation
    Cooperative
Jackson, WY 83001

DOUGLAS CLARK
School of Environment and
    Sustainability
University of Saskatchewan
Saskatoon, SK S7N 5C8

SUSAN G. CLARK
Yale School of Forestry and
    Environmental Studies
Yale University
New Haven, CT 06437

MICHAEL L. GIBEAU
Department of Geography
University of Calgary
Calgary, AB T2N 1N4

JAMES J. JONKEL
Grizzly Bear Management Specialist
Montana Department of Fish, Wildlife
    and Parks
Missoula, MT 59804

DAVID J. MATTSON
Lecturer and Senior Visiting Scientist
Yale School of Forestry and
    Environmental Studies
New Haven, CT 06511

CHRISTINA MILLOY
Wildlife and Sport Fish Restoration
    Program
US Fish and Wildlife Service
Arlington, VA 22203

GREGORY A. NEUDECKER
State Director
Montana Partners for Fish and Wildlife
    Program
US Fish and Wildlife Service
Ovando, MT 59854

J. DANIEL OPPENHEIMER
Restoration Coordinator
Tamarisk Coalition
Grand Junction, CO 81502

WILLIAM M. PYM
School of Resource and Environmental
    Management
Simon Fraser University
Burnaby, BC V5A 1S6

LAUREN RICHIE
School of Forestry and Environmental
    Studies
Yale University
New Haven, CT 06511

MURRAY B. RUTHERFORD
School of Resource and Environmental
    Management
Simon Fraser University
Burnaby, BC V5A 1S6

D. SCOTT SLOCOMBE
Dept. of Geography and Environmental
    Studies
Wilfrid Laurier University
Waterloo, ON N2L 3C5

REBECCA WATTERS
Yale School of Forestry and
    Environmental Studies
Yale University
New Haven, CT 06511

SETH M. WILSON
Program Coordinator
Blackfoot Challenge Wildlife Committee
and
Visiting Fellow
Yale School of Forestry and
    Environmental Studies
New Haven, CT 06511
and
People and Carnivores Program
Northern Rockies Conservation
    Cooperative
Jackson, WY 83001

LINAYA WORKMAN
Dept. of Heritage, Lands, and Resources
Champagne and Aishihik First Nations
Haines Junction, YT Y0B 1L1

Index

A page number followed by *f* refers to a figure, a page number followed by *t* indicates a table, and a page number followed by *b* refers to boxed material.

beekeepers, and Montana grizzly management, 177, 192, 193–94, 211

bighorn sheep (*Ovis canadensis*): declining populations of, 32–33, 36; desert subspecies of, 293; hunters' interest in, 32; mountain lion management and, 32–33, 36, 54, 55, 56, 327–28; tag fees for, 57n1; Yellowstone wolves and, 66

biophysical process, 342–43; queries and graphics for appraisal of, 348, 349–50t, 353–56, 354f, 355f

bison: wood bison of Yukon, 126; Yellowstone grizzlies and, 329

black bear, 293

Blackfoot Challenge (BC), 91, 165, 177–78; conclusions about, 203–4, 371; conservation groups working with, 182; decision-making functions in, 184–98; government agencies consulting with, 182; lessons for managers from, 198–201, 199b; Montana Fish, Wildlife and Parks and, 182; recommendations based on, 201–3, 376, 377; wolf management and, 197–98, 200, 210. *See also* grizzly bear management in Blackfoot watershed, Montana

Blackfoot Challenge Wildlife Committee, 184, 185, 187, 189, 190–92, 192, 196, 202

black-footed ferret, 309

Blackfoot River watershed, 179–80, 179f; overall conservation plan for, 192. *See also* grizzly bear management in Blackfoot watershed, Montana

Boitani, L., 251, 306

boneyards: grizzly bears scavenging at, 189, 194; wolves scavenging at, 143

bounded rationality, 368–69

bovine spongiform encephalopathy (BSE), 144, 150

Bow Valley Parkway, 222, 225, 230, 232, 233, 234, 235, 236, 238

Bow Valley Parkway Advisory Group, 233, 234, 235, 240

Boyce, M. S., 142, 143

Bracken, P., 309

bridge building, 371, 373–75

Bridger-Teton National Forest, 73, 262

bridging, 125, 308

Brunner, R. D., 67, 86, 123, 151, 239, 252, 272

Burch, William, 12, 308, 309, 310

bureaucratic model of wildlife management, 45; alternatives to, 309–10; history of, 304; problems with, 29–30, 46, 47, 49, 269, 270, 271, 296, 299; standardization of decisions in, 370

Bureau of Land Management, US, 182

Bush, George W., administration, 82

business model of agency management, 45, 47, 48, 49, 51

California ballot initiatives, on mountain lion hunting, 2, 33

California Rangeland Conservation Coalition, 90

Canada National Parks Act, 219, 236

Canadian federal government, 8. *See also* Parks Canada

caribou: Aishihik herd of, 126; wolf predation on, 119

carnivore conservation. *See* large carnivore conservation

carrying capacity, 48

cattle. *See* livestock

causal models, 336

Central Rockies Wolf Project (CRWP). *See* CRWP/SACC collaboration in Alberta

Chamberlain, Emily, 220, 232, 236

Chambers, N., 296

Champagne and Aishihik First Nations (CAFN), 109, 110, 111f; grizzly bear management and, 112, 117, 118, 119, 121; grizzly bear narratives and, 115; grizzly bear plan and, 121; relationship to Parks Canada, 118, 127; respect for bears among, 113

Charon, J. M., 257

Chellam, R., 264

Cilliers, P., 333

civility: better governance and, 3; decreased in United States, 39, 45, 53; in democratic processes, 310; eroded concerning large carnivores, 376; ideological vehemence and, 44; in liberal democracy, 46, 49, 53; mountain lion management and, 29, 38, 42, 49; in pragmatic problem solving, 50; rebuilding, 376; symbolic issues working against, 52. *See also* respect

civil society: conflict over carnivore conservation and, 346; grizzly bears of Yellowstone and, 331; mountain lion management and, 57; policy process and, 344; state-level wildlife management and, 51, 55; unreflective behaviors presenting problems for, 334

claims: about decision process, 302; North American Model and, 298, 305, 313–15

Clark, D., 123, 126, 253, 254

Clark, S. G., 221, 308, 309, 310, 311

Clark, T. W., 67, 239, 267, 273, 310

climate change, 295, 300, 301, 331

cognition: ACT-R model of, 335, 337; emotion and, 4–5; limits of, 331–33, 335, 346, 366; in orientation to complex cases, 337; reflexive versus reflective, 332; 7 ± 2 rule in, 331, 335, 338. *See also* rationality; symbols

collaborative processes: determining whether to use, 238; effective decision making in, 239–40. *See also* Blackfoot Challenge (BC); CRWP/SACC collaboration in Alberta; Grizzly Bear Dialogue Group (GBDG); Oldman Basin Carnivore Advisory Group (OBCAG)

collective-choice tier of decision making, 10

colonialist institutions in the Yukon, 123

co-management in Kluane region, Yukon, 108–9; institutional relationships involved in, 110–11, 111f, 113, 117–18, 121–22; lessons and recommendations about, 124, 125–26, 127, 128, 130, 133, 134; peoples involved in, 109–10. *See also entries concerning Kluane*

commission system: accountability in, 45, 46, 47; in Arizona, 35, 36–37, 40, 41, 43; hunters' interests in, 40, 49, 50, 328; problems with, 296; resistance to giving up power, 51

Committee on the Status of Endangered Wildlife (Canada), 8, 142

common ground: in American society, 53; for Blackfoot watershed grizzly management, 200; bridge building and, 374; democratic processes and, 259, 272; finding, by harmonizing interests, 271; finding, for public policy making, 259; as fundamental challenge of governance, 9; Grizzly Bear Dialogue Group and, 221–22, 227, 232; integrative approach to finding, 13, 14; mountain lion management and, 29–30, 37, 38, 40; political factors in finding, 7; quality of decision making and, 373; scientific management and, 76–77; state wildlife agencies and, 296; in

conservation (*continued*)
extraction, 254; standards for appraising cases of, 343–46. *See also* large carnivore conservation; North American Model of Wildlife Conservation

conservation easements, 182

conservation groups: Banff grizzly management and, 218, 223; Blackfoot Challenge and, 178, 182, 187, 191, 202; as Blackfoot watershed stakeholders, 181; litigating about delisting of wolves, 1. *See also* environmentalists

conservationist narrative, about Yukon grizzlies, 115–16, 127, 130–32

constitutional tier of decision making, 10

constitutive decision process, xi, 235, 263–65, 263f; common recurring weaknesses in, 279–82; context for, 261; for grizzly bear conservation in Alberta, 253, 254, 255; Grizzly Bear Dialogue Group and, 235–36; in integrative approach, 14; large carnivore conservation and, 251–52, 261, 264–65, 267–68, 272; in North American Model of Wildlife Conservation, 289, 303, 304; in policy process, 340; problems of, for large carnivores, 267–68; recommendations for, 273–74, 303; standards for, 266–67, 268; symbols in, 260

content knowledge, x, 269

contingent systems, 325, 331, 333–34

Cottonear, Leo, 71

cougar. *See* mountain lion (*Puma concolor*)

Craighead, Frank, 261

Craighead, John, 261

Crane, J., 308

Cromley, C. M., 265

CRWP/SACC collaboration in Alberta, 145–46; Alberta wolf management plan and, 143; appraisal of decision

processes, 168–72; common interest and, 151–52; compared to OBCAG, 159–60; conclusions on, 166–67; history of, 140, 141, 145; level of trust in, 163, 165; loss of key participants in, 163; problem orientation for, 147–48

cultural context, 256–57, 264, 310

culture of mind, 373–74

customer-oriented wildlife management, 29, 45–46, 47, 51. *See also* state-level wildlife agencies

Dahl, R. A., 9

Dall sheep (*Ovis dalli*) recovery, 126

debate, 17, 18. *See also* promotion function

deciding on a plan, 265. *See also* prescription function

decision activities (functions), 17–18, 184, 265, 301–2; in Alberta wolf management, 168–72; in Banff grizzly bear management, 243–47; in Blackfoot watershed grizzly bear management, 183–98, 206–12; criteria for evaluating, 3, 20–25; of Grizzly Bear Dialogue Group, 223–36; in Kluane grizzly bear management, 119–23, 132–34; in mountain lion management, 53–57; in policy process of North American Model, 313–15; in Wyoming wolf management, 93–102. *See also the seven specific functions*

decision making: boundedly rational, 368–69; common interest and (*see* common interest, tests of); common weaknesses in, 267, 272, 273, 274–75, 279–82, 302–3; conventional form of, 223; cultural context of, 256, 257, 264, 310; in democratic society, 9, 12, 18, 52; as focal lens of inquiry, 15; focusing on

human conflicts in, 121, 128–29; shared narrative recommended for, 129–30; studies on, 112, 113–15, 121, 130, 134
grizzly bear management in Wyoming: constitutive decision making about, 265; livestock kills and, 74; ordinary decision making about, 262
grizzly bears in Yellowstone ecosystem, 180, 329–31; changes in dietary behavior of, 329–30; politicized science about, 330; radar charts for case of, 354–56, 354f, 355f

308, 315; of grizzly bears in Blackfoot watershed, 186, 196; of grizzly bears in Yukon, 116, 120, 123, 131; in-group emphasis on, 40–41, 42; legitimate purposes of, 292; of mountain lions, 33, 34, 35–36, 37–38, 52, 328; North American Model and, 289, 292, 295, 298, 315; popularity of, 4; public condemnation of certain practices in, 44; public support for, 43, 44; of wolves in Alberta, 143. *See also* mortality of carnivores

Haines Junction, Yukon Territory, 109, 112, 121, 125, 128, 129, 130, 376
Hammond, J. S., 273
Hardin, G., 314
high-performance teams, 310, 311
Hill, B. J., 258
"human dimensions" school, 256–57, 372
Humane Society of the United States, 297
humanistic/moralistic worldview, 6, 38, 39, 43, 44, 49
hunters: of Alberta, 155; as customers of wildlife management agencies, 46, 47, 48, 328; declining population of, 290, 295; demographic profile of, 44; gender relations and, 44; mountain lion management and, 29, 34, 35, 36, 37–38, 40, 49; North American Model and, 297, 298, 308, 315; as revenue source, 293, 294; as special interest, 47; utilitarian/dominionistic worldview of, 39, 44; wildlife management agencies' alignment with, 38, 41, 47, 49, 50, 51, 54, 328, 330; wolf management in Wyoming and, 68, 69, 76–77
hunters' associations, 10, 40
hunting: Arizona's decline in, 42; of caribou in Yukon, 126; democracy of, 293;

Idaho: wolf management in, 1; wolf population in, 144; wolves moving to Canada from, 149; Yellowstone grizzly bears and, 329; Yellowstone wolves and, 65, 66
identities: federal versus state wildlife management and, 330; in Grizzly Bear Dialogue Group, 224; identifications and, 338, 339; necessity for attention to, 88b; North American Model and, 298; rectitude as value tied to, 81; social process and, 338, 339; supported by goals of carnivore management, 190; symbolic dynamics of, 52, 53
ideology, 53, 258, 259, 260
implementation, 17, 184; in Alberta grizzly management, 254; in Alberta wolf management, 153, 158; in Blackfoot watershed grizzly management, 193–96, 208–9; claims about, 302; common recurring weaknesses in, 280–81; in Grizzly Bear Dialogue Group, 230–31; in North American Model, 314. *See also* application function; invocation function
inclusive interests, 22, 160, 167, 168, 207, 255, 266, 274, 305

grizzly bear management and, 177, 180, 182, 184, 185, 187, 188, 189–90, 195, 197, 206–10; wolves and, 195

moose, bear predation on, 119

moralistic worldview. *See* humanistic/moralistic worldview

Morehouse, A. T., 143

mortality of carnivores: extirpated from much of former ranges, 4; grizzlies in Alberta, 254; grizzly-human conflicts in Banff and, 218, 219–20, 229, 230; grizzly-human conflicts in Blackfoot watershed and, 190, 192, 196–97, 207, 210, 211; grizzly-human conflicts in Yukon and, 128–29; human causes of, 4, 342–43; innovative approaches to reducing, xi, 91. *See also* hunting

mountain lion (*Puma concolor*): aesthetic valuations of, 329; attacks on people, 32; ballot initiatives on hunting of, 2, 33–34, 54, 55; charismatic and symbolic nature of, 43; emotional responses to, 5, 6; eradication efforts directed at, 31; hunting of, 33, 34, 35–36, 37–38, 52, 328; overview of conservation issues for, 4; population of, 31–32, 33, 37, 38; preying on mule deer and bighorn sheep, 32–33; range of, 31–32, 327; restoration of, 293

mountain lion management: ameliorative decision processes needed in, 40, 42–43, 45; changes during 1990s and 2000s, 33–35; complexity in, 327–29; decision activities for, 53–57; decision processes in Arizona, 36–38, 44, 46; fostering civility in American culture and, 52–53; governance problems in, 36–38, 366; history of, 31–33; by killing

threatening animals, 32, 33; lessons and recommendations for, 49–52, 51b; overview of issues in, 29–30; power in-group of, 40–43, 54; power out-group of, 43–44; radar charts for appraisal of, 354–56, 354f; responses to controversy over, 35–38; stridency of discourse on, 37–38; symbolic stakes in, 29, 31, 40, 43, 44, 50, 51, 52; worldviews at play in, 38–40

Mountain Livestock Cooperative (Alberta), 91, 165

Mt. Elden incident, 34–35, 43, 52

mule deer (*Odocoileus hemionus*), 32, 327

mutualism wildlife value orientation, 6

myths, 258–59, 260, 261, 264; associated with large carnivores, 367; working to understand those of others, 124; in Yukon grizzly management, 123

Nadasdy, P., 113, 126

Napolitano, Janet, 35

National Audubon Society, 297

national parks: Canadian, 4, 8; US federal control of, 7. *See also* Banff National Park; Yellowstone National Park

National Park Service, and wolf management, 67, 69

National Rifle Association, 290, 297

National Wildlife Federation, 297

Native Americans: cultural narrative of, 78; hunting by, 43; rectitude as value for, 81; respect demanded by, 81–82; wolf reintroduction and, 369, 370; wolves as symbol for, 77, 78. *See also* First Nations people; Wind River Indian Reservation

natural gas development, in Wyoming, 70, 73, 79–80

prescription function (*continued*)
ment, 54; in policy process, 341; to
reintroduce wolves to Yellowstone, 85;
in Wyoming wolf management, 86,
94–95; in Yukon grizzly management,
121–22, 133
Pretty, J. N., 223, 310
Primm, Steve, 91, 259, 308
problem definition, 14, 15–16; for Alberta wolf management, 147–51; for
Banff grizzly management, 217–22; for
Blackfoot watershed grizzly management, 179–80, 187, 199–200, 202, 207;
for complex conservation cases, 336,
337; deepening of, 312; facilitator for,
163; for large carnivore conservation,
267–72; for mountain lion management, 30, 31–38; policies and, 340; in
scientist management, 48–49; simplistic, 267; for wildlife conservation,
303–6; for Wyoming wolf management, 75–83; for Yukon grizzly management, 123, 133
procedural knowledge, x, 269
procedural test of common interest, 19,
102, 167, 252, 274, 377; Alberta wolf
management and, 151, 154–55; Wyoming wolf management and, 86
promotion function, 17, 18, 184, 265, 301,
302; in Alberta wolf management,
157, 158, 169; in Blackfoot watershed
grizzly management, 190–92, 201, 202,
207–8; claims about, 302; common recurring weaknesses in, 279–80; criteria
for evaluating, 21–22; focus on harmonizing different perspectives, 271; in
Grizzly Bear Dialogue Group, 226–28,
244; in mountain lion management,
54; in North American Model, 313–15;
in policy process, 341; in Wyoming

wolf management, 85–86, 87, 100–101;
in Yukon grizzly management, 121–22,
123, 133
prototyping, 202, 310, 311. *See also*
Grizzly Bear Dialogue Group (GBDG)
provincial governments, 7, 8. *See also*
grizzly bear management in Alberta;
wolf management in Alberta
Prukop, J., 302, 306, 307
public education: about bears in the
Yukon, 129; in traditional wildlife
management, 36
public interest. *See* common interest
public meetings: on Kluane region bear
plan, 120, 122–23; traditional practice
of, 36, 49
public participation, 310, 314
Public Trust Doctrine, 291–92, 296, 308,
312, 315
Putnam, Robert, 39, 125
Pym, William M., 141

Quinn, M. S., 145

rabies in wolves, 143
radar charts, 347–48, 353, 354f, 355–56,
355f
Radical Center, 90
Raiffa, H., 273
ranchers, collaboration with environmentalists, 90, 91. *See also* livestock
ranchers of Alberta, economic pressures
on, 144, 149, 150. *See also* wolf management in Alberta
ranchers of Blackfoot watershed: bear
habitat on property of, 181; characteristics of, 181; communication with,
200, 202, 207; cost sharing by, 192,
194, 203, 212; on Landowner Advisory Group, 191; mapping human-bear

conflicts of, 179; meetings with, 187; Montana Fish, Wildlife and Parks and, 182; not participating in program, 20, 190, 207; power sharing and, 178, 200; prescribed activities and, 192–93, 208; program implementation and, 193–96, 201; survey including, 185–87, 204–6
ranchers of Greater Yellowstone, 65–66, 67, 69, 70, 79–80, 86–87, 101
ranchers of Upper Green River Valley. *See* Upper Green River ranching community
ranching practices, alternative, 147, 151, 159. *See also* fladry experiments
ranchland development, 90
range riders, 101, 165. *See also* night riding
rationality, 333, 334, 335, 337, 347; bounded, 368–69; complexity and, 326; of decision-making functions, 21–22, 23, 24; emotional issues and, 52. *See also* cognition
recreational users, 10. *See also* nonconsumptive users
rectitude, 78; claims for, 298; in Grizzly Bear Dialogue Group, 224; Kluane grizzly bear narratives and, 117; value demands for, 313–15, 345; wolf management and, 81
Regan, R. J., 302, 306, 307
Reisman, W. M., 263
resilience-based ecosystem stewardship, 307
resilience theory, 333
resource extraction, 254, 255
resource management, and conflicting values, 12
respect: arenas for fostering, 89; as basic human value, 16, 78, 344, 345; for bears among Yukon Aboriginals, 113–14,

367; co-management regimes and, 125; Grizzly Bear Dialogue Group and, 224, 234; grizzly management in Banff and, 219; for interests of others, 46, 52–53; in localist narrative of Yukon, 116, 117, 130; in OBCAG wolf management decision process, 157; in solving public problems, 376; for wolves in Native American cultures, 72; Wyoming wolf management and, 68, 81–83, 86, 88. *See also* civility; disrespect
Rittel, H. W. J., 367, 373
Robichaud, C. B., 142
Rocky Mountain Elk Foundation, 290, 297
Röling, N., 129
Roosevelt, Franklin D., 293
Roosevelt, Theodore, 298
Roosevelt Doctrine, 292
Ruby Range Dall sheep (*Ovis dalli*) recovery, 126
rules. *See* prescription function
Ruther, E. J., 6
Rutherford, Murray B., 141, 221, 308

Sabino Canyon incident, 34–35, 43, 52
SACC (Southern Alberta Conservation Cooperative). *See* CRWP/SACC collaboration in Alberta
Sarewitz, D., 258
Schwartz, Shalom, 338
scientific management approach: versus adaptive governance, 123, 125; alternatives to, 306–8, 309; difficulties with, 11–13, 270, 271–72, 299, 306–7, 347, 366–67; of federal agency employees, 69; to grizzlies of Alberta, 262; to grizzlies of Kluane region, 116, 117, 119–20, 123, 125, 129, 132–33; Grizzly Bear Dialogue Group and, 224, 225,

management, 35, 40, 41, 44, 49, 57; of resource extraction industries in Alberta, 255; versus scientific management of endangered species, 76; tension with common interest, 18–19, 45, 160, 221. *See also* common interest

species: status of, 300–301, 304; as symbols for interest groups, 259–60. *See also* endangered species

Species at Risk Act (Canada), 8

spider graphs. *See* radar charts

Sport Fish Restoration Program (SFR), 293, 294, 300

Sporting Conservation Council, 306

stammtisch, 115, 123, 125

standards for appraising conservation cases, 343–46

standards for decision making, 3, 18, 20–25, 184, 252, 266–67, 268; in Alberta wolf management, 172; in Blackfoot watershed wolf management, 211–12; in Grizzly Bear Dialogue Group, 246–47

standpoint. *See* observational standpoint

Stanford, Jim, 258

star charts. *See* radar charts

state-level wildlife agencies: accepted norms of, 36; conflicts with federal bureaucracies, 7–8, 69, 309, 330; dynamics of conflict involving, 329; failure to protect threatened and endangered species, 295–96; funding for, 293–94, 328; harms of scientific management in, 299, 309; hunters given primacy by, 38, 41, 47, 49, 50, 51, 54, 328, 330; mountain lion management and, 29, 32, 36, 328; power of, 328; as problematic form of governance, 296; problem definition relating to, 305–6; in the wildlife institution, 294; worldviews

associated with, 38–39. *See also* North American Model of Wildlife Conservation; wildlife management agencies

State Wildlife Grants Program, 294

Steelman, T. A., 151, 252, 274

Stemler, J., 290, 295

stewardship of the land: Blackfoot Challenge and, 178; Native Americans' sense of, 78; ranchers' sense of, 69, 74, 75, 77; resilience-based ecosystem stewardship, 307; Yukon locals' faith in, 116

Stumpf-Allen, R. C. G., 253

substantive test of common interest, 19, 102, 167, 252, 274, 378; Alberta wolf management and, 151, 155–56; Wyoming wolf management and, 86

succession planning, 163. *See also* termination function

sufficiency versus optimization, 326, 327

sustainability: adjusting unsustainable perspectives for, 259; Blackfoot Challenge experience and, 202–3; constitutive decision making for, 268; continuum of local involvement and, 212; leadership for, 311–12; North American Model and, 299, 314; observational standpoint and, 14; public awareness of, 278

sustainable coexistence, x, 19, 202, 267

sustainable harvest, 48

Sustainable Resource Development (SRD) agency, Alberta, 146, 148, 156, 158, 165

sustainable use, 8

symbolic meaning of carnivores, 3, 4–6, 218, 219, 260–61, 367

symbolic stakes: of grizzly management, 123, 188, 218, 219, 330; of large carnivore management, 51, 334; of moun-

symbolic stakes (*continued*)
tain lion management, 29, 31, 40, 43,
44, 50, 51, 52; of wolf management,
77–78, 88b
symbols: as context, 259–61; imposed
upon the world, 334; in support of
claims about North American Model,
298–99

tag fees for game species, 57n1
Teel, T. L., 6
termination function, 17, 184, 265, 302; in
Alberta wolf management, 154, 158–
59, 162–63, 171–72; in Blackfoot watershed grizzly management, 197–98,
210–11; common recurring weaknesses
in, 281–82; criteria for evaluating, 24;
frequent inattention to, 370; in Grizzly
Bear Dialogue Group, 232–34, 236, 240,
246; in mountain lion management,
56–57; need for early attention to, 238;
in policy process, 341; in Wyoming
wolf management, 98–99; in Yukon
grizzly management, 122–23, 134
territorial government. *See* YTG (Government of the Yukon)
"three-in-three" threshold, 151, 156, 157,
165
Tocqueville, Alexis de, 53
Total Quality Management (TQM), 45, 47
tourism industries, and Banff grizzly
management, 218, 225
tragedy of the commons, 314
Tribal Fish and Game, 72, 97
Tribal Game Code, 72
trust, building, 163–65, 376; Blackfoot
Challenge and, 190, 200, 202, 207, 211;
in Grizzly Bear Dialogue Group, 226,
235

Tutchone people, Southern. *See* Southern Tutchone people

ungulates: controlling bear populations
to benefit, 118; hunters' desire to preserve, 4; large carnivores exterminated
to restore, 295; on Wind River Indian
Reservation, 72; wolf management
in Alberta and, 143. *See also* bighorn
sheep (*Ovis canadensis*); bison; deer;
elk; moose; Ruby Range Dall sheep
(*Ovis dalli*) recovery; sheep ranching
Upper Green River Cattle Association
(UGRCA), 73, 74, 95
Upper Green River ranching community,
66, 68, 69, 70, 73–75; federal reintroduction process and, 94–95, 370; hostility toward wolves in, 74–75; mineral
rights issues for, 79–80; new settlers
in, 73; power as issue in, 79–80; ranchers on Wind River Reservation and, 71,
87; rectitude as value for, 81; respect
demanded by, 81; stewards of landscape in, 69, 74, 75, 77; wolf management decision process in, 85–87, 94–
101; wolves as symbol for, 77–78
urbanization, 295, 329
US Bureau of Land Management, 182
US Department of Agriculture, Wildlife
Services, 84
US Fish and Wildlife Service: funds
controlled by, 293–94; grizzly bear
management and, 2, 180; grizzly bears
of Blackfoot watershed and, 182, 187;
new challenges asserted by, 295; ordinary decision making by, 262; public's
interest in nature and, 308; scientific
management approach in, 83; tribal
fish and game department assisted by,

71; wolf management and, 1, 8, 67, 69–70, 72, 83, 86, 87, 91, 97, 265; Wyoming Game and Fish opposition to, 69

US Forest Service: grizzly bears of Blackfoot watershed and, 182; mountain lion management and, 34; ordinary decision making by, 262; wolf management and, 67, 69

US Geological Survey, 180–81, 197

utilitarian/dominionistic worldview, 6, 38; decline of, 39; demographic profile of, 44; mountain lion management and, 39, 40, 42; of power in-group, 40, 41, 42

value deprivation, in wolf management, 78–79, 85, 88b, 96, 101

values: in Alberta wolf management conflicts, 153, 155; attitudes toward nature and, 6, 260; Banff grizzly management and, 218, 219, 242; of Blackfoot watershed interest groups, 191–92, 202, 203–4; calls for diversity of, 302; claims appealing to, 298, 299; commonwealth of human dignity and, 344; of a culture, 257; decision making and, 256, 270, 276; in Grizzly Bear Dialogue Group, 224, 225, 226, 227, 229; institutions and, 339; integrative approach and, 13–14; in interdisciplinary framework, 252; Kluane grizzly bear narratives and, 115–17, 123; Lasswell's eight basic human values, 78–79, 338, 344; in mapping social process, 16; North American Model and, 290, 297, 298, 313–15; politicization of, 79–83, 368; Schwartz's schematic of, 338; scientific management approach and, 12–13, 83, 92–93, 271–72; in social process, 337–

38, 339, 344–45, 376; universal categories of, 16; working to understand differences in, 124; in Wyoming wolf management conflict, 67, 68, 79–83, 93. *See also* worldviews

varmints, mountain lion classification as, 31, 34

volunteer burnout, 125

Washakie, Chief, 70

Washington State, and Yellowstone wolves, 66

waste management program, Blackfoot watershed, 192, 193, 195

Watson, D. O., 253

wealth: as basic human value, 78, 344; grizzly management in Alberta and, 254–55; Kluane institutions and, 117; mountain lion management and, 29, 38, 41; policy processes and, 345; wolf reintroduction and, 80

Webber, M. M., 367, 373

White House Conference on Wildlife Policy, 290

White River First Nation, 110, 111

"wicked" problems, 11, 366, 367

wildlife: attitudes toward, 6; status of, 300–301, 304; trade in, 292

wildlife conservation. *See* conservation

wildlife management: amelioration and conciliation needed in, 42–43; institution of, 293; public lack of engagement in, 43. *See also* scientific management approach

wildlife management agencies: diversification of public demands on, 39; diversifying cultures of, 42, 50; financial liability concerns of, 46–47; funding for, 41, 49, 50, 51, 328; governance role

population segment of, 265. *See also* delisting of wolves; livestock, predation on

wood bison (*Bison bison athabascae*), 126

workshops for stakeholders, 89–90

worldviews: of Aboriginal peoples, 113; decision making and, 256, 257, 259; federal versus state wildlife management and, 330; grizzly bear management in Alberta and, 254; on humans' relationship to nature, 6–7, 38–39; with inflammatory effect on conflict, 51; in interdisciplinary framework, 252; interpreting experience through, 5; about large carnivores, 6–7, 251, 257; mountain lion management and, 38–40, 44, 48, 51, 328, 329; myths and, 259; North American Model and, 294, 299; respecting differences in, 373–74; social facts created in, 258; in social process, 337, 338, 339; in wolf management conflict, 67, 68. *See also* values

Wyoming: constitutive decision making for wildlife in, 265; ordinary decision making for wildlife in, 262–63; wolf population in, 144; Yellowstone grizzly bears and, 329. *See also* wolf management in Wyoming; wolf reintroduction in United States

Wyoming Game and Fish Department (WYGF), 67, 69, 74

Yellowstone National Park, 65; elk in, 258; grizzly bears in, 180, 329–31. *See also* Greater Yellowstone Ecosystem; wolf reintroduction in United States

Young, Oran, 9

youth education, 92

YTG (Government of the Yukon), 111f; bear-human conflicts and, 118–19, 121; grizzly bear narratives and, 115; grizzly bear plan and, 121, 123; grizzly bear population survey proposal, 128, 130; grizzly planning process and, 119

Yukon. *See entries concerning Kluane*

Yukon Fish and Wildlife Management Board, 110, 121